International Review of Nuclear Physics Vol. 9

Hadronic Physics from Lattice QCD

edited by

Anthony M. Green

Helsinki Institute of Physics, Finland

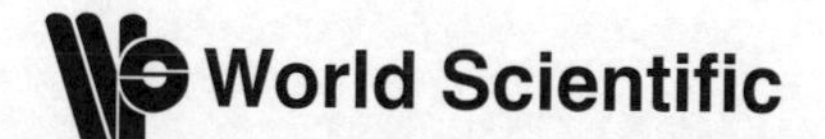

NEW JERSEY • LONDON • SINGAPORE • BEIJING • SHANGHAI • HONG KONG • TAIPEI • CHENNAI

Published by

World Scientific Publishing Co. Pte. Ltd.
5 Toh Tuck Link, Singapore 596224
USA office: 27 Warren Street, Suite 401–402, Hackensack, NJ 07601
UK office: 57 Shelton Street, Covent Garden, London WC2H 9HE

British Library Cataloguing-in-Publication Data
A catalogue record for this book is available from the British Library.

HADRONIC PHYSICS FROM LATTICE QCD

ISBN 981-256-022-X

Printed in Singapore by World Scientific Printers (S) Pte Ltd

INTERNATIONAL REVIEW OF NUCLEAR PHYSICS

Vol. 1: QUARKS AND NUCLEI
edited by W Weise

Vol. 2: COLLECTIVE PHENOMENA IN ATOMIC NUCLEI
Proceedings of the Nordic Winter School on Nuclear Physics
edited by T Engeland, J Rekstad & J S Vaagen

Vol. 3: FEW-BODY PROBLEMS
by E Hadjimichael & W Oelert

Vol. 4: CLUSTER MODELS AND OTHER TOPICS
by Y Akaishi, S A Chin, H Horiuchi & K Ikeda

Vols. 5 & 6: RELATIVISTIC HEAVY ION PHYSICS
edited by L P Csernai & D D Strottman

Vol. 7: ELECTRIC & MAGNETIC GIANT RESONANCES IN NUCLEI
edited by J Speth

Vol. 8: NUCLEAR METHODS AND THE NUCLEAR EQUATION OF STATE
by M Baldo

Vol. 9: HADRONIC PHYSICS FROM LATTICE QCD
edited by Anthony M Green

International Review of Nuclear Physics Vol. 9

Hadronic Physics from Lattice QCD

Preface

The aim of this volume is to provide an introduction to Lattice Quantum Chromodynamics (LQCD) for readers that are non-specialists in the field. In particular, we have in mind advanced undergraduates and graduate students — both theorists and experimentalists — who have a background in nuclear and/or particle physics. This leads to a very limited selection being made from the many topics that are covered by LQCD. The particular topics chosen have, in most cases, analogies in more conventional nuclear and particle physics — good examples being the chapters on the interquark potential and the interactions between hadrons, both of which can be compared with the nucleon-nucleon potential and the interactions between nucleon clusters.

Why should a nuclear/particle physicist ever be interested in LQCD? This is certainly a valid question for someone with little or no knowledge of the field. For why should such a person spend their time reading about some subject that could be dead within a few years? However, in the case of LQCD, it is firmly believed that this is indeed a subject which is here to stay and, also, is expected to be continuously developed for many years to come. The reason for being so confident about this prediction can be summarised by the following phrase — "QCD is thought to be the theory of strong interactions" — words that have appeared in the literature many times over the last 30 years. Unfortunately, to prove this in its full generality has yet to be achieved. In practice, the theory can only be checked either in limiting cases — such as perturbation theory — or by some purely numerical approach — such as LQCD.

The perturbation theory approach exploits the fact that the basic QCD interaction exhibits asymptotic freedom. This states that the strength (α) of the fundamental interaction between quarks and gluons becomes *weaker* as the momentum involved becomes *larger*. Therefore, in principle, high energy processes can be described by evaluating contributions to sufficiently high order in (α). This procedure has been applied to many cases with great success and is one of the main reasons for the above optimistic phrase about QCD being the theory of strong interactions. Unfortunately, in most situations the above perturbation limit is not applicable, since the coupling (α) is of the order unity.

In these non-perturbative regimes, attempts to deal directly with the basic QCD Lagrangian [$\mathcal{L}$(QCD)] lead to singular integrals. There are ways to regulate these using, for example, cut-offs. But the outcome is an infinite number of such integrals rendering the problem as essentially still intractable. This was overcome by Wilson in 1974 through replacing $\mathcal{L}$(QCD) by $\mathcal{L}$(Lattice) — a discretized form on a 4-dimensional space-time lattice. The effect of the lattice is essentially to introduce, in a systematic manner, a high momentum cut-off of π/a, where a is the lattice spacing. This now results in a finite number of well defined integrals — but they need to be evaluated numerically. In $\mathcal{L}$(Lattice) the quark fields only exist on the lattice sites, whereas the gluon field only exists on the links between these sites. Of course, to extract observable physical quantities, the limit $a \to 0$ must eventually be taken. However, it should be emphasized that these final results are *exact* solutions of QCD — a point that was well summarised by Martin Lüscher (hep-ph/0211220) with the words: "In general numerical simulations have the reputation of being an approximate method that mainly serves to obtain qualitative information on the behaviour of complex systems. This is, however, not so in lattice QCD, where the simulations produce results that are exact (on a given lattice) up to statistical errors. The systematic uncertainties related to the non-zero lattice spacing and the finite lattice volume then still need to be investigated, but these effects are theoretically well understood and can usually be brought under control". In most cases the size of these errors and uncertainties are indeed being continuously reduced by the access to computer resources that are ever increasing in speed and memory.

In the above, I say that there is an analogy between nuclear physics and LQCD, since both use the concept of potentials. However, it should be said immediately that this analogy can not be carried too far. For example, the nucleon-nucleon potential $V(NN)$ plays a much more central rôle in the development of nuclear physics than the interquark potential $V(QQ)$ does in QCD. This is clearly seen by opening books on nuclear physics, where often one of the first chapters is a detailed discussion of $V(NN)$ followed by its use in, for example, the Faddeev equations for an exact description of the three nucleon systems He^3 and H^3. Also nuclei, upto $A \approx 16$, can be understood using variational techniques. In all cases the models are expressed *directly* in terms of $V(NN)$ with all its complexities — such as tensor components and non-localities — and the results are essentially *exact.* This does not mean to say that these results agree with experiment, since there can be corrections due to three-body forces, relativistic effects, *etc.* For heavier nuclei with $A > 16$ (including nuclear and neutron matter) $V(NN)$ is first converted into an effective interaction $V_{\rm Eff}(NN)$ by means of Brueckner-like techniques. These treat the interaction between two specific nucleons exactly but with the effects of the other nucleons only being included in an approximate way. Models for these heavier nuclei are then expressed in terms of $V_{\rm Eff}(NN)$. The purpose of this digression is to emphasize that a major goal of nuclear physics is to explain the structure of all nuclei in terms of the basic interaction $V(NN)$. In contrast, this program is *not* generally considered to be a goal in QCD. There $V(QQ)$ is, indeed, a topic of active research, but it is not usually considered as a stepping stone in the understanding of multi-quark systems. Even for three-quark systems — the baryons — the starting point is the basic latticized lagrangian $\mathcal{L}$(Lattice) mentioned earlier. This approach treats the quarks alongwith *explicit* gluons — unlike the use of $V(NN)$ where the mesons only appear implicitly. Having just played down the rôle of $V(QQ)$ for the understanding of multi-quark systems, I should add that attempts have been made to bridge the gap between few- and multi-quark systems by way of interquark potentials. But, at present, these are still in their infancy with most of such models simply mimicking successful nuclear physics approaches.

This volume is made up from five separate chapters each of which has its counterpart in nuclear/particle physics. Also each chapter is essentially self-contained so that there is unavoidably a certain amount of repetition of, in particular, the basics of LQCD. However, this I consider to be a positive feature, since each author emphasizes different aspects that are more relevant for their particular topic.

- Chapter 1 discusses the spectroscopy of mesons and baryons — the 2 and 3 quark analogies to the deuteron and the He^3, H^3 nuclei. Unlike nuclear/particle physics, these few-quark systems are treated directly and not in terms of $V(QQ)$. Also they exhibit a much richer spectrum, since they can be constructed from combinations of 5 different types of quark, the u, d, s, c and b quarks, whereas nuclei in general are constructed from only protons and neutrons. An exception to the latter are a few hypernuclei involving strange baryons. In contrast, there are seen charmed and bottom baryons, such as the $\Omega_c^0 = ssc$ and $\Xi_b^- = dsb$. Furthermore, these are expected to exist not only in S-wave states but also in higher partial waves — again a difference to the few-baryon systems.
- Chapter 2 discusses exotic few-quark states. In Chapter 1, the states are characterised as different combinations of the 5 types of quark with no reference to the gluon field that generates the interaction between the quarks — it being assumed to be in its ground state. In this chapter the states are, in addition, characterised by the excitation of the gluon field. This explicit manifestation of the gluon field is unavoidable in QCD, but has so far not been convincingly seen experimentally. Again this is very different to the nuclear/particle physics situation where the meson fields generating $V(NN)$, for example the π and the ρ plus their many excitations, *e.g.* the $\pi(1300)$ and the $\rho(1450)$, are well documented. In this chapter, since the term "exotic" is taken to mean states not included in the naïve quark model involving simply 2 or 3 quarks, the subject of the molecular states of two mesons is also reviewed.
- Chapter 3 discusses the basic interquark potential $V(QQ)$. This is in analogy with many text books in nuclear physics that start with the nucleon-nucleon potential $V(NN)$. Also the terminology involving spin dependence and radial forms is quite similar. However, the analogy essentially ends at that point, since — unlike the nuclear/particle physics counterpart — the study of multiquark systems generally does not build directly upon this knowledge of $V(QQ)$.
- Chapter 4 discusses the interactions between few-quark systems — the analogies to deuteron-deuteron or He^3-He^3 scattering. Since this involves so many quarks, direct LQCD calculations, with a realistic vacuum containing quark-antiquark pairs, can not yet be

carried out using sufficiently light quarks and so, as the authors themselves say, the field is still "in an exploratory phase". Even so it is in the direction that is necessary for bridging the gap between QCD and nuclear/particle physics.

- Chapter 5 has a different goal compared with the earlier chapters. Here the main aim is to generate lattice data that can help in the construction of models and not necessarily for direct comparison with experiment. In a sense, it is a continuation of Chapter 4 by discussing more generally the quest for bridges between QCD and nuclear/particle physics. This requires lattice calculations with multiquark systems — mainly with 4 quarks. There is no question that models are needed for understanding multiquark systems, where the number of quarks is much larger than four, since LQCD calculations are not feasible for general multiquark systems. In fact, there are on–the–market many multiquark models that are "inspired" by QCD. Unfortunately, few are actually "based" on QCD — the inspirations mainly coming from successful multinucleon models.

Naturally, as with most projects involving authors distributed world-wide, there have been some "communication" problems. Fortunately, these were eventually resolved but — in one case — not without considerable cajoling. Several of the contributors actually expressed their pleasure for the opportunity to review their work for non-experts. But, of course, hopefully the volume will prove most useful for the audience it is primarily intended — the non-experts.

A. M. Green

Contents

Chapter 2 Exotics 103

C. Michael

Chapter 3 Two Quark Potentials 127

G. Bali

Chapter 1

Meson and Baryon Spectroscopy on a Lattice

C. McNeile

Theoretical Physics Division, Dept. of Mathematical Sciences
University of Liverpool, Liverpool L69 7ZL UK
E-mail:mcneile@sune.amtp.liv.ac.uk

I review the results of hadron spectroscopy calculations from lattice Quantum Chromodynamics (QCD) for an intended audience of low energy hadronic physicists. I briefly introduce the ideas of numerical lattice QCD. The various systematic errors, such as the lattice spacing and volume dependence, in lattice QCD calculations are discussed. In addition to the discussion of the properties of ground state hadrons, I also review the small amount of work done on the spectroscopy of excited hadrons and the effect of electromagnetic fields on hadron masses. I also discuss the attempts to understand the physical mechanisms behind hadron mass splittings.

1.1 Introduction

QCD at low energies is hard to solve, perhaps too hard for mere mortals to solve, even when assisted with the latest supercomputers. QCD is the theory that describes the interactions of quarks and gluons. QCD has been well tested in high energy scattering experiments where perturbation theory is valid. However, QCD should also describe nuclear physics and the mass spectrum of hadrons. Hadron masses depend on the coupling (g) like $M \sim e^{-1/g^2}$ hence perturbation theory can not be used to compute the masses of hadrons such as the proton.

The only technique that offers any prospect of computing masses and matrix elements non-perturbatively, from first principles, is lattice QCD. In lattice QCD, QCD is transcribed to a lattice and the resulting equations are solved numerically on a computer. The computation of the hadron spectrum using lattice QCD started in the early 80's [1, 2]. The modern era in lattice QCD calculations of the hadron spectrum started with the results of the GF11 group [3, 4]. The GF11 group were the first to try to quantify the systematic errors in taking the continuum and infinite volume limits.

The goal of a "numerical solution" to QCD is not some kind of weird and misguided reductionist quest. Our inability to solve QCD has many profound consequences. A major goal of particle physics is to look for evidence for physics beyond the standard model of particle physics. One way of doing this is to extract the basic parameters of the standard model and look for relations between them that suggest deeper structure. To test the quark sector of the standard model requires that matrix elements are computed from QCD [5]. The problem of solving QCD is symbolically summarised by the errors on the quark masses. For example, the allowed range on the strange quark mass in the particle data table [6] is 80 to 155 MeV; a range of almost 100%. The value of the top quark mass, quoted in the particle data table, is 174.3 ± 5.1 GeV. As the mass of the quark increases its relative error decreases. The dynamics of QCD becomes simpler as the mass of the quarks gets heavier. Wittig has reviewed the latest results for the light quark masses from lattice QCD [7]. Irrespective of applications of solutions to QCD to searches for physics beyond the standard model, QCD is a fascinating theory in its own right. QCD does allow us to test our meagre tools for extracting non-perturbative physics from a field theory.

In this review I will focus on the results from lattice gauge theory for the masses of the light mesons and baryons. I will not discuss flavour singlet mesons as these have been reviewed by Michael [8, 9] — see Chapter 2. There has been much work on the spectroscopy of hadrons that include heavy quarks [10, 11, 12], however I will not discuss this work. The treatment of heavy quarks (charm and bottom) on the lattice has a different set of problems and opportunities over those for light quarks. Although the spectroscopy of hadrons with heavy quarks in them can naturally be reviewed separately from light quark spectroscopy, the physics of heavy hadrons does depend on the light quarks in the sea. In particular the hyperfine splittings are known to have an important dependence on the sea quarks [12].

Until recently, the computation of the light hadron spectrum used to be just a test of the calculational tools of lattice QCD. The light hadron spectrum was only really good for providing the quark masses and estimates of the systematic errors. However, the experimental program at places such as the Jefferson lab [13, 14, 15] has asked for a new set of quantities from lattice QCD. In particular the computation of the spectrum of the $N^{\star}$'s is now a goal of lattice QCD calculations.

As the aim of the review is to focus more on the results of lattice calculations, I shall mostly treat lattice calculations as a black box that produces physical numbers. However, errors are "the kings" of lattice QCD calculations because the quality and usefulness of a result usually depends on the size of its error bar, hence I will discuss the systematic errors in lattice calculations. Most of the systematic errors in lattice QCD calculations can be understood using standard field theory techniques. I have also included an Appendix on some of the "technical tricks" that are important for lattice QCD insiders, but of limited interest to consumers of lattice results. However, it is useful to know some of the jargon and issues, as they do effect the quality of the final results.

There are a number of text books on lattice QCD. For example the books by Montvay and Munster [16], Rothe [17], Smit [18] and Creutz [19] provide important background information. The large review articles by Gupta [20], Davies [11] and Kronfeld [21] also contain pertinent information. The annual lattice conference is a snap-shot of what is happening in the lattice field every year. The contents of the proceedings of the lattice conference have been put on the hep-lat archive for the past couple of years [22, 23, 24]. The reviews of the baryon spectroscopy from lattice QCD by Bali [25] and Edwards [26] describe a different perspective on the field to mine. There used to be a plenary review specifically on hadron spectroscopy at the lattice conference [27, 28, 29, 30, 31]. The subject of hadron spectroscopy has now been split into a number of smaller topics, such as quark masses.

If the reader wants to play a bit with some lattice QCD code, then the papers by Di Pierro [32, 33, 34], contain some exercises and pointers to source code. The MILC collaboration also make their code publicly available (try putting "MILC collaboration" into a search engine).

1.2 Basic Lattice Gauge Theory

In this section, I briefly describe the main elements of numerical lattice QCD calculations. Quantum Chromodynamics (QCD) is the quantum field theory that describes the interactions of elementary particles called quarks and gluons. The key aspect of a quantum field theory is the creation and destruction of particles. This type of dynamics is crucial to QCD and one of the reasons that it is a hard theory to solve.

In principle, because we know the Lagrangian for QCD, the quantum field theory formalism should allow us to compute any quantity. The best starting point for solving QCD on the computer is the path integral formalism. The problem of computing bound state properties from QCD is reduced to evaluating Eq. 1.1

$$\langle B \rangle = \frac{1}{\mathcal{Z}} \int dU \int d\psi \int d\overline{\psi}\, B e^{-S_F - S_G} \tag{1.1}$$

$$\mathcal{Z} = \int dU \int d\psi \int d\overline{\psi} e^{-S_F - S_G}, \tag{1.2}$$

where S_F and S_G are the actions for the fermion (ψ) and gauge (U) fields respectively and B — defined in terms of the (ψ) and (U) — is an appropriate operator for studying the bound state property of interest. The path integral is defined in Euclidean space for the convergence of the measure.

The fields in Eq. 1.1 fluctuate on all distance scales. The short distance fluctuations need to be regulated. For computations of non-perturbative quantities a lattice is introduced with a lattice spacing that regulates short distance fluctuations. The lattice regulator is useful both for numerical calculations, as well for formal work [35] (theorem proving), because it provides a specific representation of the path integral in Eq. 1.1.

A four dimensional grid of space-time points is introduced. A typical size in lattice QCD calculations is 24^3 48. The introduction of a hyper-cubic lattice breaks Lorentz invariance, however this is restored as the continuum limit is taken. The lattice actions do have a well defined hyper-cubic symmetry group [36, 37]. The continuum QCD Lagrangian is transcribed to the lattice using "clever" finite difference techniques.

In the standard lattice QCD formulation, the decision has been made to keep gauge invariance explicit. The quark fields are put on the sites of the lattice. The gauge fields connect adjacent lattice points. The connection between the gauge fields in lattice QCD and the fields used in perturbative calculations is made via

$$U_\mu(x) = e^{-giA_\mu(x)}. \tag{1.3}$$

The gauge invariant objects are either products of gauge links between quark and anti-quark fields, or products of gauge links that form closed paths. All gauge invariant operators in numerical lattice QCD calculations are built out of such objects. For example, the lattice version of the gauge action is constructed from simple products of links called plaquettes, which are defined as

$$U_P(x;\mu\nu) = U_\mu(x)U_\nu(x+\hat{\mu})U_\mu^\dagger(x+\hat{\nu})U_\nu^\dagger(x). \tag{1.4}$$

The Wilson gauge action

$$S_G = -\beta\left[\sum_p \frac{1}{2N_C}\mathrm{Tr}(U_P + U_P^\dagger) - 1\right] \tag{1.5}$$

is written in terms of plaquettes. Here N_C is the number of colours. The action in Eq. 1.5 can be expanded in the lattice spacing a using Eq. 1.3, to get the continuum gauge action

$$S_G = a^4 \frac{\beta}{4N_C} \sum_x \mathrm{Tr}(F_{\mu\nu}F^{\mu\nu}) + O(a^6). \tag{1.6}$$

The coupling is related to β via

$$\beta = \frac{2N_C}{g^2}. \tag{1.7}$$

The coupling in Eq. 1.7 is known as the bare coupling. More physical definitions of the coupling [38] are typically used in perturbative calculations.

The fermion action is generically written as

$$S_F = \overline{\psi} M \psi, \tag{1.8}$$

where M, the so-called the fermion operator, is a lattice approximation to the Dirac operator.

One approximation to the Dirac operator on the lattice is the Wilson operator. There are many new lattice fermion actions, however the basic ideas can still be seen from the Wilson fermion operator

$$S_f^W = \sum_x [\kappa \sum_\mu \{\overline{\psi}_x(\gamma_\mu - 1)U_\mu(x)\psi_{x+\mu} - \overline{\psi}_{x+\mu}(\gamma_\mu + 1)U_\mu^\dagger(x)\psi_x\} + \overline{\psi}_x\psi_x]. \tag{1.9}$$

The Wilson action can be expanded in the lattice spacing to obtain the continuum Dirac action with lattice spacing corrections to the Dirac Lagrangian. The κ parameter is called the hopping parameter. It is a simple rescaling factor that is related to the quark mass via

$$\kappa = \frac{1}{2(4+m)} \tag{1.10}$$

at the tree level. An expansion in κ is not useful for light quarks, because of problems with convergence. However, for a few specialised applications, it is convenient to expand in terms of κ [39]. The fermion operator in Eq. 1.9 contains a term called the Wilson term that is required to remove fermion doubling [16, 17, 18]. The Wilson term explicitly breaks chiral symmetry so Eq. 1.10 gets renormalised. Currently, there is a lot of research effort in designing lattice QCD actions for fermions with better theoretical properties. I briefly describe some of these developments in the Appendix.

Most lattice QCD calculations obtain hadron masses from the time sliced correlator

$$c(t) = \langle \sum_{\underline{x}} e^{ipx} O(\underline{x}, t) O^\dagger(0,0) \rangle, \tag{1.11}$$

where the average is defined in Eq. 1.1.

Any gauge invariant combination of quark fields and gauge links can be used as interpolating operators ($O(\underline{x}, t)$) in Eq. 1.11. An example for an interpolating operator in Eq. 1.11 for the ρ meson would be

$$O(\underline{x}, t)_i = \overline{\psi_1}(\underline{x}, t)\gamma_i\psi_2(\underline{x}, t). \tag{1.12}$$

The interpolating operator in Eq. 1.12 has the same J^{PC} quantum numbers as the ρ.

The operator in Eq. 1.12 is local, as the quark and anti-quark are at the same location. It has been found better to use operators that build in some kind of "wave function" between the quark and anti-quark. In lattice-QCD-speak we talk about "smearing" the operator. An extended operator such as

$$O(\underline{x}, 0) = \sum_{\underline{r}} f(\underline{r}) \overline{\psi}_1(\underline{x}, 0) \gamma_i \psi_2(\underline{x} + \underline{r}, 0) \tag{1.13}$$

might have a better overlap to the ρ meson than the local operator in Eq. 1.12. The $f(\underline{r})$ function is a wave-function like function. The function f is designed to give a better signal, but the final results should be independent of the choice of f. Typical choices for f might be hydrogenic wave functions. Unfortunately Eq. 1.13 is not gauge invariant, hence it vanishes by Elitzur's theorem [40]. One way to use an operator such as Eq. 1.13 is to fix the gauge. A popular choice for spectroscopy calculations is Coulomb's gauge. On the lattice Coulomb's gauge is implemented by maximising

$$F = \sum_{x} \sum_{i=\hat{x}}^{\hat{z}} [U_i(x) + U_i^\dagger(x)]. \tag{1.14}$$

The gauge fixing conditions that are typically used in lattice QCD spectroscopy calculations are either the Coulomb or Landau gauge. Different gauges are used in other types of lattice calculations. However, it should be added that generally gauge invariant operators are used, so that gauge fixing is not a problem.

Non-local operators can be measured in lattice QCD calculations,

$$O(\underline{x}, 0) = \sum_{\underline{r}} \overline{\psi}_1(\underline{x}, 0) \gamma_i \psi_2(\underline{x} + \underline{r}, 0), \tag{1.15}$$

however these do not have any use in phenomenology [41, 42, 43, 44, 45], because experiments are usually not sensitive to single $\overline{\psi}\psi$ Fock states.

There are also gauge invariant non-local operators that are used in calculations.

$$O(\underline{x}, t) = \overline{\psi}_1(\underline{x}, t) \gamma_i F(\underline{x}, \underline{x} + \underline{r}) \psi_2(\underline{x} + \underline{r}, t), \tag{1.16}$$

where $F(\underline{x}, \underline{x} + \underline{r})$ is a product of gauge links between the quark and anti-quark.

There are many different paths between the quark and anti-quark. It has been found useful to fuzz [46] gauge links by adding bended paths *i.e.*

$$U_{\text{new}} = P_{SU(3)}[cU_{\text{old}} + \sum_{1}^{4} U_{\text{u-bend}}], \tag{1.17}$$

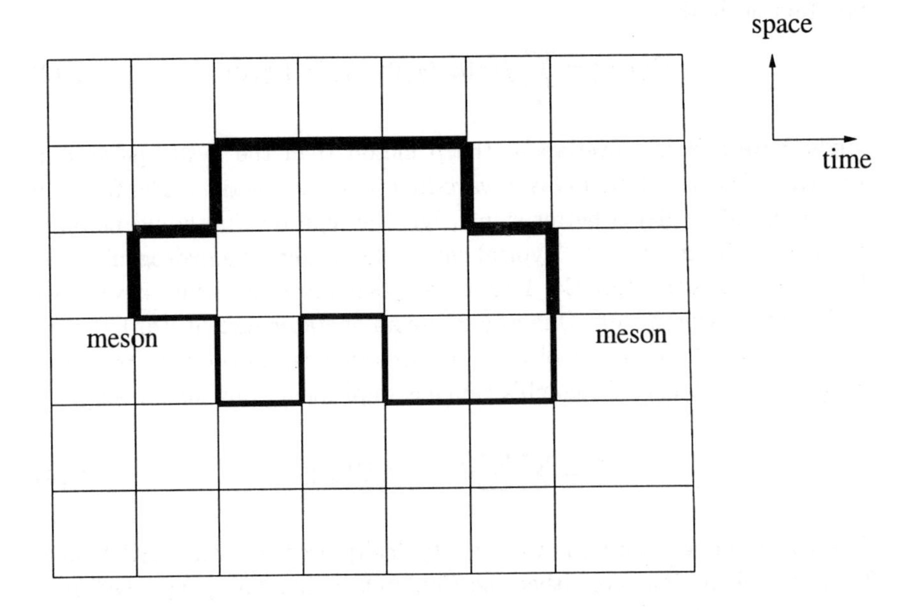

Fig. 1.1 Physical picture of a meson propagator on the lattice. The thick lines are the quark propagators.

where $P_{SU(3)}$ projects onto SU(3). There are other techniques for building up gauge fields between the quark and anti-quark fields, based on computing a scalar propagator [47, 48].

The path integral in Eq. 1.1 is evaluated using algorithms that are generalisations of the Monte Carlo methods used to compute low dimensional integrals [49]. The physical picture for Eq. 1.11 is that a hadron is created at time 0, from where it propagates to the time t, where it is destroyed. This is shown in Fig. 1.1. Equation 1.11 can be thought of as a meson propagator. Duncan *et al.* [50] have used the meson propagator representation to extract couplings in a chiral Lagrangian.

The algorithms, usually based on importance sampling, produce N samples of the gauge fields on the lattice. Each gauge field is a snapshot of the vacuum. Even though the QCD vacuum has a complicated structure, it is often treated in the so-called "quenched" approximation. This assumes the vacuum does not contain quark-antiquark pairs — in contrast to the unquenched vacuum that does contain such pairs. There is a community of people who are trying to describe the QCD vacuum in terms of objects such as a liquid of instantons (for example [51]). The lattice QCD community are starting to create publicly available gauge configurations [52, 53]. This is particularly important due to the high computational cost of unquenched calculations. An archive of gauge configurations has been running at NERSC for many years.

The correlator is a function of the light quark propagators $[M(U(i))_q^{-1}]$ averaged over the samples of the gauge fields and can be written as

$$c(t) \sim \frac{1}{N} \sum_{i}^{N} f[M(U(i))^{-1}, M(U(i))^{-1}]. \tag{1.18}$$

The quark propagator M^{-1} is the inverse of the fermion operator. The quark propagator depends on the gauge configurations from the gauge field dependence of the fermion action (see Eq. 1.9 for example).

The correlator in Eq. 1.18 is essentially computed in two stages. First the samples of the gauge configurations are generated. These are the building blocks of most lattice QCD calculations. The correlators for specific processes are computed by calculating the hadron correlator for each gauge configuration and then averaging over each gauge configuration.

To better explain the idea behind computing quark propagators in lattice QCD calculations, it is helpful to consider the computation of the propagator in perturbation theory. The starting point of perturbative calculations is when the quarks do not interact with gluons. This corresponds to quarks moving in gauge potentials that are gauge transforms from the unit configuration

$$A_\mu(x) = 0 \Rightarrow U_\mu(x) = 1. \tag{1.19}$$

Under these conditions the quark propagator can be computed analytically from the fermion operator using Fourier transforms (see [17] for derivation)

$$M^{unit} = \frac{1}{\sum_{\mu=0}^{4} \gamma_\mu \sin(p_\mu) + M(p)} \tag{1.20}$$

$$M(p) = m + \frac{2}{a} \sum_{\mu} \sin^2(ap_\mu/L), \tag{1.21}$$

where L is the size of the lattice. Although in principle the propagator in Eq. 1.20 could be used as a basis of a perturbative expansion of Eq. 1.18, the physical masses depend on the coupling like $M \sim e^{-1/g^2}$ (see Sec. 1.3.2), so hadron masses can not be computed using this approach. The quark propagator on its own is not gauge invariant. Quark propagators are typically built into gauge invariant hadron operators. However, in a fixed gauge, quark propagators can be computed and studied. This is useful for the attempts to calculate hadron spectroscopy using Dyson-Schwinger Eqs. [54, 55].

In lattice QCD calculations the gauge fields have complicated space-time dependence so the quark propagator is inverted numerically using variants of the conjugate gradient algorithms [56]. Weak coupling perturbation theory on the lattice is important for determining weak matrix elements, quark masses and the strong coupling.

The sum over $\underline{x}$ in the time slice correlator (Eq. 1.11) projects onto a specific momentum at the sink. Traditionally, for computational reasons, the spatial origin had to be fixed either at a point, or with a specific wave function distribution between the quarks. Physically, it would be clearly better to project out onto a specific momentum at the origin. The number of spatial positions at the origin is related to the cost of the calculation. There are new lattice techniques called "all-to-all" that can be used to compute a quark propagator from any point to another without spending prohibitive amounts of computer time [57, 58, 59].

There have been some studies of the point to point correlator [60, 61, 62]

$$\langle O(\underline{x}, t) O^\dagger(0,0) \rangle. \tag{1.22}$$

It was pointed out by Shuryak [63] that the correlator in Eq. 1.22 may be more easily compared to experiment. Also, this correlator could give some information on the density of states. In practice, the point to point correlators were quite noisy.

The physics from the time sliced correlator is extracted using a fit model [16]

$$c(t) = a_0 \exp(-m_0 t) + a_1 \exp(-m_1 t) + \cdots, \tag{1.23}$$

where m_0 (m_1) is the ground (first excited) state mass and the dots represent higher excitations. There are simple corrections to Eq. 1.23 for the finite size of the lattice in the time direction. In practice, as recently emphasised by Lepage *et al.* [64], fitting masses from Eq. 1.23 is nontrivial. The fit region in time has to be selected as well as the number of exponentials. The situation is roughly analogous to the choice of "cuts" in experimental particle physics. The correlators are correlated in time, hence a correlated χ^2 should be minimised. With limited statistics it can be hard to estimate the underlying covariance matrix making the χ^2/dof test nontrivial [65, 66]. As ensemble sizes have increased the problems resulting from poorly estimated covariance matrices have decreased.

It is usually better to measure a matrix of correlators

$$c(t)_{A\ B} = \langle \sum_{\underline{x}} e^{ipx} O(\underline{x}, t)_A O^\dagger(0,0)_B \rangle \tag{1.24}$$

so that variational techniques can be used to extract the masses and amplitudes [48, 67, 68]. The correlator in Eq. 1.24 is then analysed using the fit model

$$c(t)_{AB} = \sum_{N=1,..,N_0} X_{AN} e^{-m_N t} X_{NB}. \tag{1.25}$$

The matrix X is independent of time and in general has no special structure. The masses can be extracted from Eq. 1.25 either by fitting or by reorganising the problem as a generalised eigenvalue problem [69]. Obtaining the matrix structure on the right hand of Eq. 1.25 is a non-trivial test of the multi-exponential fit. One piece of "folk wisdom" is that the largest mass extracted from the fit is some average over the truncated states, and hence is unphysical. This may explain why the excited state masses obtained from [48] in calculations with only two basis states were higher than any physical state.

The main (minor) disadvantage of variational techniques is that they require more computer time, because additional quark propagators have to be computed, if the basis functions are "smearing functions". If different local interpolating operators are used as basis states, as is done for studies of the Roper resonance (see Sec. 1.7.1), then there is no additional cost. The amount of computer time depends linearly on the number of basis states. Also the basis states have to be "significantly" different to gain any benefit. Apart from a few specialised applications [70] the efficacy of a basis state is not obvious until the calculation is done.

Although in principle excited state masses can be extracted from a multiple exponential fit, in practice this is a numerically non-trivial task, because of the noise in the data from the calculation. There is a physical argument [71, 72] that explains the signal to noise ratio. The variance of the correlator in Eq. 1.11 is:

$$\sigma_O^2 = \frac{1}{N}\left[\langle (O(t)O^\dagger(0))^2\rangle - c(t)^2\right]. \tag{1.26}$$

The square of the operator will couple to two (three) pions for a meson (baryon). Hence the noise to signal ratio for mesons is

$$\frac{\sigma_M(t)}{c(t)} \sim e^{(m_M - m_\pi)t} \tag{1.27}$$

and for baryons

$$\frac{\sigma_B(t)}{c(t)} \sim e^{(m_B - 3/2 m_\pi)t}. \tag{1.28}$$

The signal to noise ratio will get worse as the mass of the hadron increases.

In Fig. 1.2 I show some data for a proton correlator. The traditional way to plot a correlator is to use the effective mass plot

$$m_{\text{eff}} = \log\left[\frac{c(t)}{c(t+1)}\right]. \tag{1.29}$$

When only the first term in Eq. 1.23 dominates, then Eq. 1.29 is a constant. There is a simple generalisation of Eq. 1.29 for periodic boundary conditions in time.

In the past the strategy used to be to fit far enough out in time so that only one exponential contributed to Eq. 1.23. One disadvantage of doing this is that the noise increases at larger times, hence the errors on the final physical parameters become larger [64].

Many of the most interesting questions in hadronic physics involve the hadrons that are not the ground state with a given set of quantum numbers, hence novel techniques that can extract masses of excited states are very important.

Recently, there has been a new set of tools developed to extract physical parameters from the correlator $c(t)$ in Eq. 1.23. The new techniques are all based on the spectral representation of the correlator in Eq. 1.11

$$c(t) = \int_0^\infty \rho(s) e^{-st} ds. \tag{1.30}$$

The fit model in Eq. 1.23 corresponds to a spectral density of

$$\rho(s) = a_0 \delta(s - m_0) + a_1 \delta(s - m_1). \tag{1.31}$$

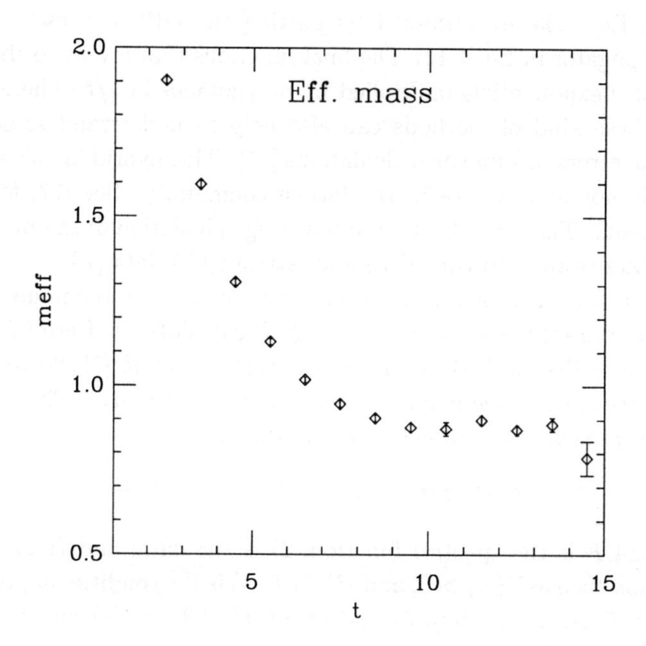

Fig. 1.2 Effective mass plot for proton correlators. Unquenched data from UKQCD [73] with κ=0.1350 , $\beta = 5.2$.

Spectral densities are rarely a sum of poles, except when the particles can not decay such as in the $N_C \to \infty$ limit. Shifman reviews some of the physics in the spectral density [74]. As the energy increases it is more realistic to represent the spectral function as a continuous function. Leinweber investigated a sum rule inspired approach to studying hadron correlators [75] on the lattice. Allton and Capitani [76] generalised the sum rule analysis to mesons. The sum rule inspired spectral density is

$$\rho(s) = \frac{Z}{2M}\delta(s - M) + \theta(s - s_0)\rho^{\mathrm{OPE}}(s). \tag{1.32}$$

The operator product expansion (OPE) is used to obtain the spectral density

$$\rho^{\mathrm{OPE}}(s) = \sum_{n=1}^{n_0} \frac{s^{n-1}}{(n-1)!} C_n O_n(m_q, \langle : \overline{q}q : \rangle), \tag{1.33}$$

where m_q is the quark mass and $: \overline{q}q :$ is the quark condensate. The C_n

factors in Eq. 1.33 are obtained by putting the OPE expression for the quark propagator in Eq. 1.11. The final functions that are fit to the lattice data involve exponentials multiplied by polynomials in $1/t$ (where t is the time). These kind of methods can also help to understand some of the systematic errors in sum rule calculations [75]. This hybrid lattice sum rule analysis is not in wide use in the lattice community. See [77] for recent developments. There has been an interesting calculation of the mass of the charm quark using sum rule ideas and lattice QCD data [78].

There is currently a lot of research into using the maximum entropy method to extract masses from lattice QCD calculations. Here I follow the description of the method developed by Asakawa *et al.* [79, 80, 81]. There are other variants of the maximum entropy method in use. The maximum entropy approach [80] is based on Bayes theorem

$$P(F \mid DH) \propto P(D \mid FH) P(F \mid H). \tag{1.34}$$

In Eq. 1.34 F is the spectral function, D is the data, and H is the prior information such as $f(w) > 0$, and $P(F \mid DH)$ is the conditional probability of getting F given the data D and priors H. The replacement of the fit model in Eq. 1.23 is

$$c_f(t) = \int_0^\infty K(w,t) f(w) dw, \tag{1.35}$$

where the kernel K is defined by

$$K(w,t) = e^{-wt} + e^{-w(T-t)} \tag{1.36}$$

with T the number of time slices in the time direction.

$$P(D \mid FH) = \frac{1}{Z_L} e^{-L} \tag{1.37}$$

$$L = \frac{1}{2} \sum_{i,j}^{N_D} (c(t_i) - c_f(t_i)) D_{ij}^{-1} (c(t_j) - c_f(t_j)), \tag{1.38}$$

where Z_L is a known normalisation constant and N_D is the number of time slices to use. The correlation matrix is defined by

$$D_{ij} = \frac{1}{N(N-1)} \sum_{n=1}^{N} (c(t_i) - c^n(t_i))(c(t_j) - c^n(t_j)) \tag{1.39}$$

$$c(t) = \frac{1}{N}\sum_{n=1}^{N} c^n(t_i), \tag{1.40}$$

where N is the number of configurations and $c^n(t)$ is the correlator for the n-th configuration. Equation 1.37 is the standard expression for the likelihood used in the least square analysis.

The prior probability depends on a function m and a parameter α.

$$P(F \mid Hm\alpha) = \frac{1}{Z_s(\alpha)} e^{\alpha S} \tag{1.41}$$

with $Z_s(\alpha)$ a known constant. The entropy S is defined by:

$$S \to \sum_{i-1}^{N_w} \left[f_i - m_i - f_i \log\left(\frac{f_i}{m_i}\right)\right]. \tag{1.42}$$

The α parameter, that determines the weight between the entropy and L, is obtained by a statistical procedure. The physics is built into the m function. Asakawa *et al.* [79, 80, 81] build some constraints from perturbation theory into the function m. The required function f is obtained from the condition

$$\frac{\delta(\alpha S(f) - L)}{\delta f} |_{f=f_w} = 0. \tag{1.43}$$

In principle the maximum entropy method offers the prospect of computing the masses of excited states in a more stable way than fitting multi-exponentials to data. It will take some time to gain more experience with these types of methods. I review some of the results from these techniques in Secs. 1.6.1 and 1.7.1.

A more pragmatic approach to constraining the fit parameters was proposed by Lepage *et al.* [64]. The basic idea is to constrain the parameters of the fit model into physically motivated ranges. The standard χ^2 is modified to the $\chi^2_{\rm aug}$

$$\chi^2_{\rm aug} = \chi^2 + \chi^2_{\rm prior} \tag{1.44}$$

with the prior $\chi^2_{\rm prior}$ defined by

$$\chi^2_{\rm prior} \equiv \sum_n \frac{(a_n - \hat{a_n})^2}{\sigma_{\hat{a_n}}} + \sum_n \frac{(m_n - \hat{m_n})^2}{\sigma_{\hat{m_n}}}. \tag{1.45}$$

The priors constrain the parameters to $m_n = \hat{m_n} \pm \sigma_{\hat{m_n}}$ and $a_n = \hat{a_n} \pm \sigma_{\hat{a_n}}$. The derivation of Eq. 1.45 comes from Bayes theorem plus the assumption that the distribution of the prior distribution is Gaussian. The idea is to "teach" the fitting algorithm what the reasonable ranges for the parameters are. Lepage *et al.* [64] do test the sensitivity of the final results against the priors.

It is difficult to understand why certain things are done in lattice calculations without an appreciation of the computational costs of lattice calculations. The SESAM collaboration [82] estimated that the number of floating point operations (N_{flop}) needed for n_f =2 full QCD calculations as

$$N_{\text{flop}} \propto \left(\frac{L}{a}\right)^5 \left(\frac{1}{am_\pi}\right)^{2.8} \tag{1.46}$$

for a box sizes of L, lattice spacing a, and N_{sample} is the number of samples of the gauge fields in Eq. 1.11. A flop is a floating point operation such as multiplication. With appropriate normalisation factors Eq. 1.46 shows how "big" a computer is required and how long the programs should run for. Equation 1.46 shows that it is easy to reduce the statistical errors on the calculation as they go like $\sim 1/\sqrt{N_{\text{sample}}}$, but more expensive to change the lattice spacing or quark mass.

In some sense Eq. 1.46 (or some variant of it) is the most important equation in numerical lattice QCD. To half the value of the pion mass used in the calculations requires essentially a computer that is seven times faster. Equation 1.46 is not a hard physical limit. Improved algorithms or techniques may be cheaper. There are disagreements between the various collaborations that do dynamical calculations as to the exact cost of the simulations. In particular, a formulation of quarks on the lattice called improved staggered fermions [83] has a much less pessimistic computational cost. The reason for this is not currently understood.

1.3 Systematic Errors

The big selling point of lattice QCD calculations is that the systematic errors can in principle be controlled. To appreciate lattice QCD calculations and to correctly use the results, the inherent systematic errors must be understood.

For computational reasons, an individual lattice calculation is done at a finite lattice spacing in a box of finite size with quarks that are heavy relative to the physical quarks. The final results from several calculations are then extrapolated to the continuum and infinite volume limit. It is clearly important that the functional forms of the extrapolations are understood. Most of the systematic errors in lattice QCD calculations can be understood in terms of effective field theories [21]. A particularly good discussion of the systematic errors from lattice QCD calculations in light hadron mass calculations is the review by Gottlieb [27].

The MILC collaboration make an amusing observation pertinent to error analysis [84]. In the heavy quark limit, the ratio of the nucleon mass to rho mass is 3/2, requiring one flop to calculate. This is within 30 % of the physical number (1.22). This suggests for a meaningful comparison at the 3σ level, the error bars should be at the 10% level.

1.3.1 *Unquenching*

The high computational cost of the fermion determinant led to the development of quenched QCD, where the determinant is not included in Eq. 1.1 and hence the dynamics of the sea quarks is omitted. Until recently the majority of lattice QCD calculations were done in quenched QCD. I will call a lattice calculation unquenched, when the dynamics of the sea quarks are included.

The integration over the quark fields in Eq. 1.1 can be done exactly using Grassmann integration. The determinant is nonlocal and the cause of most of the computational expense in Eq. 1.46. The determinant describes the dynamics of the sea quarks. Quenched QCD can be thought of as corresponding to using infinitely heavy sea quarks.

Chen [85] has made some interesting observations on the connection between quenched QCD and the large N_C limit of QCD. Figure 1.3 shows some graphs of the pion two point function. In unquenched QCD all three diagrams in Fig. 1.3 contribute to the two point function. In quenched QCD, only diagrams (a) and (c) contribute. In the large N_C (number of colours) limit only graphs of type (a) in Fig. 1.3 contribute. This argument suggests that in the large N_C limit quenched QCD and unquenched QCD should agree, hence for the real world $SU(3)$ case, quenched QCD and unquenched QCD should differ by $O(1/N_C) \sim 30\%$. Chen [85] firms up this heuristic argument by power counting factors of N_C and discusses the effect of chiral logs. This analysis suggests that the quenching error should

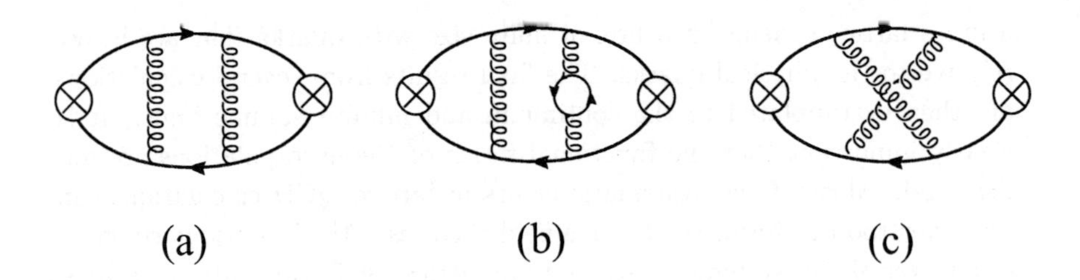

Fig. 1.3 Different contributions to the pion correlator from Chen [85]

be roughly 30%, unless there is some cancellation such that the leading $O(1/N_C)$ corrections cancel.

Quenched QCD is not a consistent theory, because omitting the fermion loops causes problems with unitarity. Bardeen *et al.* [86] have shown that there is a problem with the non-singlet 0^{++} correlator in quenched QCD. The problem can be understood using quenched chiral perturbation theory. The non-singlet 0^{++} propagator contains an intermediate state of $\eta' - \pi$. The removal of fermion loops in quenched QCD has a big effect on the η' propagator. The result is that a ghost state contributes to the scalar correlator, that makes the expression in Eq. 1.23 inappropriate to extract masses from the calculation. Bardeen *et al.* [86] predict that the ghost state will make the a_0 mass increase as the quark mass is reduced below a certain point. This behaviour was observed by Weingarten and Lee [87] for small box sizes ($L \leq 1.6$ fm). The negative scalar correlator was also seen by DeGrand [62] in a study of point to point correlators. Damgaard *et al.* [88] also discuss the scalar correlator in quenched QCD.

One major problem with quenched QCD is that it does not suppress zero eigenvalues of the fermion operator [89]. A quark propagator is the inverse of the fermion operator, so eigenvalues of the fermion operator that are zero, or close to zero, cause problems with the calculation of the quark propagator. In unquenched QCD, gauge configurations that produce zero eigenvalues in the fermion operator are suppressed by the determinant in the measure. Gauge configurations in quenched QCD that produce an eigenvalue spectrum that cause problems for the computation of the propagator are known as "exceptional configurations". Zero modes of the fermion operator can be caused by topology structures in the gauge configuration. The problem with exceptional configurations gets worse as the quark mass is reduced. The new class of actions, described in the Appendix, that have better chiral symmetry properties, do not have problems with exceptional configurations.

Please note that this section should not be taken as an apology for quenched QCD. As computers and algorithms get faster the parameters of unquenched lattice QCD are getting "closer" to their physical values. Hopefully quenched QCD calculations will fade away.

1.3.2 *Lattice spacing errors*

Lattice QCD calculations produce results in units of the lattice spacing. One experimental number must be used to calculate the lattice spacing from

$$a = am^X_{\text{latt}}/m^X_{\text{expt}}. \tag{1.47}$$

As the lattice spacing goes to zero any choice of m^X_{expt} should produce the same lattice spacing – this is known as scaling. Unfortunately, no calculations are in this regime yet. The recent unquenched calculations by the MILC collaboration [90, 91, 92] may be close.

Popular choices to set the scale are the mass of the rho, mass splitting between the S and P wave mesons in charmonium, and a quantity defined from the lattice potential called r_0. The quantity r_0 [93, 94] is defined in terms of the static potential $V(r)$ measured on the lattice as

$$r_0^2 \frac{dV}{dr} \mid_{r_0} = 1.65. \tag{1.48}$$

Many potential models [93] predict $r_0 \sim 0.5$ fm. The value of r_0 can not be measured experimentally, but is "easy" to measure on the lattice. The value of r_0 is a modern generalisation of the string tension — although it may seem a little strange to use r_0 to calculate the lattice spacing, when it is not directly known from experiment. There are problems with all methods to set the lattice spacing. For example, to set the scale from the mass of the rho meson requires a long extrapolation in light quark mass. Also it is not clear how to deal with the decay width of the rho meson in Euclidean space.

The physics results from lattice calculations should be independent of the lattice spacing. A new lattice spacing is obtained by running at a different value of the coupling in Eq. 1.7. In principle, the dependence of quantities on the coupling can be determined from renormalisation group equations,

$$\beta(g_0) = a\frac{d}{da}g_0(a). \tag{1.49}$$

The renormalisation group equations can be solved to give the dependence of the lattice spacing on the coupling [95]

$$a = \frac{1}{\Lambda}\left(g_0^2\gamma_0\right)^{-\gamma_1/(2\gamma_0^2)} e^{-1/(2\gamma_0 g_0^2)}[1 + O(g_0^2)]. \tag{1.50}$$

The $e^{-1/(2\gamma_0 g_0^2)}$ term in Eq. 1.50 prevents any weak coupling expansion converging for masses.

Equation 1.50 is not often used in lattice QCD calculations. The bare coupling g_0 does not produce very convergent series. If quantities, such as Wilson loops, are computed in perturbation theory and from numerical lattice calculations, then the agreement between the two methods is very poor. Typically couplings defined in terms of more "physical" quantities, such as the plaquette, are used in lattice perturbative calculations [38].

Allton [96] tried to use variants of Eq. 1.50 with some of the improved couplings. He also tried to model the effect of the irrelevant operators to be discussed below. An example of Allton's result for the lattice spacing determined from the ρ mass as a function of the inverse coupling is in Fig. 1.4.

There are corrections to Eq. 1.50 from lattice spacing errors. For example the Wilson fermion action in Eq. 1.9 differs from the continuum Dirac action by errors that are $O(a)$. The lattice spacing corrections can be written in terms of operators in a Lagrangian. These operators are known as irrelevant. Ratios of dimensional quantities are extrapolated to the continuum limit using a simple polynomial

$$\frac{am_1}{am_2} = \frac{am_1^{\text{cont}}}{am_2^{\text{cont}}} + xa + O(a^2). \tag{1.51}$$

The improvement program discussed in the Appendix designs fermion actions to reduce the lattice spacing dependence of ratios of dimensional quantities. When computationally feasible, calculations are done with (at least) three lattice spacings and the results are extrapolated to the continuum. This was the strategy of the large scale GF11 [3] and CP-PACS calculations [97].

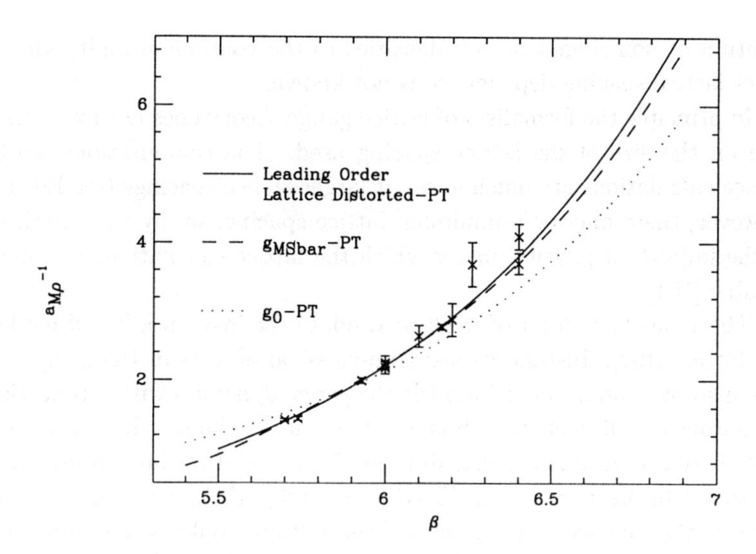

Fig. 1.4 The lattice spacing determined from the mass of the ρ meson as a function of the coupling ($\beta = \frac{6}{g^2}$) [96]

It is very important to know the exact functional form of the lattice spacing dependence of the lattice results for a reliable extrapolation to the continuum. The quantum field theory nature of the field renormalises the correction polynomial. There are potentially $O(a^n g^m)$ errors. This is particularly a problem for states with a mass that is comparable to the inverse lattice spacing. There is a physical argument that the Compton wavelength of a hadron should be greater than the lattice spacing. This is a problem for heavy quarks such as charm and bottom, that is usually solved by the use of effective field theories [10, 11, 21]. The masses of the excited light hadrons are large relative to the inverse lattice spacing, so there may be problems with the continuum extrapolations.

There have been a number of cases where problems with continuum extrapolations have been found. For example Morningstar and Peardon [98] found that the mass of the 0^{++} glueball had a very strange dependence on the lattice spacing. Later [99] they had to modify the gauge action to obtain results that allowed a controlled continuum extrapolation. The ALPHA collaboration [100] discuss the problems of extrapolating the renormalisation constant associated with the operator corresponding to the average momentum of non-singlet parton densities to the continuum limit, when the exact lattice spacing dependence is not known.

In principle the formalism of lattice gauge theory does not put a restriction on the size of the lattice spacing used. The computational costs of lattice calculations are much lower at larger lattice spacings (see Eq. 1.46). However, there may be a minimum lattice spacing, set by the length scale of the important physics, under which the lattice calculations become unreliable [101].

There has been a lot of work to validate the instanton liquid model on the lattice [102]. Instantons are semi-classical objects in the gauge fields. The instanton liquid model models the gauge dynamics with a collection of instantons of different sizes. Some lattice studies claim to have determined that there is a peak in the distribution of the size of the instantons between 0.2 to 0.3 fm in quenched QCD [103, 104, 105]. If the above estimates are correct, then lattice spacings of at least 0.2 fm would be required to correctly include the dynamics of the instanton liquid on the lattice. However, determining the instanton content of a gauge configuration is non-trivial, so estimates of size distributions are controversial. At least one group claims to see evidence in gauge configurations against the instanton liquid model [106, 107].

There have been a few calculations of the hadron spectrum on lattices with lattice spacings as coarse as 0.4 fm [98, 108, 109, 110, 111]. There were no problems reported in these coarse lattice calculations, however only a few quantities were calculated.

Another complication for unquenched calculations is that the lattice spacing depends on the value of the sea quark mass [112] as well as the coupling. This is shown in Fig. 1.5. The dependence of the lattice spacing on the sea quark mass is a complication, because for example the physical box size now depends on the quark mass.

Some groups prefer to tune the input parameters in their calculations so that the lattice spacing [90, 113] is fixed with varying sea quark mass. Other groups prefer to work with a fixed bare coupling [29].

1.3.3 *Quark mass dependence*

The cost of lattice QCD calculations (see Eq. 1.46) forces the calculations to be done at unphysical large quark masses. The results of lattice calculations are extrapolated to physical quark masses using some functional form for

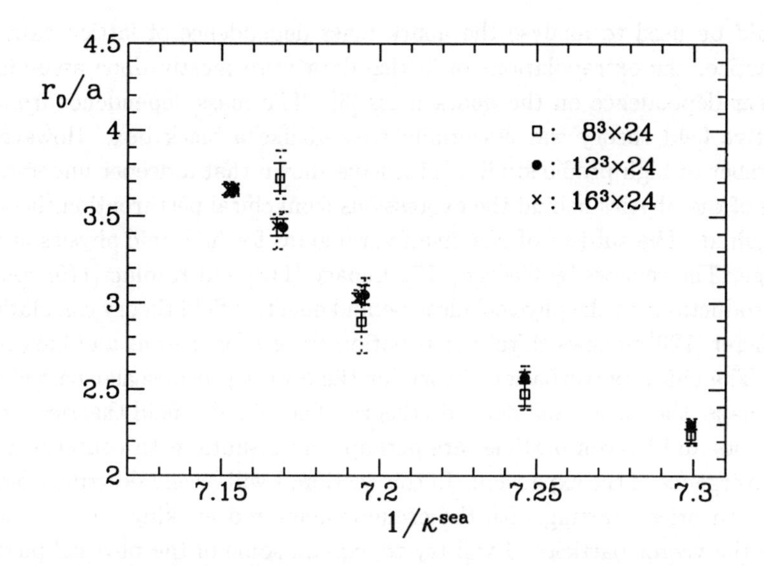

Fig. 1.5 Lattice spacing dependence on the sea quark mass $1/\kappa^{sea}$ from [112].

the quark mass dependence. The dependence on the light quark masses is motivated from effective field theories such as chiral perturbation theory. The modern view is that the lattice QCD calculations do not need to be done at exactly the masses of the physical up and down quarks, but the results can be matched onto an effective theory [114]. In this section I briefly discuss the chiral extrapolation fit models used to extrapolate hadron masses from lattice QCD data to the physical points.

The relationship between effective theories and lattice QCD is symbiotic, because the results from lattice QCD calculations can also test effective field theory methods. The extrapolation of lattice QCD data with light quark masses is currently a "hot" and controversial topic in the lattice QCD community. The controversy is over the different ways to improve the convergence of the effective theory. There was a discussion panel on the different perspectives on using effective field theories to analyse lattice QCD data, at the Lattice 2002 conference [115]. In Sec. 1.5, I show that the masses of the quarks used in lattice calculations are getting lighter and that will help alleviate many problems.

In the past chiral perturbation theory provided functional forms that could be used to analyse the quark mass dependence of lattice data. In practice, the extrapolations of lattice data were mostly done assuming a linear dependence on the quark mass [3]. The mass dependence from effective field theory was essentially treated like a black box. However, a number of high profile studies [116] have shown that a deeper understanding of the physics behind the expressions from chiral perturbation theory is required. The subject of effective Lagrangians for hadronic physics is very large. The reviews by Georgi [117], Lepage [118] and Kaplan [119] contain introductions to the physical ideas behind effective field theory calculations. Scherer [120] reviews chiral perturbation theory for mesons and baryons.

The chiral perturbation theory for the lowest pseudoscalar particles is, perhaps, the most "well defined" theory. The effective field theories for the vectors and baryon particles are perhaps more subtle with concerns about convergence of the expansion. In this section, I will briefly describe effective field theories, starting with the pseudoscalars and working up to nucleons via the vector particles. I will try to explain some of the physical pictures behind the functional dependence from effective field theories, rather than provide an exhaustive list of expressions to fit to.

In Georgi's review [117] of the basic ideas behind effective field theories, he quotes the essential principles behind such a theory as

- A local Lagrangian is used.
- In the Lagrangian density, to describe the interaction there are a finite number of parameters each of dimension $(k - d)$ for a d-dimensional space-time.
- The coefficients of the interaction terms of dimension $(k - d)$ are less than or of the order of $(\frac{1}{M})^k$, where $E < M$ for some mass independent k.

The above principles allow physical processes to be calculated to an accuracy of $(\frac{E}{M})^k$ for a process of energy E. The energy scale M helps to organise the power counting. The accuracy of the results can be improved in a systematic way if more parameters are included. This is one of the main appeals of the formalism. As the energy of the process approaches M, then in the strict effective field theory paradigm, a new effective field theory should be used with new degrees of freedom [117]. For example, Fermi's theory of weak interactions breaks down as the propagation of the W becomes important. For these process the electroweak theory can be used.

The effective field theory idea can be applied to hadronic physics. Indeed historically, the origin of the effective field theory idea was in hadronic physics. A Lagrangian is written down in terms of hadron fields, as hadrons are the more appropriate degrees of freedom at low energies. The Lagrangian is chosen to have the same symmetries as QCD, hence in its domain of validity it should give the same physics results as QCD. This is known as Weinberg's theorem [121].

For "large" momentum scales the quark and gluon degrees of freedom will become evident, so the theory based on the chiral Lagrangian with hadron fields must break down. Unfortunately, it is not clear what the next effective field theory is beyond the hadronic effective Lagrangian, because it is hard to compute anything in low energy QCD.

The lowest order chiral Lagrangian for pseudoscalars is

$$\mathcal{L}_2 = \frac{F_0^2}{4} \mathrm{Tr}(\sum_\mu D_\mu V^\dagger D_\mu V - \chi^\dagger V - \chi V), \tag{1.52}$$

$$V(x) = e^{\phi(x)/F_0}, \tag{1.53}$$

$$\phi(x) = \begin{pmatrix} \pi^0 + \frac{1}{\sqrt{3}}\eta & \sqrt{2}\pi^+ & \sqrt{2}K^+ \\ \sqrt{2}\pi^- & -\pi^0 + \frac{1}{\sqrt{3}}\eta & \sqrt{2}K^0 \\ \sqrt{2}K^- & \sqrt{2}\,\overline{K}^0 & -\frac{2}{\sqrt{3}}\eta \end{pmatrix}, \tag{1.54}$$

where F_0 is the pion decay constant at tree level, which has a value of ≈ 88 MeV. The normalization on the physical pion decay constant (F_π) is $F_\pi \approx 93$ MeV. The expansion parameter [122] for the chiral Lagrangian of pseudoscalar mesons is

$$\lambda_\chi \sim m_\rho \sim 4\pi F_\pi \sim 1\ GeV. \tag{1.55}$$

The power counting theorem of Weinberg [121] guarantees that the tree level Lagrangian in Eq. 1.52 will generate the E^4 Lagrangian and non-analytic functions.

The next order terms in the Lagrangian for the pseudoscalars are

$$\mathcal{L}_4 = \sum_{i=1}^{10} \alpha_i O_i. \tag{1.56}$$

The coefficients α_i are independent of the pion mass. They represent the high momentum behaviour. The normalisation used in lattice calculations is connected to the one in Gasser and Leutwyler [123] by $\alpha_i = 8(4\pi)^2 L_i$. For example, one of the terms in the $O(p^4)$ Lagrangian is

$$O_1 = [\mathrm{Tr}(D_\mu V D^\mu V^\dagger)]^2. \tag{1.57}$$

It can be convenient to have quarks with different masses in the sea than those used in the valence correlators. In lattice QCD jargon, this is known as "partial quenching". These partially quenched theories can provide information about the real unquenched world [124, 125, 126]. The chiral Lagrangian predicts [126, 127] the mass of the pion to depend on the sea and valence quark masses as

$$\begin{aligned} M_{\mathrm{PS}}^2 = C\Big[1 \; &+ \; \frac{1}{n_f}\left(\frac{y_{11}(y_{SS}-y_{11})\log y_{11} - y_{22}(y_{SS}-y_{22})\log y_{22}}{y_{22}-y_{11}}\right) \\ &+ \; y_{12}(2\alpha_8 - \alpha_5) + y_{SS} n_f (2\alpha_6 - \alpha_4)\Big], \end{aligned} \tag{1.58}$$

where $C = y_{12}(4\pi F_0)^2$. The function y_{ij} is related to the quark masses via

$$y_{ij} = \frac{B_0(m_i + m_j)}{(4\pi F_0)^2}, \tag{1.59}$$

where the mass of the sea quark is m_S, the masses of the valence quarks are m_1 and m_2, F_0 is the pion decay constant and B_0 is given, for example, by Eq. 1.62. The leading order term in Eq. 1.58 represents the PCAC relation.

The terms like $\log y_{11}$ are known as "chiral logs" and are caused by one loop diagrams. As the chiral logs are independent of the Gasser-Leutwyler coefficients (α_4, α_5, α_6, α_8), they are generic predictions of the chiral perturbation theory. A major goal of lattice QCD calculations is to detect the presence of these chiral logs. This will give confidence that the lattice QCD calculations are at masses where chiral perturbation theory is applicable (see [7] for a recent review of how close lattice calculations are to this goal).

There have been some attempts to use Eq. 1.58 to determine the Gasser-Leutwyler coefficients from lattice QCD [127, 128, 129, 130]. Table 1.1 contains some results for the Gasser-Leutwyler coefficients from a two flavour lattice QCD calculation at a fixed lattice spacing of around 0.1 fm [131], compared to some non-lattice estimates [132].

It is claimed that resonance exchange is largely responsible for the values of the Gasser-Leutwyler coefficients [132, 133]. For example, in this

GL coeff	continuum	$n_f = 2$ [131]
$\alpha_5(4\pi F_\pi)$	0.5 ± 0.6	$1.22^{+11}_{-0.16}(\text{stat})^{+23}_{-26}$
$\alpha_8(4\pi F_\pi)$	0.76 ± 0.4	$0.79^{+5}_{-7}(\text{stat})^{+21}_{-21}$

Table 1.1 Comparison of the results for the Gasser-Leutwyler coefficients from lattice QCD (with two flavours) [131], to non-lattice estimates [132].

approach the value of α_5 is related to the properties of the scalar mesons

$$\alpha_5^{\text{scalar}} = 8(4\pi)^2 \frac{c_d c_m}{M_S^2}, \tag{1.60}$$

where M_S is the mass of the non-singlet scalar and the parameters c_d and c_m are related to the decay of the scalar into two mesons.

The α_5 and α_8 coefficients are important to pin down the Kaplan-Manohar ambiguity [134] in the masses in the chiral Lagrangian at order $O(p^4)$. The latest results on estimating Gasser-Leutwyler coefficients using lattice QCD are reviewed by Wittig [7].

Most lattice QCD calculations do not calculate at light enough quarks so that the *log* terms in Eq. 1.58 are apparent. In practice, many groups extrapolate the squares of the light pseudoscalar meson mass as either linear or quadratic functions of the light quark masses. For example the CP-PACS collaboration [135] used fit models of the type

$$m_{\text{PS}}^2 = b_S^{\text{PS}} m_S + b_V^{\text{PS}} m_V + c_S^{\text{PS}} m_S^2 + c_V^{\text{PS}} m_V^2 + c_{SV}^{\text{PS}} m_V m_S \tag{1.61}$$

for the mass of the pseudoscalar (m_{PS}) where m_S is the mass of the sea quark and m_V is the mass of a valence quark.

In traditional chiral perturbation theory (see [120] for a review) framework B_0 in Eq. 1.59 is

$$B_0 = -\frac{1}{3F_0^2}\langle \overline{q} q \rangle. \tag{1.62}$$

It is possible that B_0 is quite small [136, 137], say of the order of the pion decay constant. In this scenario the higher order terms in the quark mass become important. There is a formalism called generalised chiral perturbation theory [137] that is general enough to work with a small B_0. The chiral condensate has been computed using quenched lattice QCD [138, 139]. The generalised chiral perturbation theory predictions have been compared to some lattice data by Ecker [140]. The next generation of experiments may be able to provide evidence for the standard picture of a large B_0.

Morozumi, Sanda, and Soni [141] used a linear sigma model to study lattice QCD data. Their motivation was that the quark masses in lattice QCD calculations may be too large for traditional chiral perturbation to be appropriate.

There is also a version of chiral perturbation theory developed especially for quenched QCD, by Morel [142], Sharpe [143], Bernard and Golterman [144]. Quenched QCD is considered as QCD with scalar ghost quarks. The determinant of the ghost quarks cancels the determinant of the quarks. The relevant symmetry for this theory is $U(3 \mid 3) \times U(3 \mid 3)$. This is a graded symmetry, because it mixes fermions and bosons. The chiral Lagrangian is written down in terms of a unitary field that transform under the graded symmetry group.

In this section, I follow the review by Golterman [145]. The problems with quenched QCD can be seen by looking at the Lagrangian for the η' and $\overline{\eta}'$ (the ghost partner of the η').

$$\begin{aligned} \mathcal{L}_{\eta'} &= \frac{1}{2}(\partial_\mu \eta')^2 - \frac{1}{2}(\partial_\mu \overline{\eta}')^2 + \frac{1}{2}m_\pi^2((\eta')^2 - (\overline{\eta}')^2) \\ &+ \frac{1}{2}\mu^2(\eta' - \overline{\eta}')^2 + \frac{1}{2}\alpha^2(\partial_\mu(\eta' - \overline{\eta}'))^2. \end{aligned} \tag{1.63}$$

The propagator for the η' can be derived from Eq. 1.63 as

$$S_{\eta'}(p) = \frac{1}{p^2 + m_\pi^2} - \frac{\mu^2 + \alpha p^2}{(p^2 + m_\pi^2)^2}. \tag{1.64}$$

The double pole in Eq. 1.64 stops the η' decoupling in quenched chiral perturbation theory as μ gets large. This has a dramatic effect on the dependence of the meson masses on the light quark mass. For example the dependence of the square of the pion mass on the light quark mass m_q is

$$m_\pi^2 = A m_q [1 + \delta(\log(B m_q)) + C m_q], \tag{1.65}$$

where δ is defined by

$$\delta = \frac{\mu^2}{48\pi^2 F_0^2} \tag{1.66}$$

and A, B and C are functions of the parameters in the Lagrangian. In standard chiral perturbation theory, the δ term in Eq. 1.65 is replaced by

$m_\pi^2/(32\pi^2 F_0^2)$. As the quark mass is reduced, Eq. 1.65 predicts that the pion mass will diverge. Values of δ from 0.05 to 0.30 have been obtained [7] from lattice data. It has been suggested [7] that the wide range in δ might be caused by finite volume effects in some calculations.

The effective Lagrangians, encountered so far, assume that the hadron masses are in the continuum limit. In practice most lattice calculations use the quark mass dependence at a fixed lattice spacing. The CP-PACS collaboration [97] have compared doing a chiral extrapolation at finite lattice spacing and then extrapolating to the continuum, versus taking the continuum limit, and then doing the chiral extrapolation. In quenched QCD, the CP-PACS collaboration [97] found that the two methods only differed at the 1.5σ level.

It is very expensive to generate unquenched gauge configurations with very different lattice spacings, so it would be very useful to have a formalism that allowed chiral extrapolations at a fixed lattice spacing. Rupak and Shoresh [146] have developed a chiral perturbation theory formalism that includes $O(a)$ lattice spacing errors for the Wilson action. This type of formalism had already been used to study the phase of lattice QCD with Wilson fermions [147]. An additional parameter ρ is introduced as

$$\rho \equiv 2W_0 \, a \, c_{SW}, \tag{1.67}$$

where W_0 is an unknown parameter and c_{SW} is defined in the Appendix. This makes the expansion a double expansion in p^2/Λ_χ^2 and ρ/Λ_χ, *i.e.*

$$\mathcal{L}_2 = \frac{F_0^2}{4}\mathrm{Tr}(\sum_\mu D_\mu V^\dagger D_\mu V) - \frac{F_0^2}{4}\mathrm{Tr}\left[(\chi + \rho)V + V(\chi^\dagger + \rho^\dagger)\right]. \tag{1.68}$$

The Lagrangian in Eq. 1.68 is starting to be used to analyse the results from lattice QCD calculations [148].

For vector mesons it less clear how to write down an appropriate Lagrangian than for the pseudoscalar mesons. There are a variety of different Lagrangians for vector mesons [149, 150, 151, 152], most of which are equivalent. A relativistic effective Lagrangian for the vector mesons [151] is

$$\mathcal{L} = -\frac{1}{4}\mathrm{Tr}(D_\mu V_\nu - D_\nu V_\mu)^2 + \frac{1}{2}M_V^2 \mathrm{Tr}(V_\nu V^\nu), \tag{1.69}$$

where V_ν contains the vector mesons. As discussed earlier in this section, a crucial ingredient of the effective field theory formalism is that a power counting scheme can be set up. The large mass of the vector mesons complicates the power counting, hence other formalisms have been developed. Jenkins *et al.* [152] wrote down a heavy meson effective field theory for vector mesons.

In the heavy meson formalism the velocity v with $v^2 = 1$ is introduced. Only the residual momentum (p) of vector mesons, defined as,

$$k_V = M_V v + p \tag{1.70}$$

enters the effective theory. In the large N_C limit the meson fields live inside N_μ

$$N_\mu = \begin{pmatrix} \rho^0_\mu + \frac{1}{\sqrt{2}}\omega_\mu & \rho^+_\mu & K^{\star +}_\mu \\ \rho^-_\mu & -\frac{1}{\sqrt{2}}\rho^0_\mu + \frac{1}{\sqrt{2}}\omega^0_\mu & K^{\star 0}_\mu \\ K^{\star -} & \overline{K}^{\star 0}_\mu & \phi_\mu \end{pmatrix}. \tag{1.71}$$

The Lagrangian for the heavy vector mesons (in the large N_C and massless limit) is

$$\mathcal{L} = -i\mathrm{Tr}\left[N^\dagger_\mu (v.\mathcal{D}) N^\mu\right] - ig_2 \mathrm{Tr}\left[\{N^\dagger_\mu, N_\nu\} A_\lambda v_\sigma \epsilon^{\mu\nu\lambda\sigma}\right]. \tag{1.72}$$

The connection between the Lagrangian in Eq. 1.69 and the Lagrangian in Eq. 1.72 is discussed by the Bijnens *et al.* [151].

At one loop the correction to the ρ mass [152] is

$$\Delta m_\rho = -\frac{1}{24\pi^2 F^2_\pi}\left[g^2_2\left(\frac{2}{3}m^3_\pi + 2m^3_k + \frac{2}{3}m^3_\eta\right) + g^2_1(m^3_\pi)\right], \tag{1.73}$$

where g_1 and g_2 are parameters (related to meson decay) in the heavy vector Lagrangian and F_π is the pion decay constant. The next order in the expansion of the masses of vector mesons is in the paper by Bijnens *et al.* [151]. The equivalent expression in quenched and partially quenched chiral perturbation theory has been computed by Booth *et al.* [153] and Chow and Ray [154]. The heavy vector formalism suggests that for degenerate unquenched quarks the mass of the vector mass should depend on the quark mass m_q like

$$M^{\mathrm{Vec}} = M^{\mathrm{Vec}}_0 + M^{\mathrm{Vec}}_1 m_q + M^{\mathrm{Vec}}_2 m^{3/2}_q + O(m^2_q). \tag{1.74}$$

In practice, it has been found to be difficult to detect the presence of the $m^{3/2}_q$ term in Eq. 1.74 from recent lattice calculations. The mass of the light vector particle from lattice QCD calculations is usually extrapolated to the physical point using a function that is linear or quadratic in the quark mass. For example the CP-PACS collaboration [135] used the fit model

$$M_{\text{Vec}} = A^{\text{Vec}} + B_S^{\text{Vec}} m_S + B_V^{\text{Vec}} m_V + C_S^{\text{Vec}} m_S^2 + C_V^{\text{Vec}} m_V^2 + C_V^{\text{Vec}} m_V m_S \tag{1.75}$$

to extrapolate the mass of the vector meson (M_{Vec}) in terms of the sea (m_S) and valence (m_V) quark masses in their unquenched calculations. CP-PACS [135] also investigated the inclusion of terms from the one loop calculation of the correction to the rho mass in Eq. 1.73.

There is an added complication for the functional dependence of the mass of the ρ meson on the light quark mass, because in principle the ρ can decay into two pions (see Sec. 1.10). This decay threshold complicates the chiral extrapolation model. The first person to do an analysis of this problem for the ρ meson in lattice QCD was DeGrand [155]. There was further work done by Leinweber amd Cohen [156]. The effect of decay thresholds on hadron masses is also a problem for the quark model [157, 158].

The Adelaide group [159] has studied the issue of the effect of the ρ decay on the chiral extrapolation model of the ρ-meson mass in more detail. The physical motivation behind the Adelaide group's program in the extrapolation of hadron masses in the light quark mass has been reviewed by Thomas [160].

The mass of the ρ is shifted by $\pi - \pi$ and $\pi - \omega$ intermediate states. The effect of the two meson intermediate states can be found by computing the Feynman diagrams in Fig. 1.6 from an effective field Lagrangian. The self energies from the $\pi - \pi$ ($\Sigma^{\rho}_{\pi\pi}$) and the $\pi - \omega$ ($\Sigma^{\rho}_{\pi\omega}$) intermediate states renormalize the mass of the ρ meson, which becomes

$$m_\rho = \sqrt{m_0^2 + \Sigma^{\rho}_{\pi\pi} + \Sigma^{\rho}_{\pi\omega}}. \tag{1.76}$$

To explain the idea, I will consider the self energy correction from $\pi - \omega$ intermediate states in more detail. This correction is

$$\Sigma^{\rho}_{\pi\omega} = -\frac{g^2_{\omega\rho\pi}\mu_\rho}{12\pi^2}\int_0^\infty \frac{dk k^4 u^2_{\omega\pi}(k)}{w_\pi(k)^2}, \tag{1.77}$$

where

$$w_\pi(k)^2 = k^2 + m_\pi^2 \tag{1.78}$$

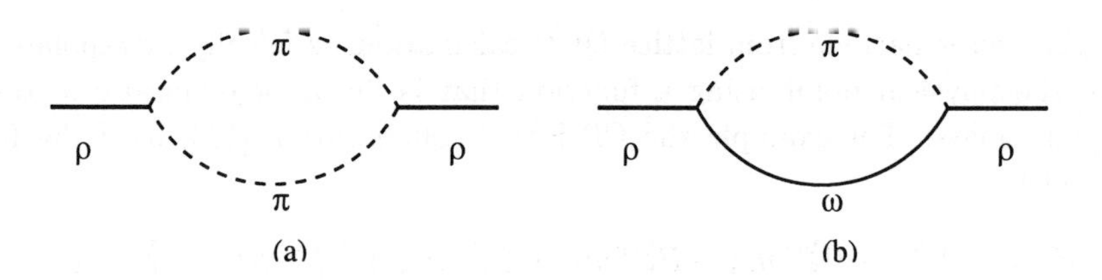

Fig. 1.6 Self energy contributions to the mass of the rho meson.

and μ_ρ is the physical mass of the ρ meson. The integral for the $\Sigma^\rho_{\pi\pi}$ is similar, but the algebra is more complicated.

As the momentum increases, the effective field theory description of the physics in terms of meson fields breaks down. Adelaide prefer to parameterise the breakdown of the effective field theory by introducing a form factor $u_{\omega\pi}(k)$ at the interaction between the two pseudoscalar mesons and the vector meson. A dipole form factor used in Eq. 1.77 has the form

$$u_{\omega\pi}(k) = \left(\frac{\Lambda^2_{\pi\omega} - \mu^2_\pi}{\Lambda^2_{\pi\omega} + k^2} \right)^2 . \tag{1.79}$$

The $\Lambda_{\pi\omega}$ parameter is a (energy) scale associated with the finite extent of the hadrons. This is a fit parameter that the Adelaide group [159] determine from lattice data. They obtain $\Lambda_{\pi\omega} = 630$ MeV.

It is instructive to look at the self energy $\Sigma^\rho_{\pi\omega}$ with a sharp cut off, *i.e.*

$$\Sigma^\rho_{\pi\omega} = -\frac{g_{\omega\rho\pi}\mu_\rho}{12\pi^2} \left[m^3_\pi \arctan\left(\frac{\Lambda}{m_\pi}\right) + \Lambda^3 - \Lambda m^2_\pi \right] . \tag{1.80}$$

There is a strong dependence on Λ in Eq. 1.80. The $\Sigma^\rho_{\pi\omega}$ term contains the m^3_π term of chiral perturbation theory.

In the Adelaide approach [159] the fit model used to extrapolate the mass of the ρ-meson in terms of the quark mass is

$$m_\rho = c_0 + c_1 m^2_\pi + \frac{\Sigma^\rho_{\pi\omega}(\Lambda_{\pi\omega}, m_\pi) + \Sigma^\rho_{\pi\pi}(\Lambda_{\pi\pi}, m_\pi)}{2(c_0 + c_1 m^2_\pi)} . \tag{1.81}$$

This is a replacement for the chiral extrapolation model in Eq. 1.74.

The Adelaide group [159] note that the coefficient of the m^3_π term in Eq. 1.73 is known, hence this is a constraint on the fits from the lattice data. However, the actual fits to the mass of the ρ particle from lattice QCD do not reproduce the known coefficient in Eq. 1.73. When the Adelaide group fit the expression in Eq. 1.81 to the mass of the ρ at the coarse lattice spacings from calculations by UKQCD [112] and CP-PACS, the correct coefficient is obtained.

It is interesting to compare the Adelaide group's approach to the chiral extrapolation of the mass of the ρ-meson to a more "traditional" effective field theory calculation. Equation 1.77 looks similar to a perturbative calculation with a cut off. In an effective field theory calculation, terms with powers of Λ would be absorbed into the counter terms. In a field theory with a cut off, the actual cut off should have no effect on the dynamics in the effective theory. A strong dependence on the cut off would signify the breakdown of the effective theory. The lecture notes by Lepage discuss the connection between a cut off field theory and renormalisation in nuclear physics [118].

In his review on effective field theories Georgi [117] quotes Sidney Coleman as asking "What's wrong with form factors?" Georgi's answer is "Nothing". My translation of Georgi's more detailed answer to Coleman's question is in the next paragraph.

Both the Adelaide group's approach and effective field theory agree for low momentum scales, hence both formalisms can reproduce the non-analytic correction in Eq. 1.73 to the mass of the ρ. However, the two formalisms differ in the treatment of the large momentum behaviour. In the effective field theory paradigm the large momentum behaviour is parameterised by a local Lagrangian with terms that are ordered with a power counting scheme. In the Adelaide group's approach the long distance physics is parameterised (presumably with Coleman's blessing) by a form factor. In an ideal world, the use of an effective field theory is clearly superior to the use of form factors as the accuracy of the approximation is controlled. For the ρ-meson and nucleon it is not obvious how to set up a power counting scheme in the energy (although people are trying). Also an effective field theory based on hadrons will no longer describe the physics at large momentum when the quark and gluon degrees of freedom become important. Hence the use of a form factor may be a more pragmatic way to control the long distance behaviour of hadronic graphs. It may be easier to introduce decay thresholds in a form factor based approach. The hard part of a form factor based approach is in controlling the errors from the approximate nature of the form factor. The Adelaide group [161] do check the sensitivity of their final results by using different form factors. For the light pseudoscalars, the standard effective field theory formalism is clearly superior.

There is a "tradition" of not using "strict" effective field theory techniques for the ρ meson. For example DeGrand [155] used a twice subtracted dispersion relation to regulate the graph in Fig. 1.6.

Nucleons have also been incorporated into the Chiral Lagrangian approach (see [120] for a review) as

$$\Psi = \begin{pmatrix} p \\ n \end{pmatrix} , \tag{1.82}$$

$$\mathcal{L}_{\pi N} = \overline{\Psi} i \not{D} - \hat{m_N} + \frac{\hat{g_A}}{2} \gamma_\mu \gamma_5 u_\mu \Psi, \tag{1.83}$$

where $\hat{m_N}$ and $\hat{g_A}$ are parameters in the chiral limit and

$$u^2 = U, \tag{1.84}$$

$$u_\mu = i u^\dagger \partial_\mu U u^\dagger, \tag{1.85}$$

$$\Gamma_\mu = \frac{1}{2}[u^\dagger, \partial_\mu u], \tag{1.86}$$

$$D_\mu = \partial_\mu + \Gamma_\mu, \tag{1.87}$$

where U is the $SU(2)$ matrix containing the pion fields.

There are a number of complications with baryon chiral perturbation theory based on the relativistic action over meson perturbation theory. As discussed earlier, the power counting is the key principle in an effective theory as it allows an estimate of the errors from the neglected terms. However, the nucleon mass in baryon chiral perturbation theory is the same order as $4\pi F_0$. This complicates the power counting. Also the expansion is linear in the small momentum.

Most modern baryon chiral perturbation theory calculations are done using ideas motivated from heavy quark effective field theory [162, 163]. The four momentum is factored into a velocity dependent part and a small residual momentum part, *i.e.*

$$p_\mu = m v_\mu + k_\mu . \tag{1.88}$$

The baryon field is split into "large" and "small" fields that can be written as

$$\mathcal{N}_v \equiv e^{imv.x} P_{v+} \Psi \tag{1.89}$$

$$\mathcal{H}_v \equiv e^{imv.x} P_{v-} \Psi, \tag{1.90}$$

where the projection operator is defined by

$$P_{v+/-} = \frac{1 + v_\nu \gamma^\nu}{2}. \tag{1.91}$$

The leading order Lagrangian for heavy baryon chiral perturbation theory (HBChPT) is

$$\mathcal{L}_{\pi N} = \overline{\mathcal{N}}_v (iv.D + \frac{g_A}{2} S_v.u) \mathcal{N}, \tag{1.92}$$

where

$$S_v^\mu = \frac{i}{2} \gamma_5 \sigma^{\mu\nu} v_\nu. \tag{1.93}$$

In principle there are $1/M_Q$ corrections to Eq. 1.92. See the review article by Scherer [120] for a detailed comparison of the relativistic and heavy Lagrangians.

The final extrapolation formula for the nucleon mass M_N as a function of the quark mass m_q is

$$M^N = M_0^N + m_q B + m_q^{3/2} C + ... \tag{1.94}$$

The coefficient C is negative and is a prediction of the formalism.

The convergence of baryonic chiral perturbation theory is very poor even in the continuum. On the lattice the pion masses are even larger, hence there are additional concerns about the convergence of the predictions. Using chiral perturbation theory, Borasoy and Meissner [164] computed the nucleon mass (and other quantities) using heavy baryonic chiral perturbation theory, including all quark mass terms up to and including the quadratic order. The result for the nucleon mass for each order of the quark mass is

$$M^N = \hat{m}(1 + 0.34 - 0.35 + 0.24), \tag{1.95}$$

where $\hat{m} = 767$ MeV. Note that this is not a lattice calculation. Although, when all the terms are summed up, the correction is small to the nucleon mass, there are clearly problems with the convergence of the series. The corrections to the masses of the Λ, Σ and Ξ baryons were also sizeable.

Donoghue *et al.* [165] blame the poor convergence of baryonic chiral perturbation theory on the use of dimensional regularisation. They argue that, for distances below the size of the baryon, the effective field theory description breaks down. However, the dimensionally regulated graphs include all length scales. The incorrect physics from the graphs is compensated by the counter terms in the Lagrangian. Unfortunately, this "compensation" makes the expansion poorly convergent. Lepage [118] also gives an example (from [166]) from nuclear physics where using a cut off gives a better representation of the physics than using minimal subtraction with dimensional regularisation.

The integral in Eq. 1.96 occurs at the one loop level for the baryon masses

$$\int \frac{d^4k}{(2\pi)^4} \frac{k_i k_j}{(k_0 - i\epsilon)(k^2 - m^2 + i\epsilon)} = -i\delta_{ij} \frac{I(m)}{24\pi}. \tag{1.96}$$

In dimensional regularisation [167] the integral gives

$$I_{\mathrm{dim-reg}}(m) = m^3. \tag{1.97}$$

The graph in Eq. 1.96 contributes the lowest non-analytic term in the octet masses ($M_N \propto m_\pi^3$). Donoghue *et al.* [165] point out that it is a bit suspicious that the result in Eq. 1.97 is finite when the integral in Eq. 1.96 is cubicly divergent.

Consider now Eq. 1.96 regulated with a dipole regulator, *i.e.*

$$\int \frac{d^4k}{(2\pi)^4} \frac{k_i k_j}{(k_0 - i\epsilon)(k^2 - m^2 + i\epsilon)} \left(\frac{\Lambda^2}{\Lambda^2 - k^2} \right) = -i\delta_{ij} \frac{I_\Lambda(m)}{24\pi}, \tag{1.98}$$

where

$$I_\Lambda(m) = \frac{1}{2}\Lambda^4 \frac{2m + \Lambda^2}{m + \Lambda}. \tag{1.99}$$

In the limit $m < \Lambda$ limit

$$I_\Lambda(m) \longrightarrow \frac{1}{2}\Lambda^3 - \frac{1}{2}\Lambda m^2 + m^3 + \ldots \tag{1.100}$$

Hence for small m the result from dimensional regularisation is reproduced.

Up to this point the treatment of the integrals looks very similar to the approach originally advocated by the Adelaide group [161, 168]. However,

Donoghue *et al.* [165] treat Λ as a cut off. Strong dependence on Λ is removed by renormalising the mass as

$$M_0^r = M_0 - d\Lambda^3, \tag{1.101}$$

where d is a function of the other renormalised parameters in the Lagrangian (such as the pion decay constant). In the original work by the Adelaide group [161, 168] the parameter Λ was a physical number that could be extracted from the lattice data. In the formalism of Donoghue *et al.* [165], the physical results should not depend on Λ, although a weak dependence on Λ may remain because the calculations are only done to one loop. The results for the mass of the nucleon as a function of the order of the expansion are

$$M^N = 1.143 - 0.237 + 0.034 = 0.940 \text{ GeV} \tag{1.102}$$

with the cut off $\Lambda = 400$ MeV. See [169] for a brief critique of the formalism of Donoghue *et al.* [165].

The Adelaide group [161, 168] consider the one loop pion self energy to the nucleon and delta propagators. The method is essentially the same as the one applied to the chiral extrapolation of the ρ mass. Their fit model for the nucleon mass is

$$M_N = \alpha_N + \beta_N m_\pi^2 + \sigma_{NN}(m_\pi, \Lambda) + \sigma_{N\Delta}(m_\pi, \Lambda), \tag{1.103}$$

where the σ's come from one loop graphs. Equation 1.103 is a fit model with three free parameters: α_N, β_N and Λ.

The Adelaide group have applied the formalism of Donoghue *et al.* [165] to the analysis of lattice QCD data from CP-PACS and have compared it with their previous formalism [170, 171].

Lewis and collaborators have studied the lattice regularisation of chiral perturbation theory [172, 173]. This type of calculation can not be used to quantify the lattice spacing dependence of lattice results [173], as the lattice spacing dependence of the two theories could be very different. However, it is interesting to explore different regularisation schemes. There is renewed interest in the relativistic baryon Lagrangian, because a method [174] called "infrared regularisation" allows a power counting scheme to be introduced (see [120] for a review).

All of the above analyses of effective Lagrangians relied on perturbation theory to study the theory. Hoch and Horgan [175] used a numerical lattice calculation to study the $SU(2) \times SU(2)$ non-linear model for pions and the nucleon. There the action

$$\begin{aligned} S(V) &= -\frac{1}{4}F_0^2 \sum_x \mathrm{Tr}\left[\sum_{j=1}^{4} V(x)V^\dagger(x+\mu_j) + V(x+\mu_j)V^\dagger(x)\right] \\ &\quad -\frac{1}{4}F_0^2 m_0^2 \sum_x \mathrm{Tr}\left[U(x) + U^\dagger(x)\right] \end{aligned} \tag{1.104}$$

is expressed in terms of the unitary matrix V

$$V = \frac{1}{F_0}(\sigma.1 + i\pi_a \tau_a \gamma_5) \text{ and } F_0^2 = \sigma^2 + \pi^2. \tag{1.105}$$

The numerical calculation was done with a small volume 8^4, and no finite size study was done. The comparison of the lattice results with the perturbative results is complicated by the effect of the unknown parameters in the next order Lagrangian (Eq. 1.56 for example). Hoch and Horgan [175] found that the lattice calculation disagreed with the predictions of one loop perturbation theory for *log* divergent quantities.

A more conservative (some might say cowardly) approach to chiral extrapolations is to only interpolate the appropriate hadron masses to the mass of the strange quark, in an attempt to try to minimise the dependence of any results on an uncontrolled extrapolation to the light quark masses. One formalism [176] for doing this is called the "method of planes". Similar methods have been used by other groups (see for example [177, 178]). Obviously, this type of technique is not useful to get the nucleon mass. In unquenched QCD, the sea quark masses should be extrapolated to their physical values, so there is no way to avoid a chiral extrapolation even for heavy hadrons.

It is traditional to plot the hadron masses before any chiral extrapolations have been done, so as not to contaminate the raw lattice data from the computer with any theory. In an "Edinburgh plot" the ratio of the nucleon to ρ mass is plotted against the ratio of the π to ρ mass [179]. If there were no systematic errors, such as lattice spacing dependence, then the data should fall on a universal curve. It is also common to use $1/r_0$ [93] (see section 1.3.2) as a replacement for the mass of the ρ. There is also an APE plot that plots the ratio of the nucleon to ρ mass against the square of the π to ρ mass ratio [180]. This parameterisation is meant to have a smoother mass dependence than the Edinburgh plot.

1.3.4 *Finite size effects*

The physical size of the lattice represents an obvious and important systematic error. One simple way to estimate the size of a hadron is to consider its charge radius. For example, the proton's charge radius is quoted as 0.870(8) fm in the particle data table [6]. The sizes of all recent unquenched lattice QCD calculations are all above 1.6 fm (see Sec. 1.5). The fact that all the physical lattice sizes were much bigger than the charge radius does not rule out finite size effects. In this section I will discuss some of the mechanisms thought to be behind finite size effects in lattice data.

A nice physical explanation of the origin of finite size effects for hadron masses has been presented by Fukugita *et al.* [181]. Consider a hadron in a box with length L and periodic boundary conditions. Most lattice QCD calculations use periodic boundary conditions in space. The path integral formalism requires that the boundary conditions in time are anti-periodic [182], although many groups use periodic boundary condition in time as well as space. The finite size of the box will mean that a hadron will interact with periodic images a distance L away. The origin of finite size effects is closely related to nuclear forces. The MILC collaboration have described a qualitative model for finite size effects based on nuclear density [84]. The nucleon is considered as part of a nucleus comprising of the periodic images of the nucleon.

The self energy of the hadron (δE) will be

$$\delta E = \sum_{\underline{n}} V(\underline{n}L), \tag{1.106}$$

where $V(x)$ is the potential between two hadrons a distance x apart. Various approximations to $V(x)$ give different functional forms for the dependence of the hadron mass on the volume.

In the large L limit the potential is approximated by one particle exchange ($e^{-m_\pi r}/r$). The interaction energy goes like $V(0) + 6V(L)$, so the dependence of the hadron mass will be $\sim e^{-m_\pi L}$. This argument can be made rigorous [183, 184].

It is useful to consider Eq. 1.106 in momentum space using the Poisson resummation formulae

$$\delta E = \frac{1}{L^3} \sum_{\underline{n}} \hat{V}\left(\underline{n}\frac{2\pi}{L}\right). \tag{1.107}$$

A more general expression for the potential V between two hadrons can be derived if the spatial size of the hadron is modelled with a form factor $F(k)$ as

$$\hat{V}(k) = \frac{F(k)^2}{k^2 + m^2}. \tag{1.108}$$

Momentum is quantised on the lattice in quanta of $\frac{2\pi a}{L}$. In physical units the value of a quantum of momentum is around 1 GeV, hence the $\underline{k} = \underline{0}$ term in Eq. 1.107 should dominate the sum. Therefore this model predicts that the masses should depend on the box size like

$$M = M_\infty + cL^{-3}. \tag{1.109}$$

This model is physically plausible but not a rigorous consequence of QCD. Fukugita *et al.* [181] noted that their data could also be fit to the functional form

$$m^2 = m_\infty^2 + cL^{-3} \tag{1.110}$$

rather then Eq. 1.109. Unfortunately only the theory for the regime of point particles interacting at large distances is really rigorous [183, 184].

Chiral perturbation theory can be used to compute finite volume corrections. For example the ALPHA/UKQCD collaboration have used [185] a chiral perturbation theory calculation by Gasser and Leutwyler [186] to estimate the dependence of the pion mass $m_{PS}(L)$ on the box size L to be

$$\frac{m_{PS}(L)}{m_{PS}(\infty)} = 1 + \frac{1}{N_f} \frac{m_{PS}^2}{F_{PS}^2} g(m_{PS}L), \tag{1.111}$$

where

$$g(z) = \frac{1}{8\pi^2 z^2} \int_0^\infty \frac{dx}{x^2} e^{-z^2 x - 1/(4x)}. \tag{1.112}$$

Garden *et al.* [185] used Eqs. 1.111 and 1.112 to show that the error in m_{PS} was 0.1 % for $m_{PS}L > 4.3$. This is an example of the rule of thumb that finite size effects in hadron masses become a concern for $m_{PS}L < 4$. Colangelo *et al.* [187] are extending Eq. 1.111 to the next order. Ali Khan *et al.* [188] have started to use chiral perturbation theory to study the finite size effects on the nucleon.

It would be useful if some insight could be gained from finite size effects in quenched QCD that could be applied to unquenched QCD where the volumes are smaller. Unfortunately, there are theoretical arguments [189] that show that finite size effects in the unquenched QCD will be larger than in the quenched QCD. A propagator for a meson can formally be written in terms of gauge invariant paths. Conceptually this can be understood using the hopping parameter expansion. The quark propagators are expanded in terms of the κ parameter. The hopping parameter expansion [190] was used in early numerical lattice QCD calculations, but was found not to be very convergent for light quarks. The correlator can be written as

$$c(t) = \sum_{C} \kappa_{\text{val}}^{L(C)} \langle W(C) \rangle + \sum_{C'} \kappa_{\text{val}}^{L(C')} \sigma_{\text{val}} \langle P(C') \rangle, \tag{1.113}$$

where $W(C)$ are closed Wilson loops inside the lattice of length $L(C)$ and $P(C')$ are Polyakov lines that wrap around the lattice in the space direction. The σ_{val} parameter is 1 for periodic boundary conditions and -1 for anti-periodic boundary conditions for Polyakov lines that wrap around the lattice an odd number of times.

In quenched QCD $\langle P(C') \rangle = 0$ because of a Z_3 symmetry of the pure gauge action, so only the first term in Eq. 1.113 contributes to the correlator $c(t)$. The centre of SU(3), the elements that commute with all the elements of SU(3), is the $Z(3)$ group [17]. The Wilson gauge action is invariant under the gauge links being multiplied by a member of the centre of the group on each time plane. In unquenched QCD, the Z_3 symmetry is broken by the quark action, so both Wilson loop and Polyakov lines contribute to the correlators. There is no clear connection between the arguments based on Eq. 1.113 and the nuclear force and chiral perturbation formalism for finite size effects.

In their large scale quenched QCD spectroscopy calculations the CP-PACS collaboration kept the box to be 3 fm [97], so they did not apply any corrections for the finite size of the lattice. Using the finite size estimate from previous calculations CP-PACS [97] estimated that the finite size effects were of the order 0.6 %. The GF11 group [4] did a finite size study at a coarse lattice spacing ($a = 0.17$ fm). The results for ratios of hadron masses at finer lattice spacings were then extrapolated to the infinite volume limit using the estimated mass shift at the coarse lattice spacing. Gottlieb [27] gives a detailed critique of the method used by the GF11 group to extrapolate their mass ratios to the infinite volume.

If a formalism could be developed that would predict the dependence of hadron masses on the box size, then this would help make the calculations cheaper (see Eq. 1.46). The additional savings in computer time could then be spent on reducing the size of the light quark masses used in the calculations [191]. Gottlieb [27] suggested that the only way to control finite size effects is to keep increasing the box size until the masses no longer change.

The finite box size is not always a bad thing. The size of the box has been used to advantage in lattice QCD calculations (see [192] for a review). A definition of the coupling is chosen that is proportional to $1/L$ (where L is the length of one side of the lattice). A recursive scheme is setup that studies the change in coupling as the length of the lattice side L is halved. The Femtouniverse, introduced by Bjorken [193], is a useful regime to study QCD in. There is a chiral perturbative expansion based on the limit $m_\pi \ll \frac{1}{L}$ (see [88] for a modern application). See Van Baal [194] for a discussion of the usefulness of the finite volume on QCD.

It seems possible to run calculations in a big enough box to do realistic calculations. So the prospects are good for a first principles lattice calculation of hadron masses without resort to approximations in simulations of the kind required in condensed matter systems in simulations of macroscopic size systems.

1.4 An Analysis of Some Lattice Data

To consolidate the previous material, I will work over a simple analysis of some lattice QCD data. I think it is helpful in understanding the steps in lattice QCD calculation, if the ideal case where QCD could be solved analytically is considered first. All the masses of the hadrons would then be known as a function of the parameters of QCD, namely, quark masses (m_i) and coupling g, *i.e.*

$$M_H = f_H(m_u, m_d, m_s, m_c, m_b, g). \tag{1.114}$$

As the masses and coupling are not determined by QCD, the equation for all the hadrons H would have to be solved to get the parameters. The solution would be checked for consistency that a single set of parameters could reproduce the entire hadron spectrum. For a calculation of a form factor, the master function f_H would also depend on the momentum.

The formalism of lattice QCD is, in some sense, a discrete approximation

κ	am_π	am_ρ
0.13460	0.2803^{+15}_{-10}	0.3887^{+32}_{-28}
0.13510	0.2149^{+19}_{-14}	0.3531^{+55}_{-51}
0.13530	0.1836^{+23}_{-18}	0.3414^{+72}_{-82}

Table 1.2 Lattice QCD data for the masses of the π and ρ from lattice QCD calculations with UKQCD [195].

to the function f_H. The result from calculation would be a table of numbers,

$$M_H = f_H^{\text{latt}}(m_u, m_d, m_s, m_c, m_b, g, L, a), \tag{1.115}$$

i.e. an individual lattice calculation would also depend on the lattice spacing and lattice volume. Lattice calculations have to be done at a number of different lattice spacings and volumes to extrapolate the dependence of f_H^{latt} on L and a.

By doing calculations with a number of different parameters the results can be combined to produce physical results in much the same way that could be done if the exact solution was known. In particular the lattice spacing and lattice volume have to be extrapolated away to get access to the function f_H.

To understand the procedure in more detail, I will work through a naive analysis of some lattice QCD data from the UKQCD collaboration [195]. Table 1.2 and Fig. 1.7 contain the results for the mass of the π and ρ particles in lattice units from a quenched QCD calculation [195]. The lattice volume was 24^3 48, $\beta = 6.2$, and the ensemble size was 216. The clover action using the ALPHA coefficients was used. To use the data in Table 1.2, the secret language of the lattice QCD cabal must be converted to the working jargon of the continuum physicist.

The ensemble size of 216 means that 216 snapshots of the vacuum (the value of N in Eq. 1.18) were used to compute estimates of the ρ and π correlators. A supercomputer was used to compute the correlators for each gauge configuration from quark propagators (see Eq. 1.18). The masses were calculated by fitting the correlator to a fit model of the form in Eq. 1.23, using a χ^2 minimiser such as MINUIT [196]. The error bars in Table 1.2 come from a statistical procedure called the "bootstrap" method [197].

A table of numbers of hadron masses (or even a graph) is not too useful.

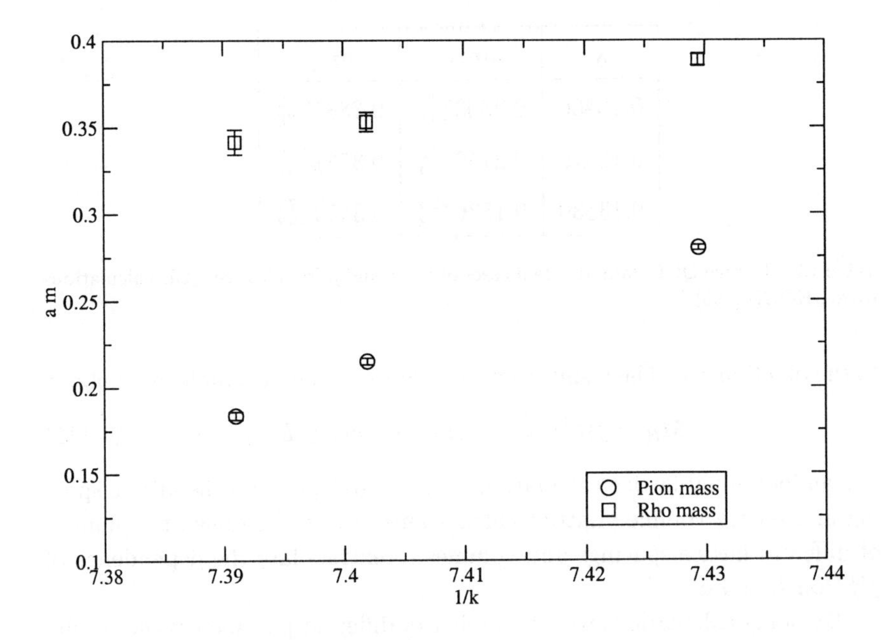

Fig. 1.7 ρ and π masses in lattice units as a function of inverse κ value. The data set had $\beta = 6.2$, and a volume of $24^3 48$. The fermion operator was the non-perturbatively improved clover operator.

A better way to encapsulate the hadron masses as a function of quark mass is to use them to tune effective Lagrangians, as discussed in Sec. 1.3.3. To plot the data in Table 1.2 in a more physical form, I convert from the κ value to the quark mass by using the formula

$$m_q = \frac{1}{2}\left(\frac{1}{\kappa} - \frac{1}{\kappa^{\text{crit}}}\right). \tag{1.116}$$

There are additional $O(a)$ corrections to Eq. 1.116 [192]. The κ^{crit} parameter is required because clover fermions break chiral symmetry. The value of κ^{crit} is the chosen to give a zero pion mass. Equation 1.116 is the basis of the computation of the masses of quarks from lattice QCD. However, perturbative factors are required to convert the quark mass to a standard scheme and scale. This perturbative "matching" can be involved, so the value of the quark mass is rarely used to indicate how light a lattice calculation is.

The simplest thing to do is to use a fit model such as

$$m_{\mathrm{PS}}^2 = S_{\mathrm{PS}} m_q. \tag{1.117}$$

There are classes of fermion actions (see the Appendix), such as staggered or Ginsparg-Wilson actions, that do not have an additive renormalisation.

A simple fit to the data in Table 1.2 with the fit model in Eqs. 1.116 and 1.117 gives $\kappa^{\mathrm{crit}} = 0.135828(6)$, to be compared to $\kappa^{\mathrm{crit}} = 0.135818^{+18}_{-17}$ from the explicit analysis from UKQCD that included a $O(a)$ correction term.

To simplify the analysis, I will assume that the physical mass of the light quark is zero. I fit the data for the mass of the ρ in Table 1.2 to the linear model

$$m_V = A_V + S_V m_q. \tag{1.118}$$

The result for the A_V parameter is 0.304(2). If the mass of the light quark is assumed to be zero, then $am_\rho = 0.304$, hence the inverse lattice spacing is 2530 MeV (using m_ρ = 770 MeV). This can be compared against the a^{-1} = 2963 MeV from r_0 [198].

As the masses of quarks get lighter more sophisticated fit models based on the ideas in Sec. 1.3.3 can be used. Although the basic ideas behind the analysis outlined in this section are correct, there are many improvements that can be made, particularly if the ρ and π correlators for each configuration are available.

1.5 Parameter Values of Lattice QCD Calculations

The results from unquenched lattice QCD calculations, with the lattice spacing and finite size effects accounted for, are the results from QCD at the physical parameters of the calculation. Hence a key issue in unquenched calculations is how close the parameters are to the physical parameters. For example, as discussed in Sec. 1.3.3, ideally the masses of the quarks must be light enough to match the lattice results to chiral perturbation theory. The parameters used in a lattice calculation are usually dictated by the amount of computer time available, or what gauge configurations are publicly available.

In this section, I will describe the current state of the art in the parameters used in lattice QCD calculations of the hadron spectrum. It is

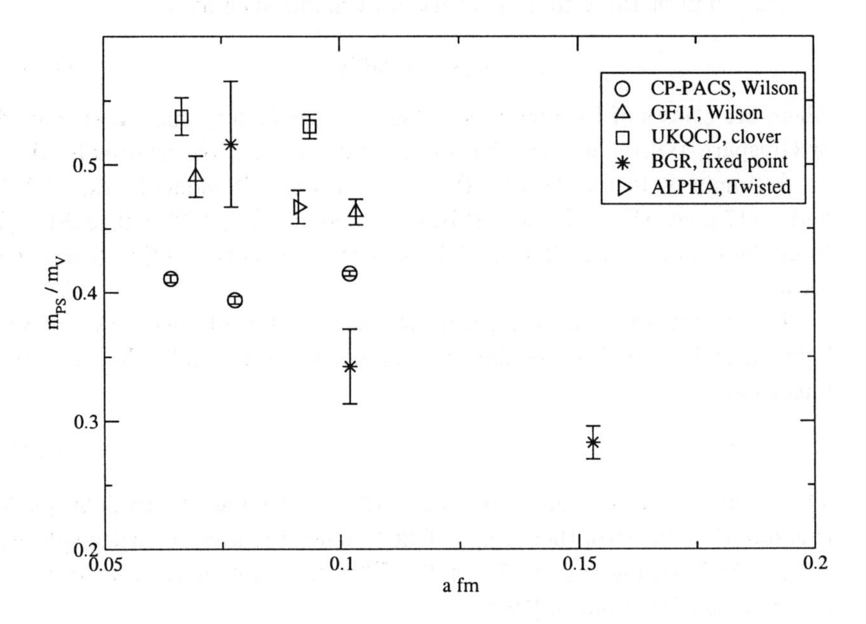

Fig. 1.8 Smallest ratio of pseudoscalar mass to vector mass as a function of lattice spacing for a number of quenched QCD calculations. The labels have the collaboration name followed by the type of fermion operator.

not entirely obvious which parameters to use to characterise a lattice calculation. The obvious choice of using quark masses is complicated by the need for renormalisation and running. To show how light the quarks are in a calculation, I plot the ratio of the pseudoscalar mass to the vector mass as a function of lattice spacing. The danger of this type of plot is that it says nothing about finite size effects. I usually just show the ratio for the lightest quark mass, as this is computationally the most expensive point. The error bars on the ratio gives some indication on the statistical sample size. I have always used a lattice spacing defined by $r_0 = 0.49$ fm (see Sec. 1.3.2).

In Fig. 1.8, I plot the smallest pseudoscalar mass to vector mass ratio as a function of lattice spacing for some recent large quenched calculations from the following collaborations: GF11 [3, 4], CP-PACS [199], UKQCD [195], BGR [200] and ALPHA [201]. The improvements in parameters between the GF11 calculation [3, 4], at the start of 1990's, and the CP-PACS calculation [199], at the end of 1990's, can clearly be seen in the figure.

The two main benchmark calculations by CP-PACS [199] and GF11 [3, 4] used both the Wilson fermion and gauge actions. As I discuss in the Appendix, lattice QCD actions with better properties have been developed. For example, the point in Fig. 1.8 from UKQCD used the clover fermion action. The clover action is designed to have smaller lattice spacing errors compared to the Wilson action. Unfortunately, the calculations by UKQCD [195] and QCDSF [202] that used the non-perturbative improved clover action reported problems with exceptional configurations, hence for those lattice spacings the light quark masses can not be reduced further at these lattice spacings. There is a formalism called twisted mass QCD [201], which is a natural extension of the clover action, that can be used to explore the light quark mass regime [203].

To reach lighter quark masses, it seems likely, that new fermion actions such as the overlap-Dirac, Domain wall, fixed point [204] and twisted mass QCD [201] will be required to study the light quark mass region of quenched QCD, and hence improve on the results of the CP-PACS calculation [199].

The first results from these types of calculations are shown in the Fig. 1.8. Unfortunately, the fermion operators that obey the Ginsparg-Wilson relation are computationally more expensive than Wilson like actions. At the moment the error bars are too large to be competitive with those from actions that use Wilson type fermions. However, the control of the systematic errors in the calculations that use Ginsparg-Wilson operators is rapidly improving [200].

Table 1.3 shows the parameters of some recent large scale unquenched calculations. I have also included the mass of the lightest pion in the calcu-

Collaboration	n_f	a fm	L fm	$\frac{M_{PS}}{M_V}$	m_{PS} MeV
MILC [91]	2+1	0.09	2.5	0.4	340
CP-PACS [205]	2	0.11	2.5	0.6	900
UKQCD [73]	2	0.1	1.6	0.58	600
SESAM [206, 207]	2	0.074	1.8	0.57	530

Table 1.3 Typical parameters in recent unquenched lattice QCD calculations.

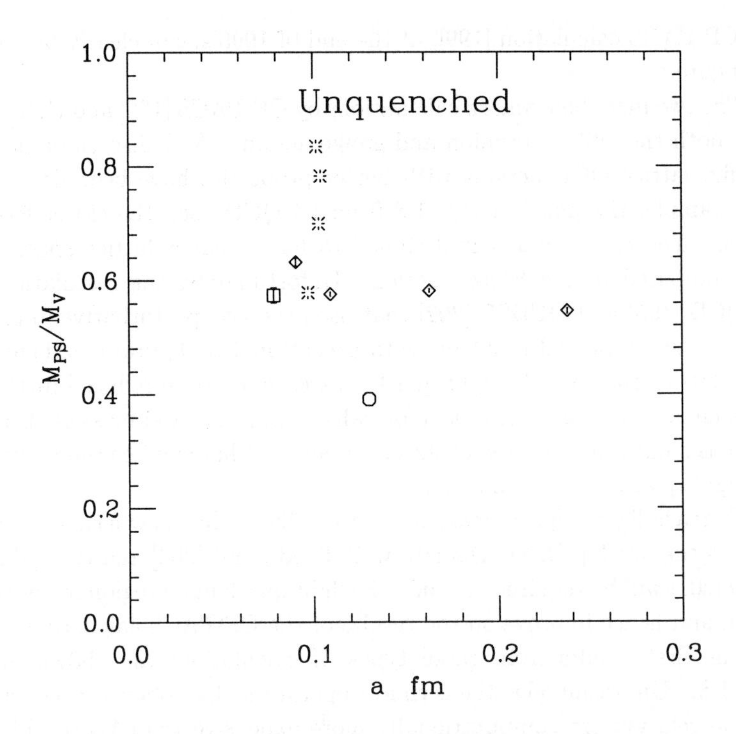

Fig. 1.9 Smallest ratio of pseudoscalar mass to vector mass as a function of lattice spacing for a number of full QCD calculations. The data is from CP-PACS [205] (diamonds), UKQCD [73] (bursts), MILC [91] (octagon) and SESAM [206] (squares).

lations. Although the choice of the lattice spacing in a specific calculation can have large uncertainties, I feel that the mass of the pion in physical units gives a more immediate measure of the "lightness" of the quarks in a calculation. For the table I just used the lattice spacing quoted by the collaboration. The review article by Kaneko [31] contains a more thorough survey of the parameters of some recent unquenched calculations.

In Fig. 1.9 the ratio of the masses of the lightest pseudoscalar to vector for some recent full QCD calculations is plotted. The effect of the cost for full QCD from formula 1.46 can be clearly seen, as the finest lattice spacing used is around 0.1 fm. Only the CP-PACS calculation had a number of different lattice spacings. The non-perturbative improved clover action was constrained to lie around 0.1 fm, because the computation of the clover coefficient was too expensive at coarser lattice spacings [208].

Operator	J^{PC}	Lightest particle
$\overline{\psi}_1 \gamma_5 \psi_2$	0^{-+}	π
$\overline{\psi}_1 \gamma_4 \gamma_5 \psi_2$	0^{-+}	π
$\overline{\psi}_1 \gamma_i \psi_2$	1^{--}	ρ
$\overline{\psi}_1 \gamma_4 \gamma_i \psi_2$	1^{--}	ρ
$\overline{\psi}_1 \gamma_i \gamma_j \psi_2$	1^{+-}	b_1
$\overline{\psi}_1 \gamma_i \gamma_5 \psi_2$	1^{++}	a_1
$\overline{\psi}_1 \psi_2$	0^{++}	a_0

Table 1.4 Interpolating operators for light mesons. The 1 and 2 subscripts label flavour and show that the mesons are non-singlet.

The most interesting point is from the calculations done by the MILC collaboration [90]. The MILC collaboration are currently running with a lattice spacing of 0.09 fm.

1.6 The Masses of Light Mesons

Light mesons have a number of important uses in lattice QCD calculations. In calculations that use Wilson like fermions, the mass of the pion is used to calculate the additive mass renormalisation (κ^{crit} in Eq. 1.116). The ρ is sometimes used to set the lattice spacing. The mass of one of the mesons — kaon, $K^\star$ or ϕ — is used to calculate the strange quark mass. After the light quark masses are calculated any remaining masses are used as a consistency check. These masses of the quarks are then used in any further calculations, such as the computation of matrix elements.

The interpolating operators for mesons to be used in Eq. 1.11 are in Table 1.4. The J^{PC} quantum numbers of the meson operators can be derived using the standard representation of the parity (P) and charge conjugation (C) operators from the Dirac theory. The meson interpolating operators are usually extended in space using one of the prescriptions in Sec. 1.2. Most calculations concentrate on the S-wave mesons as the signal to noise ratio is better for these mesons, than for P-wave states. I discuss

P-wave states in Sec. 1.6.1.

The QCD field strength tensor (F_{jk}) has also been used with the fermion bilinears in Table 1.4. The QCD field strength tensor has specific J^{PC} quantum numbers that can be used to obtain fermion bilinear operators with different J^{PC} quantum numbers. For example the MILC collaboration [209] used an interpolating operator of the form

$$\epsilon_{ijk}\overline{\psi}_1\gamma_i F_{jk}\psi_2 \tag{1.119}$$

with $J^{PC} = 0^{-+}$. They obtained the same mass for the pion using the operator in Eq. 1.119 as the pion operator in Table 1.4. The operator in Eq. 1.119 is more "gluey" than a $\overline{\psi}\gamma_5\psi$ operator, so might be expected to couple to hybrid (a quark-antiquark pair with excited glue) pions. However, the pion is still the lightest state that couples to the operator in Eq. 1.119, hence it is the state extracted from the fits.

In Fig. 1.10, I plot the mass of the π and ρ mesons as a function of the physical box size for Wilson fermions at a fixed lattice spacing ($\beta = 6.0$, $a \sim 0.1$ fm). The data seem to show that a linear box size of 3 fm is big enough for "small" finite size errors. The errors on the masses of the mesons with the lighter quark masses are too big to draw any conclusions.

To get a more quantitative estimate of the volume dependence UKQCD [195] studied finite size effects in quenched QCD at $\beta = 6.0$ with non-perturbatively improved clover fermions. Two volumes were used with sides 1.5 fm and 3 fm. For the pseudoscalar channel there was a 2σ difference between the mass in the two volumes. There were no statistically significant difference between the masses for the vector particle between the two lattice volumes.

To give some idea of the size of lattice spacing errors in the masses of the light mesons, I plot the mass of the $K^\star$ meson as a function of the lattice spacing in Fig. 1.11 from the CP-PACS collaboration [97]. The figure shows the difference between using the kaon or ϕ meson to determine the strange quark mass. The clear difference between the mass of the $K^\star$ when the strange quark mass is determined in two different ways is caused by quenched QCD not being the theory of nature. The results for meson masses in the continuum limit are in Table 1.5 [97].

The physical summary of the results in Table 1.5 is that the hyperfine splitting is too low from quenched lattice QCD. To isolate the reduction of the hyperfine splitting, Michael and Lacock [177, 178] introduced the J

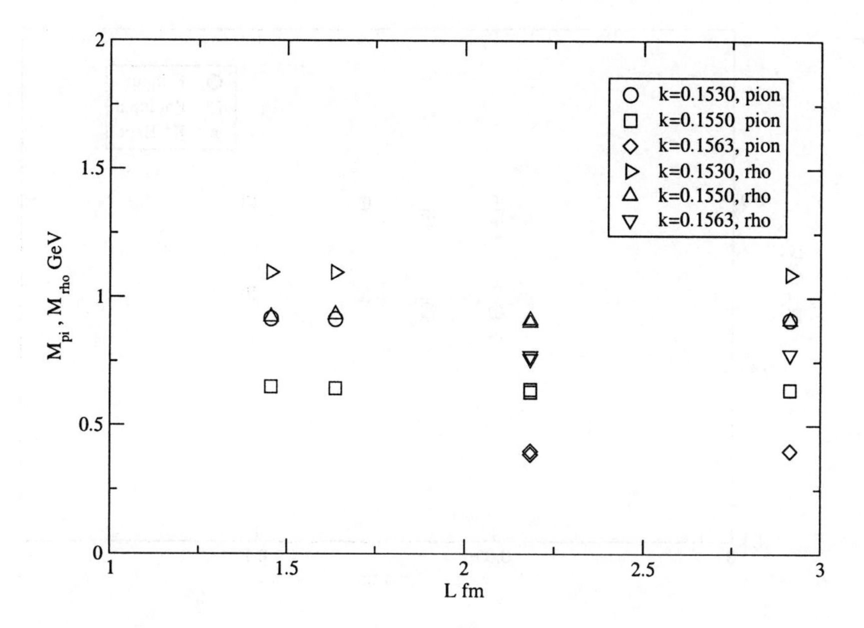

Fig. 1.10 Dependence of the mass of the π and ρ mesons on the box size in quenched QCD for Wilson fermions at $\beta = 6.0$. The data was taken from the compendium of World data in [202].

Mass	Result (m_K) GeV	Result (m_ϕ) GeV	Expt. GeV
m_K	-	0.546(06)	0.496
m_{K^*}	0.846(07)	0.891(05)	0.892
m_ϕ	0.970(06)	-	1.020

Table 1.5 Masses of light S-wave mesons in quenched QCD from CP-PACS [97]. The different analyses depend on which meson is used to determine the strange quark mass.

parameter

$$J = M_V \frac{dM_V}{dM_P^2} |_{M_V = 1.8 M_P}, \tag{1.120}$$

where M_V and M_P are the vector and pseudoscalar masses respectively.

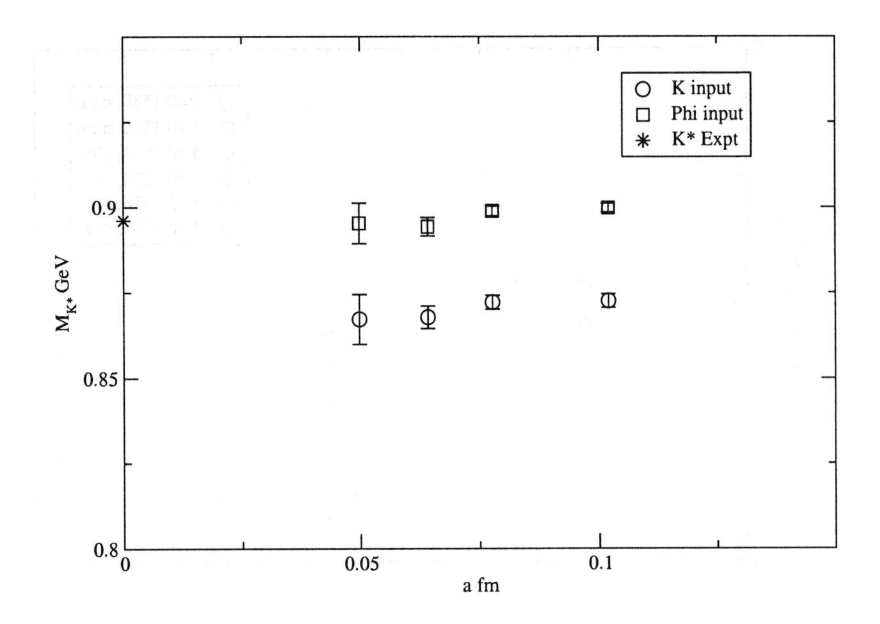

Fig. 1.11 Lattice spacing dependence of the K^* meson in quenched QCD from the CP-PACS collaboration [97].

The condition $M_V = 1.8 M_P$ corresponds to the experimental ratio of $K^\star$ and K masses. This mass ratio was chosen so that an extrapolation to quark masses below strange was not required. Some theoretical problems with the value of J defined at a light quark reference point are discussed by Leinweber *et al.* [159].

The J parameter has been chosen to be independent of the lattice spacing and an explicit definition of the quark mass. One experimental estimate for J is obtained from

$$J = m_{K^\star} \frac{m_{K^\star} - m_\rho}{m_K^2 - m_\pi^2} = 0.48. \tag{1.121}$$

Including the uncertainty from using the $\phi - K^\star$ difference rather than the $K^\star - \rho$ mass difference, Michael and Lacock [177] estimate the experimental value of J to be 0.48(2).

In quenched QCD, Michael and Lacock [177] obtained $J = 0.37(2)(4)$ in disagreement with the experimental estimate. Using their bigger data set, the CP-PACS collaboration [97] obtained J = 0.346(23) from a continuum

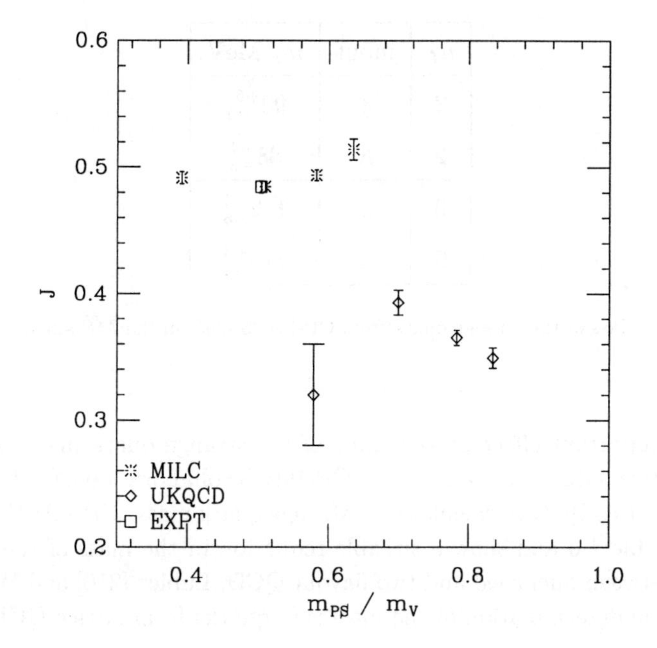

Fig. 1.12 Comparison of the J parameter from UKQCD [73] (diamonds) and MILC [90] (bursts). The experimental point is the square.

extrapolation in quenched QCD.

The clear disagreement of the J parameter from quenched QCD with experiment, in principle makes J a good quantity to measure the effect of the sea quarks. The actual definition of J in Eq. 1.120 for unquenched QCD is not trivial because, as discussed in Sec. 1.3.2, the lattice spacing does depend on the mass of the sea quarks. The issues in defining J for unquenched QCD are discussed by the UKQCD collaboration [73].

Figure 1.12 compares the value of J from the unquenched lattice QCD from UKQCD [73] and MILC [90]. Both calculations done by the UKQCD and MILC collaborations were at a fixed lattice spacing. The calculation by MILC was done at a lighter sea quark mass then UKQCD, and that presumably explains why the value of J from MILC agrees better with the experimental value. The MILC result needs to be confirmed by a study of the lattice spacing dependence. Kaneko [31] has recently reviewed the status of calculations of J from unquenched calculations.

The effect of unquenching on the hyperfine splitting in light mesons

n_f	input	m_s MeV
2	ϕ	90^{+5}_{-11}
2	K	88^{+4}_{-6}
0	ϕ	132^{+4}_{-6}
0	K	110^{+3}_{-4}

Table 1.6 Mass of the strange quark from CP-PACS [205] in the $\overline{MS}$ scheme at a scale of 2 GeV.

has an important effect on the value of the strange quark mass extracted from lattice data. For example, the CP-PACS collaboration [205] have used $n_f = 2$ lattice QCD to calculate the strange quark mass. CP-PACS's results are in Table 1.6 and show a sizeable reduction in the mass of the strange quark between quenched and two flavour QCD. Lubicz [210] and Wittig [7] review the determination of the masses of quarks from lattice QCD.

1.6.1 *P-wave mesons and higher excitations*

The calculation of the masses of ground state mesons and baryons is essentially done to validate lattice QCD methods and to compute quark masses and the strong coupling. For excited S-wave and P-wave states, the issues are to determine the quark and glue content of the state as well as to check that the masses agree with experiment. The signal to noise ratio is worse for P-wave mesons than for S-wave mesons [72], so the calculations are harder. These type of calculations tend to be more exploratory, so usually no attempt is made to take the continuum or infinite volume limit. As the underlying lattice techniques mature, there will be attempts to quantify all the systematic errors.

The interpolating operators in Table 1.4 can be used to create the P-wave a_0, b_1, and a_1 states. However, in the quark model these states have zero wave function at the origin, so non-local interpolating operators with a node at the origin have been tried [211, 212, 213]. Reasonable results can be obtained using the operators in Table 1.4. Spin 2 states are not accessible from local interpolating operators, so must be created using non-local operators.

Continuum representation	cubic rep.
J=0	A_1
J=1	T_1
J=2	E,T_2
J=3	A_2,T_1,T_2
J=4	A_1,E,T_1,T_2

Table 1.7 Representation of $SU(2)$ in terms of representations of the cubic group.

A more general approach [214] is to consider a generic non-local meson operator at a specific time t (as is required for an interpolating operator for a time correlation function)

$$O(\underline{r}) = \overline{\psi}(\underline{x}, t)\Gamma \prod_{\underline{x},\ \underline{x}+\underline{r}} U\psi(\underline{x} + \underline{r}, t), \tag{1.122}$$

where the quark and anti-quark are separated by a distance $\underline{r}$ and Γ is an arbitrary gamma matrix. The $\prod_{\underline{x},\underline{x}+\underline{r}} U$ is a set of gauge links that connects the quark to the anti-quark in a gauge invariant way. In general $\prod_{\underline{x},\underline{x}+\underline{r}} U$ is not unique and will effect the transformation properties of $O(\underline{r})$ under the cubic group. Operators are designed to transform under specific representations of the cubic group.

The connection between the representations of the cubic group and the $SU(2)$ rotation group are in Table 1.7. The dimensions of the representations A_1, A_2, E, T_1 and T_2 are 1, 1, 2, 3 and 3 respectively. Hence the dimensions of the representations match between the cubic group and the $SU(2)$ rotation group with representation of $2J + 1$ in Table 1.7.

If the gauge configurations are fixed to the Coulomb gauge then non-local interpolating operators based on spherical harmonics can be used. This approach was studied by DeGrand and Hecht [72, 215]. As hadron masses are gauge invariant quantities, gauge fixing at an intermediate stage should not effect the final results. Meyer and Teper discuss how to construct higher spin glueball operators [216].

There are a number of interesting puzzles with the phenomenology of "P-wave" mesons. For example, there are speculations that the $a_0(980)$ particle is potentially not well described by a $\overline{q}q$ state in the quark model.

Its mass is very close to the threshold for two kaon decay. There are many models that treat this state as a $\overline{K}K$ molecule [217] or $\overline{qq}qq$ state. As lattice QCD calculations use the non-singlet $\overline{q}q$ operator to create this state, it may not couple strongly to a molecular $\overline{qq}qq$ state. Therefore I would expect that the lightest state in the $\overline{q}q$ channel to be the $a_0(1450)$. However, this speculation should, and will be [218] tested in unquenched lattice calculations.

As discussed in Sec. 1.3.1, the interpretation of the a_0 state is complicated by a quenched chiral artifact [86] in quenched QCD. The largest systematic study of the a_0 in quenched QCD was done by Lee and Weingarten [87]. Unfortunately, they could only find the mass of the a_0 at the mass of the strange quark, possibly because of problems with the artifact in this channel. Alford and Jaffe [110] review some of the earlier results for the mass of the a_0 from quenched QCD. In a lattice calculation that used domain wall fermions and modelled the quenched chiral artifact, Prelovsek and Orginos in Ref. [219] obtained the lightest state in the a_0 channel to be 1.04(7) GeV. At one coarse lattice spacing, Bardeen *et al.* [86] obtained a value of 1.34(9) GeV for the lightest state in the a_0 channel, also with an analysis that was aware of the quenched chiral artifact. The existing quenched lattice QCD data does not determine the mass of the lightest non-singlet 0^{++} state.

In a two flavour unquenched calculation the UKQCD collaboration [218] quote the preliminary result for the mass of the a_0 to be 1.0(2) GeV at one lattice spacing. The MILC collaboration [90] have computed the a_0 mass with 2+1 flavours of sea quarks. The lightest mass of the a_0 in MILC's calculation is 0.81 GeV [90]. The MILC collaboration [90] also claimed to see evidence for the open decay channels of the a_0 state. Hence, more work is required in unquenched QCD to determine the mass of the lightest non-singlet 0^{++} state.

The only values of the masses of the a_1 and b_1 mesons from unquenched QCD, I could find, were from the MILC collaboration [90]. MILC obtained masses of the a_1 and b_1 of 1.23(2) and 1.30(2) GeV compared to the experimental values of 1.23(4) GeV and 1.2295(32) GeV respectively [6]. There does not seem to be any interesting experimental issues with the a_1 and b_1 mesons. As noted by MILC [90], according to the quark model the masses of the a_1 and b_1 mesons should be similar to the mass of the a_0. Hence, any large splitting between the masses of a_0 and a_1 and b_1 is indicative of dynamics beyond the simple quark model.

The most surprising aspect of hadron spectroscopy is the existence of Regge trajectories. Empirically, the square of the mass of a meson with mass $M(l)$ is linearly related to the spin l by

$$l = \alpha' M^2(l) + \alpha(0). \tag{1.123}$$

Although the Regge trajectories can be explained by models (see [220] for a review), in particular the string model, the existence of Regge trajectories has not been shown to be a rigorous consequence of QCD. Equation 1.123 is a very useful tool in classifying baryon states. Historically some of the spin assignments of baryons were guessed from Eq. 1.123 [221]. Equation 1.123 is also useful in scattering experiments (see [222] for a discussion).

The simple relation in Eq. 1.123 has recently been challenged by a number of authors. The improved precision of the experimental data on hadron masses has allowed fits to the spectrum that seem to show some nonlinearity in the relation between the square of the meson masses and their spin [223]. Brisudova *et al.* [224] discuss various hadronic models that can reproduce linear and nonlinear Regge trajectories. The nonlinearities in the Regge trajectories were related to flux tube breaking [220], hence the study of Regge trajectories is complimentary to the study of string breaking [225]. Brisudova *et al.* [220] make the prediction that there are no light quark quarkonia beyond 3.2 GeV due to the termination of the Regge trajectories. If this conjecture is correct, then this might simplify searches for hybrid and glueballs in the region from 3 to 5 GeV.

The problem for lattice QCD calculations is that it is hard to construct interpolating operators with spin higher than 2 on the lattice. As higher orbital states have larger masses than ground state mesons, they have a worse signal to noise ratio, so it is hard to extract a signal from the calculations.

There have been some calculations of spin 3 masses from lattice QCD. The UKQCD collaboration [214] studied spin 2 and spin 3 states in a quenched lattice QCD calculation. In Fig. 1.13, I plot their results in a Chew-Frautschi plot [226] with experimental data from the particle data table.

There have been a few attempts to study the spectroscopy of excited mesons using lattice QCD. As discussed in Sec. 1.2 the computation of the masses of excited states requires a multi-exponential fit to the lattice correlators that can be unstable.

The CP-PACS collaboration in Ref. [81] used the Maximum Entropy

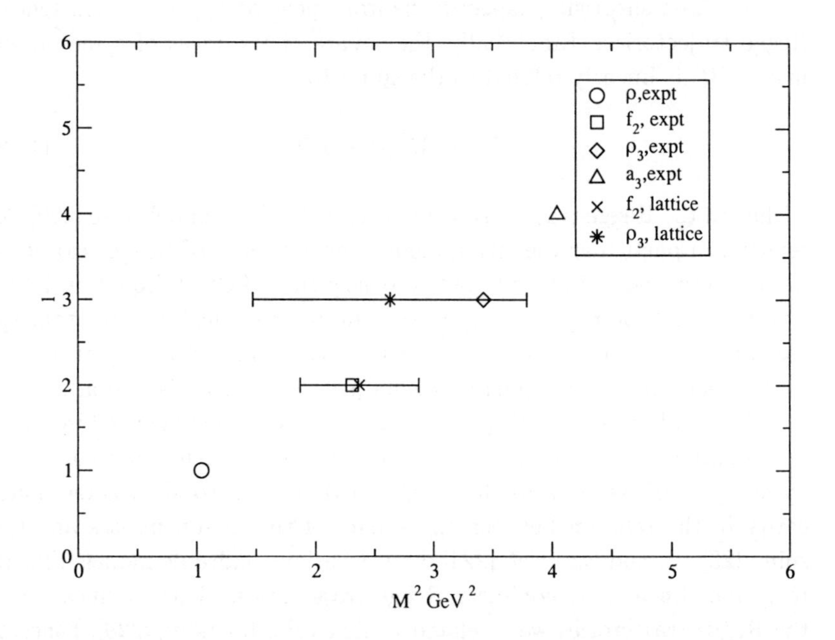

Fig. 1.13 Square of the mass of the meson mass versus its spin l. The octagons are the experimental numbers and the bursts are the lattice QCD data from UKQCD [214]. The quarks have the mass of the strange quark. There are only error bars on the lattice data.

Method (MEM) — briefly revived in Sec. 1.2 — to study the excited spectrum of the ρ- and π-mesons. The calculations were done in quenched QCD, using the same data set that was used for their calculation of the ground state masses [199].

A spectral density from CP-PACS [81] is plotted in Fig. 1.14. The masses of the states are obtained from the peaks in Fig. 1.14. Figure 1.15 from CP-PACS [81] shows the masses of the excited ρ meson as a function of lattice spacing. The diverging graph in Fig. 1.15 is thought to be a bound state of fermion doublers.

The final results from CP-PACS [81] were that the first excited state of the π had a mass of 660(590) MeV and the mass of the first excited ρ-meson was 1540 (570) MeV from quenched QCD. The errors also include an estimate of the error from taking the continuum limit.

Experimentally [6], the first excited π is the $\pi(1300)$ with the mass of

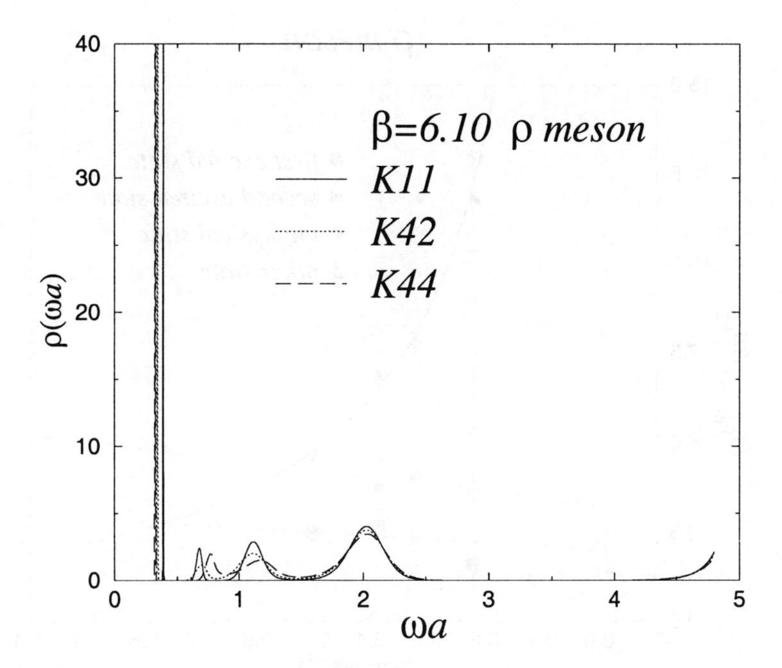

Fig. 1.14 Spectral function (from MEM) obtained by the CP-PACS collaboration for the ρ correlators. The lattice spacing is $a^{-1} = 2.58$ GeV. The $K11$, $K42$ and $K44$ keys are the plots for mesons with different quark masses.

1300 ± 100 MeV. The excited states of the ρ meson are more interesting. There are two states: $\rho(1459)$ with a mass of 1465 ± 25 and the $\rho(1700)$ with a mass of 1700 ± 20 MeV. Donnachie [227] reviews the evidence for a hybrid state in the 1^{--} channel.

The errors on the excited states from CP-PACS are really too big for a meaningful comparison with experiment. The results from CP-PACS [81] need confirmation from other, less Bayesian based fitting methods, such as variational methods.

1.7 The Masses of Light Baryons

An important, but perhaps slightly boring, goal of lattice QCD is to compute the mass of the nucleon with reliable errors from first principles. The nucleon is the most important hadron for the real world of the general pub-

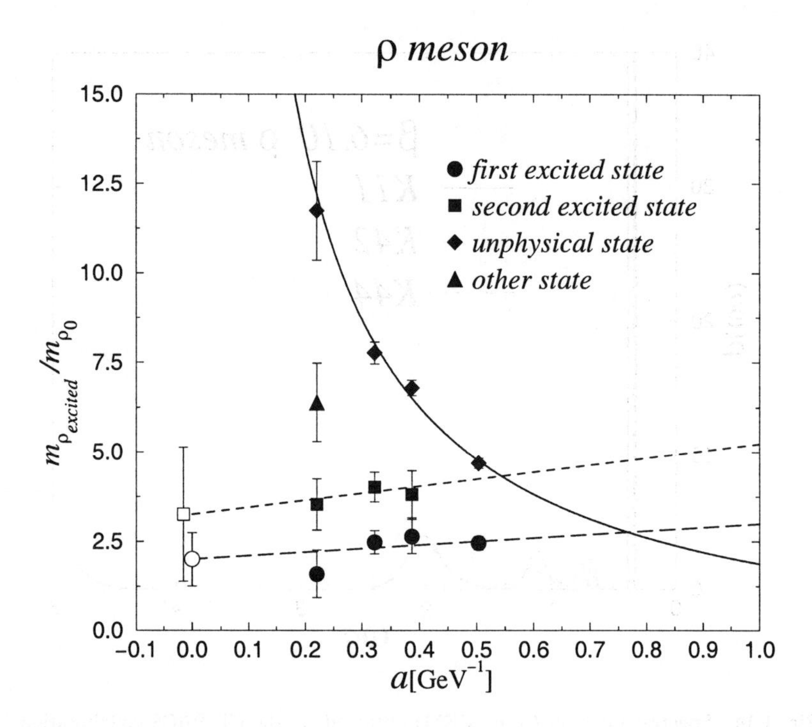

Fig. 1.15 Continuum extrapolation of the masses in the vector channel from CP-PACS.

lic, but the nucleon's rôle in the esoteric domain of particle physics is as "background" to the more interesting stuff. In this section, I will discuss the highlights of some recent large scale quenched and unquenched QCD calculations. As the aim of lattice QCD calculations for these baryons is to validate lattice QCD before the spectroscopy of more interesting hadrons is attempted, the focus will be on the error analysis.

In principle, the mass of the nucleon should be an ideal quantity to compute on the lattice, because it is stable within QCD, hence there are no concerns with the formalism due to decay widths. The main complication with getting an accurate value of the nucleon mass is the large chiral extrapolation required (see Sec. 1.3.3) caused by the large quark masses used in the calculations. In the majority of lattice QCD calculations, electromagnetism is ignored, so I refer to the generic nucleon rather than the neutron or proton.

The "standard" interpolating operators for the nucleon [16, 228] are

$$\begin{aligned} N_1^{1/2+} &= \epsilon_{ijk}(u_i^T C\gamma_5 d_j)u_k \\ N_2^{1/2+} &= \epsilon_{ijk}(u_i^T C d_j)\gamma_5 u_k, \end{aligned}$$

where u and d are operators that create the up and down quark respectively and C is the charge conjugation matrix. In the non-relativistic limit (keeping the upper components) the N_2 operator vanishes, so it does not couple strongly to the nucleon. Empirically the N_2 operator has been found useful for $N^\star$ states. Leinweber [229] discusses the connection between the interpolating operators for the nucleon used on the lattice and those used in QCD sum rules.

The nucleon correlator is constructed, in the standard way, by creating a nucleon at the origin, and then destroying it at time t later *i.e.*

$$C_\pm(t) = \sum_{\underline{x}} \langle 0 \mid N(\underline{x}, t)(1 \pm \gamma_4)\overline{N}(\underline{0}, 0) \mid 0 \rangle.$$

The specific representation of the fitting Eq. 1.23 is now slightly more subtle, namely,

$$\begin{aligned} C_+(t) &\to A^+ e^{-m_+ t} + A^- e^{-m_-(T-t)} \\ C_-(t) &\to A^- e^{-m_- t} + A^+ e^{-m_+(T-t)}, \end{aligned}$$

where m_+ and m_- are the masses of the lightest positive and negative parity states.

To illustrate the finite size effects on the mass of the nucleon in quenched QCD, I plot the nucleon mass as a function of box size for Wilson fermions at $\beta = 6.0$ in Fig. 1.16. This data set was chosen because there was data at a number of different volumes. For all the data I used the lattice spacing of a= 0.091 fm [198]. The dependence of the nucleon mass on the length of the lattice looks "mild" from Fig. 1.16. Unfortunately, the statistical errors on the lightest data are too large to make any statement about finite size effects.

For many years, the quality of a lattice QCD calculation was judged by the final value of the ratio of the nucleon mass to ρ mass. I now discuss some recent quenched QCD calculations in more detail.

There was a large scale calculation of the quenched QCD spectrum by the MILC collaboration [230] that used staggered fermions. The study

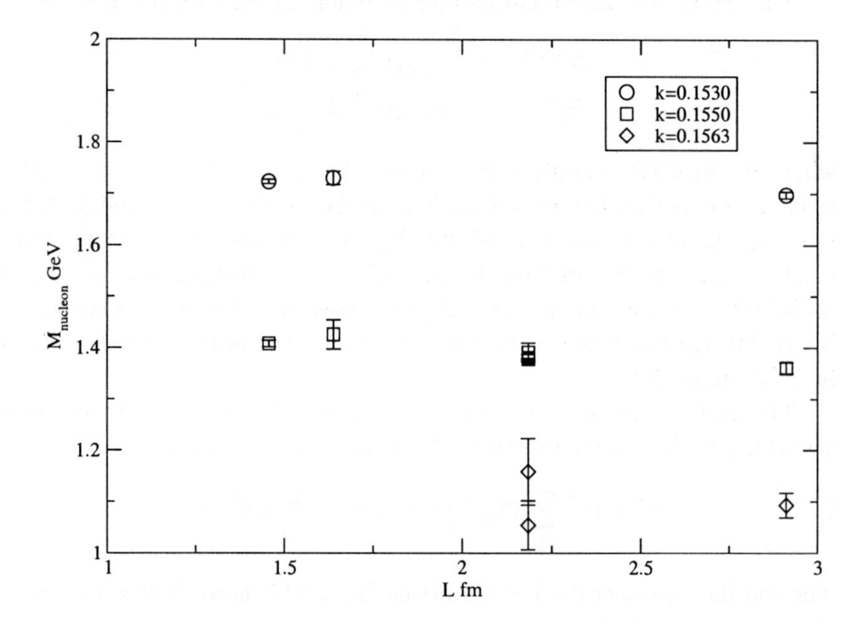

Fig. 1.16 Dependence of the mass of the nucleon on the box size in quenched QCD for Wilson fermions at $\beta = 6.0$. The data was taken from the compendium of World data in [202].

included four different lattice spacings with a^{-1} ranging from 0.63 up to 2.38 GeV and also investigated finite size effects.

This collaboration tested 12 different chiral extrapolation models based on generic models for the dependence of a hadron mass M_H on the quark mass m_q of the form

$$M_H = M + am_q^{1/2} + bm_q + cm_q^{3/2} + dm_q^2 + em_q^2 \log m_q. \tag{1.124}$$

Only three of the fit parameters (a, b, c, d, e) were varied in a fit model. The fit model in Eq. 1.124 was broad enough to include both full and quenched chiral perturbation theory. It is important to test more than the quenched chiral perturbation theory results, because it is not clear that the data is in the regime where the expressions are valid. The simple linear fit, where only b was non-zero in Eq. 1.124, was not consistent with the data [230].

The MILC collaboration found that the coefficient of the $m_q^{1/2}$ term from their fits had the wrong sign compared with the expectations from quenched chiral perturbation theory, hence this term was not included in their final fits. For the ρ extrapolation the coefficient of the $m_q^{3/2}$ term was an order of magnitude lower than expected. The analysis of the chiral extrapolations was complicated by the flavour symmetry breaking terms of staggered fermions [230]. This seems to be a perennial feature of the staggered fermion formalism [231].

The result from the MILC collaboration [230] for the ratio of the nucleon to ρ mass ratio was $m_N/m_\rho = 1.254 \pm 0.018 \pm 0.027$, where the first error is statistical and the second error is systematic (in the summary, Fig. 1.17, I have added the two errors in quadrature).

Kim and Ohta [232] studied quenched QCD using staggered fermion with a smaller lattice spacing ($a^{-1} = 3.63 \pm 0.06$ GeV) and lighter quarks than were used by the MILC collaboration [230]. The spatial length was 2.59 ± 0.05 fm and their lightest quark mass was 4.5 MeV. They fitted the same chiral extrapolation models as MILC [230] did, but also had problems unambiguously detecting the predictions from quenched chiral perturbation theory. Kim and Ohta also studied some chiral extrapolation formulae, suggested by [233], that looked for finite volume effects masquerading as quenched chiral logs. In Ref. [232] their final result was $m_N/m_\rho = 1.24 \pm 0.04(\text{stat}) \pm 0.02(\text{sys})$.

The CP-PACS collaboration [97] found $m_N/m_\rho = 1.143\ (33)(18)$. In fits of the vector particle as a function of the quark mass, CP-PACS excluded the $m_q^{3/2}$ term, but included the $m_q^{1/2}$ term. The coefficient of the $m_q^{1/2}$ term was found to be an order of magnitude less than the naive expectation.

In Fig. 1.17, I plot the results for the ratio of the nucleon and rho masses from some recent quenched lattice QCD calculations [4, 97, 195, 230]. I have selected calculations where an attempt was made to take the continuum limit. For comparison, in the strong coupling limit $g \rightarrow \infty$ the hadron spectrum can be computed analytically. In this limit the ratio of the nucleon and ρ masses is $\frac{m_N}{m_\rho} = 1.7$ [18].

The agreement with experiment for the m_N/m_ρ ratio from quenched QCD is surprisingly good. Although agreement at the 10 % level may sound quite impressive, errors of this magnitude are too large for QCD matrix elements required in determining CKM matrix elements from experiment [5]. The analysis of Booth *et al.* [153] using the non-analytical terms from quenched chiral perturbation theory estimated that the value of m_N/m_ρ from quenched QCD could be as low as 1.0. This situation is not seen in the lattice results. However, none of the quenched QCD calcu-

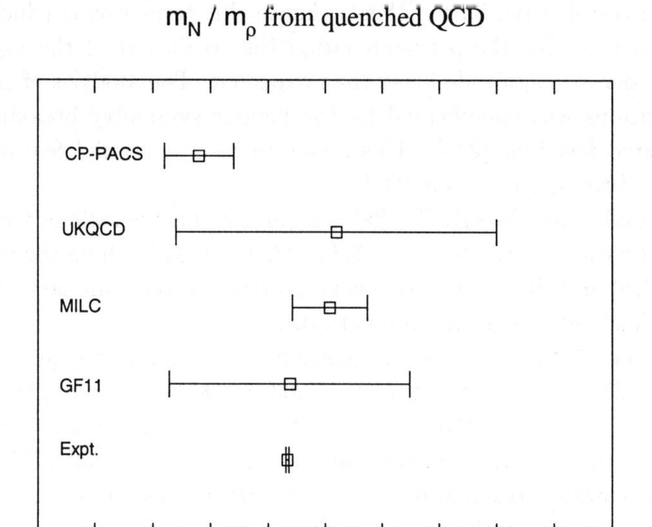

Fig. 1.17 The ratio of the nucleon mass to rho mass from several recent large scale quenched QCD calculations [4, 97, 195, 230] .

lations have detected the predicted dependence of the mass of the ρ meson on the quark mass. To confirm the quark mass dependence predicted by quenched chiral perturbation theory requires new calculations with lighter quarks. The QCDSF collaboration [234] obtained a coefficient of the $m_q^{1/2}$ term from their fits that was roughly in agreement with expectations from quenched chiral perturbation theory.

A summary of some of the hadron spectra for quenched QCD from the work by CP-PACS [97] is shown in Fig. 1.18. The ALPHA collaboration notice that the largest deviations from experiment of the quenched lattice results from CP-PACS are for resonant hadrons [185] if the lattice spacing is determined from the nucleon mass. The BGR collaboration also note a similar trend in their data [200]. This trend could be modelled using techniques similar to those employed by Leinweber and Cohen [156] to study the ρ meson. It is difficult to see from Fig. 1.18 whether the hadrons with large widths have masses in quenched QCD that differ more from experiment than those with smaller widths.

As discussed by Aoki [29] and the CP-PACS collaboration [97] the value

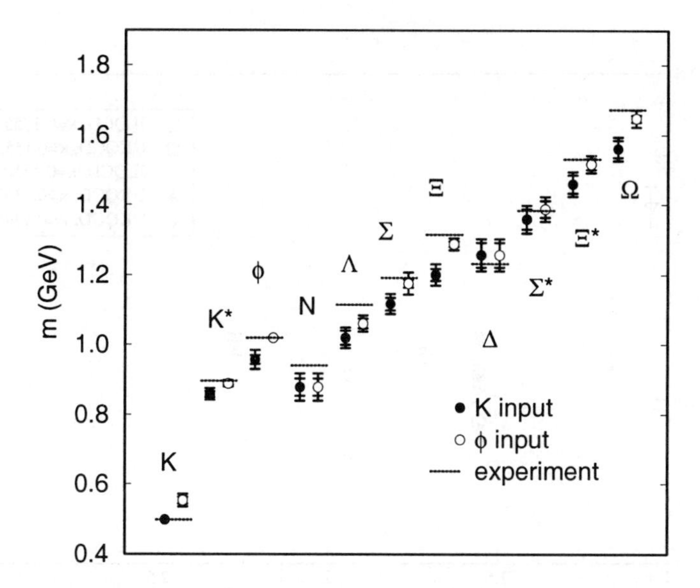

Fig. 1.18 Final results for the light hadron spectrum from CP-PACS in quenched QCD [97].

of the mass of the nucleon from CP-PACS [97] and MILC [230] differ at the 2.5σ level. The two calculations used different fermion formulations, each with a different set of potential theoretical problems, that should in principle produce the same results in the continuum limit.

The main criticisms of the CP-PACS quenched study are on their treatment of the quenched chiral perturbation theory. Wittig reviews the lattice results for detecting the quenched chiral log [7]. The new calculations that use light fermion actions with better chiral symmetry properties are disagreeing with the result from CP-PACS. These new calculations are done at a fixed (small volume) and large lattice spacings, so perhaps there are systematic errors in their results [7]. The CP-PACS and MILC collaborations used the mass of the ρ to determine the lattice spacing. It would have been interesting to see the dependence of the final results on using different quantities to determine the lattice spacing.

Although it may be possible to do a more systematic study of quenched QCD at lighter quark masses than CP-PACS [199], I am not sure that it is worth it. As discussed in Sec. 1.3.3 the crucial non-analytic terms in quenched and unquenched chiral perturbation theory are very different,

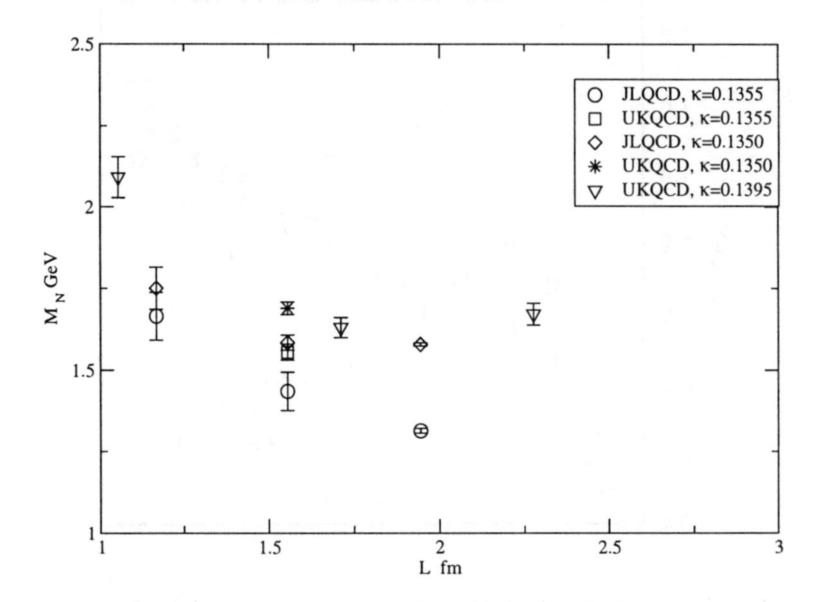

Fig. 1.19 Dependence of the mass of the nucleon on the box size from UKQCD [112] and JLQCD [235]. These are two flavour unquenched calculations.

so calculations with light quarks in quenched QCD will have very little direct relevance for unquenched calculations. As the masses of the quarks in unquenched calculations decrease we should start to see the effect of the decay of the hadrons. The effect of particle decay on the mass spectrum can not be studied in quenched QCD.

I will now describe the results from recent unquenched lattice QCD calculations. As usual the systematic errors must first be discussed. There have been two recent studies of finite size effects in the nucleon from unquenched QCD, carried out by the JLQCD [235] and UKQCD [112]. The results are plotted in Fig. 1.19. One slight concern with the nice study of finite size effects from JLQCD [235] is the large statistical errors on the two smaller volumes. Also there is some disagreement between the results from UKQCD and JLQCD on the 16^3 volume. As stressed by the MILC collaboration [84] a careful control of statistical errors is required to see definitive evidence for the effect of the box size on the masses of hadrons.

In Fig. 1.20 I plot the dependence of some hadron masses on the lattice spacing from the calculations by the CP-PACS collaboration [97]. This

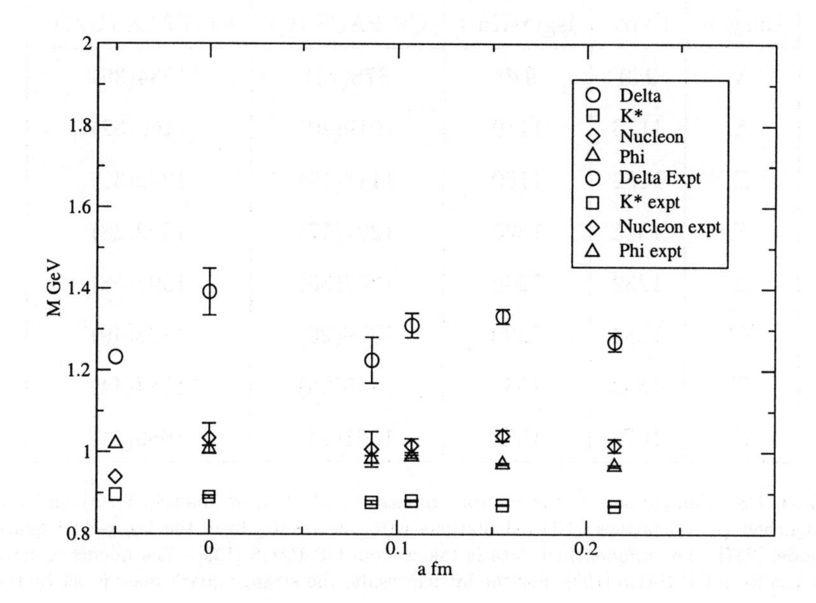

Fig. 1.20 Dependence of hadron masses on the lattice spacing, from CP-PACS [97] in 2 flavour unquenched QCD. Data from the finest lattice spacing was not used in the continuum extrapolation.

should give some idea of the size of the lattice spacing errors in unquenched QCD.

As discussed in Sec. 1.6 the main "successes" of unquenching have occurred in the meson sector. A detailed comparison of the baryon spectrum with experiment is obscured by lattice spacing and finite size effects [135, 235]. In a preliminary analysis the MILC collaboration [236] found that m_N/m_ρ from two flavour unquenched QCD in the continuum limit was higher than the value in the quenched QCD, so the quenched value agreed better with experiment than the unquenched result. The MILC collaboration [90] are now studying this issue using a better version of staggered fermions with 2+1 unquenched flavours.

In Table 1.8, I compare the results from lattice QCD with the old results from the Isgur-Karl quark model [237]. Although the Isgur-Karl model agrees better with experiment than the two lattice QCD calculations, because the lattice QCD calculations are based on the QCD Lagrangian, the hadron masses can be used to extract quark masses. This is not possible

Baryon	Expt.	Isgur-Karl	CP-PACS (Q)	CP-PACS (UnQ)
N	940	940	878(25)	1034(36)
Λ	1116	1110	1019(20)	1160(32)
Σ	1193	1190	1117(19)	1202(30)
Ξ	1315	1325	1201(17)	1302(28)
Δ	1232	1240	1257(35)	1392(58)
$\Sigma^\star$	1384	1390	1359(29)	1488(49)
$\Xi^\star$	1532	1530	1459(26)	1583(44)
Ω	1673	1675	1561(24)	1680(41)

Table 1.8 Comparison of the hadron masses, in MeV, from quenched (Q) and unquenched (UnQ) lattice QCD calculations with the results from the Isgur-Karl quark model [237]. The unquenched data is taken from CP-PACS [135]. The quenched data is also from CP-PACS [199]. For the lattice results the strange quark mass is set by the kaon mass.

from quark models where there is no way of relating the constituent quark masses to the masses in the Lagrangian. The more interesting tests of the quark model occur for excited baryons. I discuss the lattice results for these hadrons in Sec. 1.7.1.

In fact the agreement between the hadron masses from the quark model and experiment is actually too good. The quark model calculation of Isgur and Karl [237] does not include the dynamics of hadron decay. For example the Δ baryons have decay widths of around 120 MeV [6]. It might have been expected that the errors in the quark model predictions for the hadron masses would be of the order of the decay width. Isgur and Geiger [157, 158] have developed a formalism to absorb some of the effects of quark-antiquark loops into the potential. This issue is also reviewed by Capstick and Roberts [238].

1.7.1 *Excited baryon states*

Recently there has been a lot of work on the spectroscopy of excited nucleon states from lattice QCD. This research is mostly motivated by the experimental program at the Jefferson lab [13, 14, 15]. The accurate spectroscopy of the $N^\star$ states will be an accurate test of our understanding of the forces and effective degrees of freedom in hadrons [239]. Realistically, lattice QCD calculations may only be able to obtain one or two excited baryon states from a specific channel. However, this is enough to (potentially) solve some very interesting and long standing puzzles.

At the moment the identification of the Roper resonance from lattice QCD is controversial, so I will introduce some notation to prevent ambiguity. I call the nucleon the N state, the first excited nucleon with positive parity the N' state, and the first excited nucleon with negative parity the $N^\star$ state. In the particle data table, the N' state would be the $N(1440)$ and the $N^\star$ state would be the $N(1535)$.

There is a potential artifact associated with quenched QCD for baryon states [240, 241]. The mass of the $\eta' N$ state is experimentally close to the mass of the $N(1440)$ and $N(1535)$ states. In quenched QCD the η' is treated incorrectly, so the intermediate $\eta' N$ state is also incorrect. This is the analogue of the artifact in the scalar correlator found by Bardeen *et al.* [86] discussed in Sec. 1.3.1. Dong *et al.* [242] claim to have seen the correlator, that would be positive definite in unquenched QCD, for the $N^\star$ state go negative for pion masses below 248 MeV. If this artifact causes the correlator to go negative this may "confuse" fitting techniques such as the maximum entropy method that relies on a positive definite correlator. There is no chiral artifact in unquenched QCD. Work has started on studying the Roper resonance using unquenched QCD [243].

In Table 1.9 I have collected some results for the ratio of the $N^\star$ mass to the nucleon mass from some recent quenched lattice QCD calculations. Only the calculation by Gockeler *et al.* [241] took the continuum limit. The experimental number corresponds to the mass of the $N(1535)$ divided by the nucleon mass *i.e.* about 1.63. It is not clear what effect the 150 MeV width of the $N(1535)$ will be on the lattice result.

The nature of the Roper resonance $N(1440)$ is still a mystery. There is an experimental signal for this state, but it is not clear what the quark and glue composition of this hadron is. On the lattice three quark interpolating operators are used to study the Roper state. If the mass of the Roper is not reproduced, then this would be evidence that additional dynamics, beyond three valence quarks, is important for this state.

The quark model has problems reproducing the experimental mass of

Reference	Comments	$M_{N^\star}/M_N$
Blum *et al.* [244]	Domain wall	1.49(9)
Gockeler *et al.* [241]	Clover	1.50(3)
Broemmel *et al.* [245]	Chirally improved	1.77(7)
Dong *et al.* [242]	Overlap	1.67(12)
Nemoto *et al.* [246]	Anisotropic clover	1.463(51)

Table 1.9 Ratio of the mass of the parity partner of the nucleon ($N^\star$) to the nucleon mass from the quenched QCD.

the Roper resonance. Using a simple harmonic oscillator potential to study the hadron spectrum, Isgur and Karl [244, 247] used an oscillator quantum of 250 MeV. In quark model language, the $N^\star$ state would have one quantum above the ground state, and the N' state would have two quanta above the ground state. This predicted ordering is opposite to the experimental masses of the $N(1440)$ and $N(1535)$. Capstick and Roberts review the nature of the Roper resonance in the context of potential models [238]. Isgur also discusses the problems of the Roper resonance in the quark model [239]. The predictions of the quark model for the lowest excitations of the nucleon improve if a more realistic potential is used, and the mixing between states is taken into account [238, 239]. Glozman [248] has critically discussed some of the issues with the description of the Roper resonance in the standard quark model framework.

There are predictions from flux tube and bag models that the lightest hybrid baryon (three quarks with excited glue) [249, 250] is $J^P = 1^+$ with a mass in the region 1.5 to 1.9 GeV, hence close to the mass of the Roper resonance.

Sasaki, Blum, and Ohta [244] studied the first excited state of the nucleon at a fixed lattice spacing of 0.1 fm and with a physical length of the lattice as 1.7 fm. Their calculations were done in quenched QCD with quark masses in the range M_{PS}/M_V 0.59 to 0.9. The excited state masses were extracted using a variational technique with two basis states that were different interpolating operators for the nucleon. Sasaki *et al.* [244] could only obtain a signal for the N' state for pion masses above 600 MeV. The mass of the N' state was larger than the $N^\star$ state. If the variational technique was not biased by truncation of the sum of excited states, then the calculation of Sasaki [244] should be able to resolve the ground and first excited states. For the negative parity states, the masses obtained from the ground and first excited state were degenerate. This could be interpreted as the variational technique not being able to resolve the two states. Experimentally the lightest $N^\star$ states are the $N(1535)$ and $N(1650)$.

Melnitchouk *et al.* [251] have also studied the masses of the N, $N^\star$, and N' states. Their raw data for the splitting between the N and $N^\star$ states is consistent with other groups [241, 244]. The mass for the N' state was much higher than the mass of the experimental Roper resonance.

Dong *et al.* [242] claim agreement between the mass of the N' state from their calculation and the mass of the Roper resonance from experiment. The calculation was done with the lattice spacing of 0.2 fm and physical box lengths of 2.4 and 3.2 fm. The calculations used very light valence quarks (for this lattice anyway). The lightest π mass was 180 MeV. The excited state masses were extracted using the constrained curve fitting method developed from the proposal by Lepage *et al.* [64]. Dong *et al.* [242] reported that the mass of the N' state started to decrease rapidly with π masses below 400 MeV. The mass of the N' state was less than that of the $N^\star$ state for a π mass of about 220 MeV. The mass quoted for the N' state is 1462(157) MeV. The large error bars on this result mean that a 2σ statistical fluctuation would give 1776 MeV.

Sasaki [252] has reported a study of the finite size effects on the mass of the excited state of the nucleon. The excited state of the nucleon was studied at the fixed lattice spacing of 0.09 fm. Three physical box sizes were used with L = 1.5, 2.2 and 3.0 fm. The mass of the N' state was extracted using the maximum entropy method. The N' and N states had finite size effects that increased as the light quark masses were reduced. The final result for the N' state looks as though it would extrapolate to the experimental value from pion masses around 600 MeV. No dramatic decrease in the mass of the N' is required. The picture from the calculation of Sasaki [252] disagrees with that from Dong *et al.* [242].

Broemmel *et al.* [245] have tried to study the Roper resonance using lattice QCD. They used a chirally improved fermion operator with a lattice spacing of 0.15 fm and two physical lattice sizes of 1.8 and 2.4 fm. To study excited states they used a variational technique based on three interpolating operators. They could get a signal for the nucleon with pion masses as low as 220 MeV. Unfortunately, they could extract a signal for excited nucleon states only with π masses above about 550 MeV. This stopped them being able to confirm the mass dependence of the N' state claimed by Dong *et al.* [242]. This analysis did not include the effect of the quenched chiral artifact in this channel. Broemmel *et al.* [245] extracted both the ground and first excited state of the negative parity nucleon channel.

From the very interesting recent studies [242, 244, 245, 252] of the Roper resonance on the lattice it is clear that careful attention will have to be paid to the systematic errors. Pion masses below 200 MeV may be required. It would be good to have variational calculations with wider sets of interpolating operators as a check on the various Bayesian based fitting techniques. For example it would be reasonable for "fuzzed" nucleon operators [46] to couple more to hybrid baryon states because they contain more glue. It is a high priority that other groups try to reproduce the results of Dong *et al.* [242] and Sasaki [252].

One particular concern with getting the mass of the $N(1440)$ is that its width is 380 MeV. The decay of the $N(1440)$ will effect its mass. This is also a difficulty for potential model calculations. Capstick and Roberts [238] estimate that ignoring the width of the $N(1440)$ causes an uncertainty of the order of 100 MeV on the mass within the potential model framework.

Although I have focused on the excited states of the nucleon, there is a growing body of work on the parity partners of other baryons [246, 251].

1.8 Electromagnetic Effects

The majority of lattice QCD calculations do not incorporate the effect of the electromagnetic fields in hadron mass calculations. This is reasonable, because the dominant interaction for hadron masses is the strong force. I now review the work done on including electromagnetic fields in lattice QCD calculations.

The theory of QED has been studied by many groups using lattice techniques. The formalism is similar to that discussed in Sec. 1.2, except that the gauge group is U(1). However, there are some conceptual differences, because the U(1) gauge theory is not asymptotically free. Issues relating to the non-asymptotically free nature of QED have been studied in a non-perturbative way on the lattice [253, 254]. Also there are compact and non-compact versions of the lattice U(1) theory. In non-compact QED the gauge fields A_μ^{em} take to the range $-\infty$ to ∞.

To study electromagnetic effects on the hadron spectrum the electromagnetic fields have been quenched. The dynamics of the sea quarks have not been included in the generation of the U(1) gauge fields. A comparison between the use of background fields in sum rules and lattice QCD has been made by Burkardt *et al.* [255].

The most "comprehensive" study of the effect of electromagnetism on the masses of hadrons has been performed by Duncan *et al.* [256, 257]. They used a non-compact version of QED. The gauge fields were generated using the action

$$S_{em} = \frac{1}{4e^2} \sum_{x} \sum_{\nu\mu} (D_\mu A_\nu(x) - D_\nu A_\mu(x)), \tag{1.125}$$

where e is the electromagnetic charge. The $A_\mu(x)$ fields were subject to the linear Coulomb condition and then promoted to the compact fields $U(x)_\mu = e^{\pm iqA_\mu(x)}$. This field coupled to a quark field with charge $\pm qe$.

Currently, lattice QCD calculations are usually not accurate to 10 MeV, the order of magnitude of electromagnetic effects on the masses of light hadrons. To increase the mass splitting Duncan *et al.* [256, 257] used large charges (2 to 6 times the physical values) and then matched onto chiral perturbation theory that included the photon field.

The chiral extrapolation fit model used for pseudoscalars with electromagnetic fields is

$$m_P^2 = A(e_q, e_{\overline{q}}) + m_q B(e_q, e_{\overline{q}}) + m_{\overline{q}} B(e_q, e_{\overline{q}}), \tag{1.126}$$

where e_q, $e_{\overline{q}}$ (m_q, $m_{\overline{q}}$) are the charges (masses) of the quark and anti-quark.

The calculations of Duncan *et al.* [256, 257] were done with one coarse lattice spacing $a^{-1} \sim 1.15$ GeV with a box size of 2 fm. Some of their results for the mass splittings are in Table 1.10. I have included both the data and the corrected results, where theoretical expressions were used to correct for finite volume effects.

There has been some work on computing electromagnetic polarizabilities [258, 259, 260] from lattice QCD. The electric (E) and magnetic (B) polarizabilities measure the interaction of a hadron with constant electromagnetic fields. Under the electromagnetic field the mass of the hadron is shifted by

$$\delta m = -\frac{1}{2}\alpha E^2 - \frac{1}{2}\beta B^2. \tag{1.127}$$

Mass splitting	Raw Lattice QCD	Corrected	Experiment
$m_{\pi^+} - m_{\pi^0}$	4.9(3)	5.2(3)	4.594
m_n - m_p	2.83(56)	1.55(56)	1.293
$m_{\Sigma^0} - m_{\Sigma^+}$	3.43(39)	2.47(39)	3.27
$m_{\Sigma^-} - m_{\Sigma^0}$	4.04(36)	4.63(36)	4.81
$m_{\Xi^-} - m_{\Xi^0}$	4.72(24)	5.68(24)	6.48

Table 1.10 Electromagnetic mass splittings, in MeV, from Duncan *et al.* [256, 257]

The α and β quantities should be computable from QCD. Holstein [261] reviews the theory and experiments behind the nucleon polarizabilities. The experimental values for α and β are extracted from Compton scattering experiments (see [262] for example). A comparison of the results from lattice QCD to models and experiment can be found in two recent papers [259, 260].

In contrast to the work of Duncan *et al.* the formalism used for electromagnetic polarizations [258] uses static electromagnetic fields. The $SU(3)$ gauge fields are modified as

$$U_1(x) \to e^{i\alpha q E x_4} U_1(x), \tag{1.128}$$

where x_4 is the Euclidean time variable and E is the constant electric field. The phase factor in Eq. 1.128 can be linearised. Smit and Vink have described how to put a constant magnetic field on the lattice [263].

There are speculations that in very strong magnetic fields (*i.e.* for $B \geq 5 \times 10^{14}$ T), the proton will become unstable to the decay to neutrons [264]. Magnetic fields of this intensity may be realised in the universe [265]. The original estimate [264] of the instability of the proton was done in the quark model. In an attempt to remove some of the uncertainty in the hadronic calculation, Rubinstein *et al.* [266] used lattice QCD to study the dependence of the masses of the proton and neutron on the magnetic field.

Some early lattice calculations included the magnetic fields to look at the magnetic moments of hadrons [267, 268]. However, it is best to calculate magnetic moments from form factors [269, 270, 271], so electromagnetic fields are no longer used.

Electric fields were used in the first (unsuccessful) attempts to compute the electric dipole moment of the neutron [272, 273]. The calculation of the neutron electric dipole moment has recently been reformulated [274] in a way that does not require the use of electromagnetic fields.

1.9 Insight from Lattice QCD Calculations

The start of the book Numerical Methods for Scientists and Engineers by Hamming has the immortal phrase: "The purpose of computing is insight not numbers." In the previous sections I have described how lattice QCD is used to compute the masses of hadrons. This may give the impression that lattice QCD is essentially just a black box that produces the masses of hadrons without any insight into the physical mechanisms or relevant degrees of freedom. In this section I hope to show that lattice QCD can also help to explain the physical mechanisms behind the hadron mass spectrum. A very good overview of the type of insight wanted from hadronic physics, that contrasts the hadron spectroscopy approach to the study of confinement with the results from DIS type studies, is the paper by Capstick *et al.* [275]. Note, however, for the B physics experimental program, high precision numbers with reliable error bars are required to look for evidence for physics beyond the standard model of particle physics [276].

Isgur's motivation [239] for studying the N^* particles is based on trying to understand the important degrees of freedom that describe low energy QCD. Increasingly, lattice QCD calculations are being used to provide insight into the dynamics of QCD. Some of Isgur's last papers [106, 107] and [277, 278] were devoted to using lattice QCD to validate the quark model picture of hadronic physics.

At first sight the model building approach to studying QCD appears to give more insight into the dynamics of QCD. It is usually quite easy to study the effect of adding new interactions to the model. There are many models for QCD interactions: the quark model [238], the instanton liquid model [279], or the bag model [280]. The main problem with the model based approach to hadron spectroscopy is that it is very difficult to judge whether the assumptions in the models are valid. The models are only believed when they provide a reasonable description of "most" experimental data with a good χ^2, but this does not prove that they are correct. Different models of the QCD dynamics can be based on very different physical pictures, but may give equally valid descriptions of experimental data. For example the physical assumptions behind the bag model seem to be very different to the assumptions behind the instanton liquid model [279]. The major advantage of lattice QCD calculations is that the theory can be mutilated in a controllable way, so that the physical mechanism underlying a process can be studied.

For example, the question about what mechanism in QCD causes a linearly rising potential for heavy quarks at intermediate distances is not something that can be answered by the quark model. Greensite [281] reviews the work on studying "confinement" using lattice QCD.

In this section I will focus on various attempts to explain the value of the mass splitting between the nucleon and delta. According to the quark model [6], the masses of the nucleon and delta are split by a spin-spin hyperfine term

$$H_{HF} = -\alpha_S M \sum_{i>j} (\vec{\sigma}\lambda_a)_i (\vec{\sigma}\lambda_a)_j, \tag{1.129}$$

where the sum runs over the constituent quarks and λ_a is the set of SU(3) unitary spin matrices (a runs from 1 to 8). In perturbative QCD a term of the form in Eq. 1.129 is naturally generated, but it is not clear whether it is relevant for light relativistic quarks.

Lattice QCD calculations can produce more masses than experiment, so the hyperfine interaction can be tested [282, 283]. A pictorial measure of the quark mass dependence of the masses of the octet and decuplet is given by the "Edinburgh" plot from CP-PACS [97] in Fig. 1.21. The continuous curve is from a quark model by Ono [284] that uses

$$\begin{aligned} M_{\text{baryon}} &= M_b + \sum_{i=1}^{3} m_i + \xi_b \sum_{i>j} \frac{\vec{S}_i \vec{S}_j}{m_i m_j} \\ M_{\text{meson}} &= M_m + \sum_{i=q,\bar{q}} m_i + \xi_m \frac{\vec{S}_q \vec{S}_{\bar{q}}}{m_q m_{\bar{q}}}. \end{aligned} \tag{1.130}$$

The agreement between the lattice data and Ono's model is reasonable in this figure. The lattice QCD data from CP-PACS [97] is almost precise enough to show deviations from this model by Ono [284].

In the instanton liquid model [102], the vacuum is made up of a liquid of interacting instantons. There has been a lot of work on comparing the

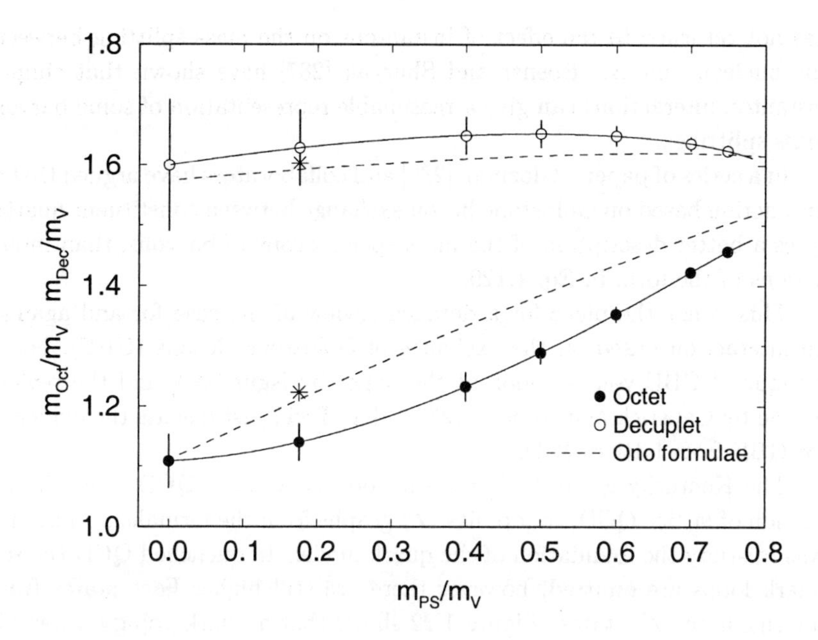

Fig. 1.21 Edinburgh plot from CP-PACS [97] from quenched QCD. The lattice data is compared against the quark model of Ono [284].

instanton liquid model against lattice QCD. The basic idea is to cool the gauge configuration [51]. This removes the perturbative part of the gauge field and leaves the classical configurations, that can be compared against the predictions of the instanton liquid model. The cooling procedure is essentially a way of smoothing the perturbative noise from the configurations. This perturbative noise presumably has something to do with the one gluon exchange term.

In the first study of this on the lattice [285], the masses of the nucleon, ρ, Δ and π were measured in the usual way. The gauge configurations were then cooled and the simple hadron spectrum was measured again. The masses for the ρ, π and nucleon particles were qualitatively the same before and after cooling. As the cooling does not effect the instantons this suggested that the mass splittings are largely due to instantons. The mass splitting between the nucleon and Δ was reduced by smoothing. Chu *et al.* [285] claimed that this was due to the problems with extracting the mass of the Δ from the lattice data. Later work on this issue [101, 286] has not returned to the effect of instantons on the mass splitting between the nucleon and Δ. Rosner and Shuryak [287] have shown that simple instanton interactions can give a reasonable representation of some baryon mass splittings.

In a series of papers, Glozman [288] and collaborators have argued that a interaction based on Goldstone boson exchange between constituent quarks gives a better description of the mass spectroscopy of baryons, than interactions of the form in Eq. 1.129.

This is not the place for a detailed review of the case for and against an interaction based on the exchange of Goldstone Bosons (GBE). For a critique of GBE you can look at the paper by Isgur [289] and the review article by Capstick and Roberts [238]. Here I will just discuss the evidence for GBE from lattice QCD.

The Kentucky group [290] have introduced valence QCD, a mutilated version of lattice QCD, that omits "Z" graphs from the formalism. The aim was to study the foundation of the quark model. In quenched QCD the sea quark loops are omitted, however there are still higher Fock states from intermediate "Z" states. Figure 1.22 shows that a quark going backwards in time can be interpreted as meson state.

The lattice version of valence QCD is the Wilson fermion action (Eq. 1.9) with the backwards hopping terms removed.

$$S_f^W = \sum_x \quad [\quad -\kappa \sum_{i=1}^{3} \{ \overline{\psi}_x (1-\gamma_i) U_i(x) \psi_{x+i} + \overline{\psi}_{x+i} (\gamma_i + 1) U_i^\dagger(x) \psi_x \} \\ + \quad \overline{\psi}_x \psi_x - \kappa \overline{\psi}_{x+\hat{t}} (\gamma_4 + 1) U_{\hat{t}}^\dagger(x) \psi_x + \overline{\psi}_x (1-\gamma_4) \psi_x] \quad (1.131)$$

The Kentucky group [290] studied valence QCD in a lattice calculation at $\beta = 6.0$ with a volume of 16^3 24, and sample size of 100. They also used valence QCD to study form factors and matrix elements, but I will just focus on their results for the hadron spectrum.

In Fig. 1.23, I show a comparison of some hadron masses (from the comment by Isgur [277]). The main conclusion from Fig. 1.23 is that the hyperfine splittings seem to have been reduced for light hadrons.

Valence QCD still has the physics of gluon exchange, so the near degeneracy of the nucleon and Δ suggests that the "Z" graphs are the important mechanism behind the nucleon-delta mass splitting. For heavy hadrons the results from vQCD are not suppressed relative to the those from quenched QCD. For example, the vQCD prediction for the mass splitting between

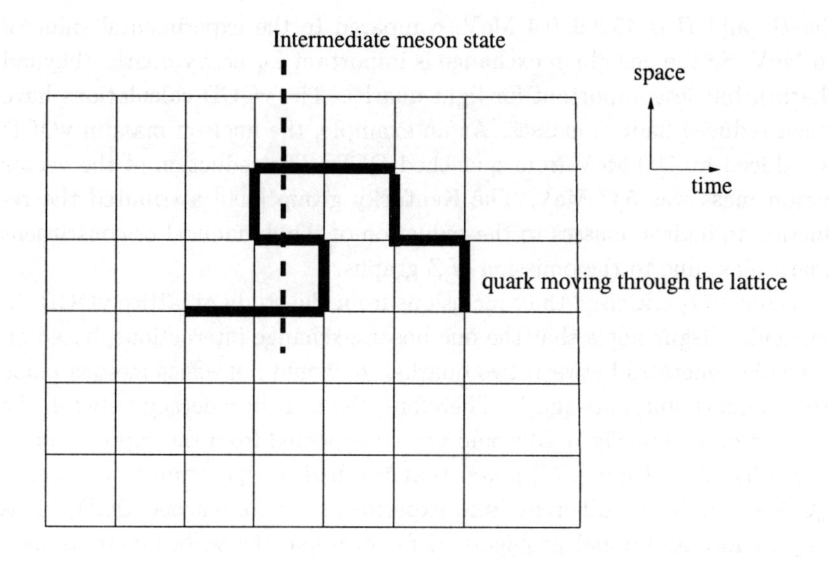

Fig. 1.22 Z graph for a quark.

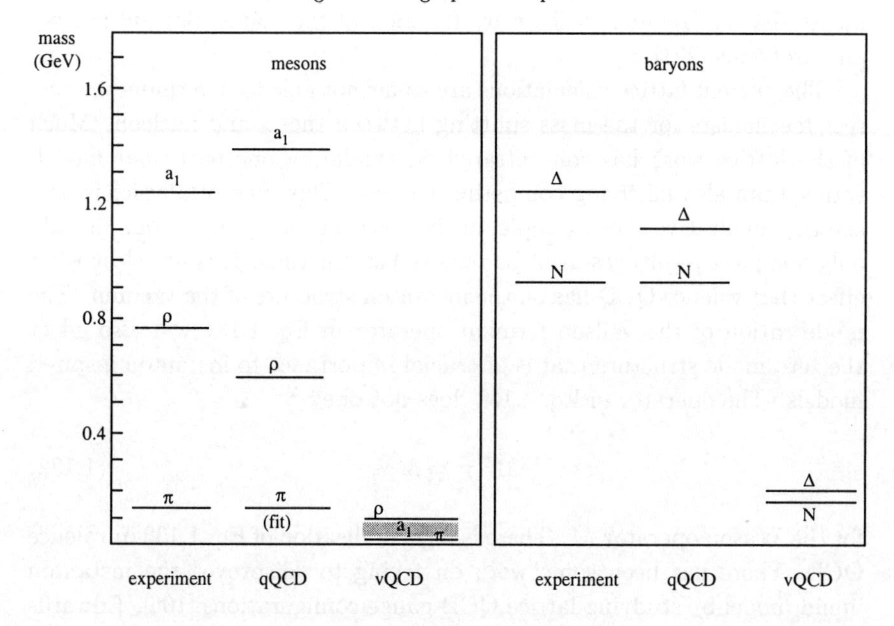

Fig. 1.23 Spectrum of valence (vQCD) and quenched (qQCD) QCD from [277].

the $B^{\star}$ and B is 45.8 ± 0.4 MeV, compared to the experimental value of 46 MeV. So the one gluon exchange is important for heavy quarks (beyond charm), but less important for light quarks. The vQCD calculations have much reduced hadron masses. As an example, the nucleon mass in vQCD is reduced by 700 MeV from quenched QCD. The reduction of the vector meson mass was 537 MeV. The Kentucky group [290] attributed the reduction in hadron masses to the reduction of the dynamical or constituent quark mass due to the omission of Z graphs.

Isgur [277] criticised the conclusions from the study of lattice vQCD. In particular, Isgur notes that the one boson exchange interactions, based on Z graphs, operates between two quarks, so should not effect mesons made from a quark and anti-quark. Therefore, the near degeneracy between the ρ and π masses in Fig. 1.23 would not be expected from the suppression of Z graphs. Also Isgur [277] noted that the hadron spectrum from valence QCD was radically different from experiment (and quenched QCD). This might cause additional problems, if for example the wave-functions were very different for valence QCD compared to the real world, then this would effect the hyperfine splittings. The Kentucky group did study matrix elements that are related to the wave function of the states, but did not see any problems [291].

The current lattice calculations are so far not able to determine the correct mechanism for the mass splitting between the Δ and nucleon. Much of the lattice work has concentrated on validating one particular model, rather than also falsifying competing models. This type of physics is necessarily qualitative. For example, in the spirit of theory mutilation, ideally only one piece of physics must be removed at one time. It is not clear what effect that valence QCD has on the instanton structure of the vacuum. The modification of the Wilson fermion operator in Eq. 1.131 will also effect the zero mode structure that is of crucial importance to instanton inspired models. The operator in Eq. 1.131 does not obey

$$M^{\dagger} = \gamma_5 M \gamma_5 \tag{1.132}$$

for the Wilson operator M. There is a generalisation of Eq. 1.132 in valence QCD. There has been some work on trying to "disprove" the instanton liquid model by studying lattice QCD gauge configurations [106]. Edwards reviews the work by many groups on this [292].

Lattice QCD simulations have been used to test other assumptions made in models of the QCD dynamics. For example, there are some models of hadronic structure that are based on diquarks [293]. A critical assumption in diquark models is that two quarks actually do cluster to form a diquark. This assumption has been tested in lattice gauge theory calculations by the Bielefeld group [294], where they found no deeply bound diquark state in the Landau gauge. Leinweber [295] has claimed that lattice QCD data on the charge radii of hadrons provides evidence against scalar diquark clustering. As a test of the MIT bag model [296] and the Skyrme model [297], density operators from the models were compared against results from quenched QCD.

The large N_C limit of QCD provides much insight into QCD (see [298] for a review). Teper and collaborators [299] have studied the glueball masses in the large N_C limit. Lattice QCD calculations can be done with any gauge group. They also studied the glueball masses for N_C = 2,3,4,5. This allowed them to estimate the size of the corrections to the $N_C \to \infty$ limit. As the $O(1/N_C^2)$ corrections are small, it was important to control both the finite volume and lattice spacing errors.

1.10 What Lattice QCD Is Not Good at

In the previous sections I have implicitly assumed that all the hadrons are stable. In the real world, most hadrons are unstable to strong decays. For example the ρ meson has a mass of 770 MeV and a decay width of about 150 MeV. Most lattice practitioners never "worry" about the ρ's decay width (in public at least). The determination of a hadron's mass from experiment is inextricably linked to the determination of the decay width. This is perhaps most dramatically demonstrated by the problem of missing baryon resonances [300]. The quark model predicted more excited baryons [247] than were actually seen in $N\pi$ reactions. It was claimed that the additional states were not seen because they coupled very weakly to the $N\pi$ channel. The quark model did predict that the missing baryons might be seen in $N\pi\pi$ reactions [300]. There are experiments at the Jefferson laboratory that are trying to detect these "missing resonances" [14].

Lattice QCD calculations have to be done in Euclidean space for convergence of the path integral [35]. This implies that the amplitudes and masses from lattice calculations are real. This makes the study of resonances non-trivial, because decay widths are inherently complex quantities. This is also a problem for calculations with a finite chemical potential, although there has been some progress in this area [301].

I do not discuss the very elegant formalism of Lüscher [302] for studying unstable particles in a finite volume. The formalism and results from the scattering formalism are reviewed by Fiebig and Markum in Chapter 4 of this volume [303]. In this section, I would like to describe the possible implications for mass determinations from standard correlation functions.

The momentum is quantised on a lattice of length L and periodic boundary conditions. The momentum of mesons can only take values

$$p_n = \frac{2\pi n}{aL}, \tag{1.133}$$

where n is an integer between 0 and $L-1$. For a typical lattice, $L = 16$ and $a^{-1} = 2.0$ GeV, so the quantum of momentum is 0.79 GeV. The quantisation of momentum makes the coupling of the state to the scattering states different to that in the continuum. This "feature" has been used to advantage by Lüscher [302] in his formalism.

The quantisation of momentum has important consequences for mesons that decay via P-wave decays such as the ρ meson. A ρ meson at rest can only decay to two pions with momentum p and $-p$. The decay threshold is $2\sqrt{m_\pi^2 + (2\pi/L)^2}$. The quantisation of momentum on the lattice does not effect the threshold for $S-$wave decays. An example of an S-wave decay is the decay of the 0^{++} meson to pairs of mesons. The quarks in current lattice calculations are almost light enough for the strong decay of the flavour singlet 0^{++} [304].

Hadron masses are extracted from lattice QCD calculations using two point correlators (Eq. 1.23). However, the use of Eq. 1.23 may not be appropriate for hadrons that can decay. The most naive modification of the lattice QCD formalism caused by the introduction of decay widths is the replacement

$$m \to m + i\frac{\Gamma}{2}, \tag{1.134}$$

where Γ is the decay width. This modifies Eq. 1.23 into

$$c(t) = a_0 e^{-m_0 t} e^{-i\frac{\Gamma t}{2}} + \cdots \tag{1.135}$$

as the form used to extract the masses from correlators. There are a number of problems with Eq. 1.135, so a more thorough derivation is required.

I review the work by Michael [305] (see also the text book by Brown [306]) on the effect of decays on two point functions. I will consider two scalar fields (σ, π) interacting with the interaction $\sigma\pi\pi$. The mass of the π particle is μ. To study the implications of particle decay on the two point correlator, consider the renormalisation of the propagator of the σ particle in Euclidean space,

$$P_B^{-1}(p) = p^2 + m^2. \tag{1.136}$$

The effect of the interaction of the π particle with the σ particle renormalises the σ propagator as

$$P^{-1}(p) = p^2 + m^2 - X(p^2), \tag{1.137}$$

where X is the self energy.

Masses are extracted from lattice QCD calculations using the time sliced propagator

$$G(t) = \frac{1}{2\pi} \int_{-\infty}^{\infty} dp_0 e^{ip_0 t} P(p_0, \underline{0}) \tag{1.138}$$

$$= \frac{1}{\pi} \int_{2\mu}^{\infty} dE e^{-Et} \rho(E), \tag{1.139}$$

where

$$\rho(E) = \frac{ImX(E)}{(m^2 - E^2 - X)^2} \ . \tag{1.140}$$

If the pole around $E \sim -m$ is neglected then the expression for the spectral density is

$$\rho(E) = \frac{1}{2m} \frac{\gamma(E)}{(m - E)^2 + \gamma(E)^2} \ . \tag{1.141}$$

In the limit $\gamma \ll (m - 2\mu)$, $(m - 2\mu)t \gg 1$

$$G(t) = \frac{1}{2m} e^{-mt} \cos(\gamma t) + \frac{\gamma e^{-2\mu t}}{2\pi m (m - 2\mu)^2 t} \ . \tag{1.142}$$

In Fig. 1.24, I plot separately the *log* of the first and second terms in Eq. 1.142 using the parameters: $(m = 0.5, \gamma = 0.05, \mu = 0.1)$. The breakdown of a pure exponential decay can be seen for the second term.

There is no evidence for a breakdown of the simple exponential fit model in current lattice calculations (apart from the effects of chiral artifacts in

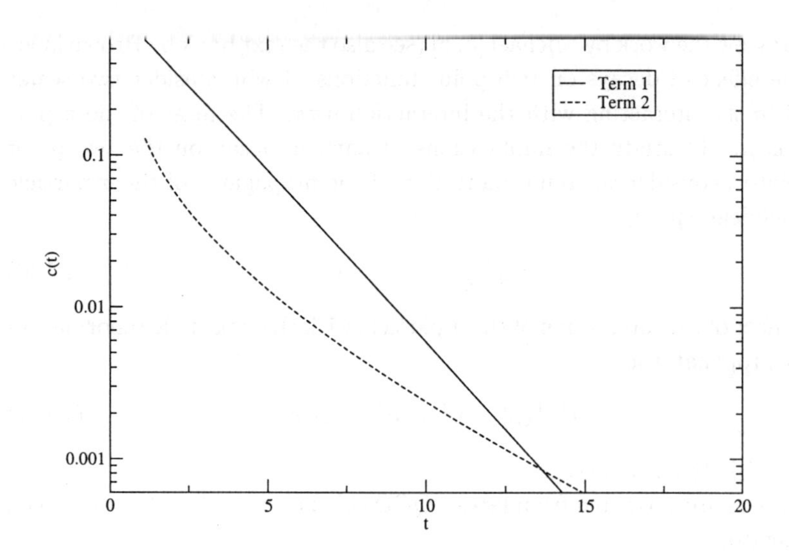

Fig. 1.24 Resonant correlators from Eq. 1.142.

quenched QCD [86]). This effect may become apparent as the sea quark masses are reduced to where particle decay is energetically allowed.

The maximum entropy approach [79, 81] to extracting masses from correlators produces an estimate of the spectral density, so in principle could be used to extract the decay widths of particles. It is not clear to me whether a decay width obtained from an analysis based on maximum entropy method would be physical. Yamazaki and Ishizuka [307] have recently compared the maximum entropy approach to studying unstable particles to the method advocated by Lüscher [302] in a model theory. Yamazaki and Ishizuka claimed good agreement between the two methods.

Another way of looking for "evidence" of resonant behaviour is to look for peculiarities in the quark mass dependence of the hadron masses. One of the first applications of this idea was by DeGrand who studied the effect of ρ decay on the quark mass dependence of the ρ mass [155]. The mixing of the ρ correlator with $\pi\pi$ states at nonzero momentum makes the mass dependence of the ρ correlator more complex than described by the theory in Sec. 1.3.3. The MILC collaboration [90] have recently claimed to see some evidence for the decay of the a_0 (non-singlet 0^{++}) meson by comparing the mass dependence of the a_0 particle in unquenched and quenched QCD.

The MILC collaboration [84] tried to look for the evidence of ρ meson decay by studying the dispersion relation of the ρ particle. The momentum can be injected into particle correlators, so that the hadron masses can be computed at nonzero momentum. The signal to noise gets worse, as the momentum increases, so typically the dispersion relation is only known for a few values of momentum. On the lattice the dispersion relation is of the form [308]

$$\sinh^2 E = \sinh^2 m + \sin^2 p. \tag{1.143}$$

The "improvement" program, described in the Appendix, aims to make the lattice dispersion relation closer to the continuum one. Some clever lattice theorists call this computing "the speed of light" [109].

The relationship in Eq. 1.143 predicts the dependence of the hadron mass as a function of momentum. Consider an interpolating operator for the ρ meson, $\overline{\psi}\gamma_i\psi$. If a unit of momentum is injected in the z direction, then the $\overline{\psi}\gamma_z\psi$ operator will couple to the ρ with one unit of momentum, as well as two pions with unit momentum. The two states will mix and the masses will be modified. However, the operators: $\overline{\psi}\gamma_y\psi$, $\overline{\psi}\gamma_x\psi$ do not couple to the two pion states, essentially because the $\rho\pi\pi$ interaction is zero for these kinematics. A signal for ρ decay would be a different mass from the ρ's polarised perpendicular and parallel to the momentum of the state. The MILC collaboration did not see this effect, perhaps because of artifacts with the type of fermions used. The UKQCD collaboration have recently claimed to see evidence for ρ decay via this mechanism [309] that was consistent with the results from other methods.

In this review I have focused on just computing the masses of hadrons. The computation of decay widths from first principles is the holy grail of hadron spectroscopy. There have been very few attempts at the calculation of the decay widths from lattice QCD. The hadronic coupling constants are computed from matrix elements, rather than trying to fit expressions like Eq. 1.142 to data. Additional correlators over the standard two point correlators are required to be calculated. There have been some attempts to compute the ρ to $\pi\pi$ coupling constant [309, 310, 311, 312]. Perhaps the most famous calculation of decay widths is that of partial widths for the decay of the 0^{++} glueball calculated by the GF11 group [313]. A clarification of the formalism for studying decay widths and mixing in lattice QCD calculations is described in [69, 305].

1.11 Conclusions

The computation of the light hadron mass spectrum is essential to checking the validity of lattice QCD techniques. For example, to reliably extract quark masses requires a precise and consistent calculation of the masses of the light mesons. A precise determination of the light hadron mass spectrum would unequivocally demonstrate that non-perturbative quantities can be extracted from a physically relevant quantum field theory.

The recent large scale quenched lattice QCD calculations of the spectrum of the lightest hadrons from CP-PACS [97, 199] are the benchmark for future calculations. After nearly twenty years of work the deviations of the predictions of quenched QCD from experiment have been quantified. The next generation of large scale quenched QCD calculations (if they are worth doing at all) will probably use the new fermion operators, such as the overlap-Dirac operator, to push to lighter quark masses where M_{PS}/M_V is below 0.4. The light quark mass region of quenched QCD is full of pathologies, such as quenched chiral logs, that will be fun to study theoretically, but are of limited or no relevance to experiment. Quenched QCD calculations will still be of some value for many interesting quantities where the uncertainty is larger than the inherent error of quenched QCD.

The main challenges in unquenched hadron spectrum calculations is determining the lattice spacing dependence and reducing the size of the quark masses used in the calculations. The unquenched calculations of the CP-PACS collaboration [135] have shown that at least three lattice spacings will be needed to obtain high "quality" results. Lattice QCD calculations with 2+1 flavours of quarks are starting to produce important results [90]. The effective field theory community is increasingly doing calculations specifically to analyse data from lattice QCD calculations.

Motivated by the "new nuclear physics" experimental programs at facilities, such as the Jefferson laboratory [13, 14, 15], lattice calculations are starting to be used to study interesting particles such as the $N^{\star}$s. The use of more sophisticated interpolating operators and more advanced statistical techniques (such as maximum entropy techniques) may allow some information to be obtained on some of the lowest excited states of hadrons. The determination of the masses of excited states from lattice QCD would be a big step forward for hadron spectroscopy, if this was indeed possible.

The theory of hadronic physics is the ultimate postmodern playground, as it sometimes seems that the use of a particular model for hadronic physics, from a mutual incompatible set of possibilities, is almost a matter of personal preference [314]. One advantage of lattice QCD calculations is that they provide qualitative information about physical mechanisms, that is not directly accessible from experiment. If as Shuryak [279] claims, the physical picture behind the bag model and the instanton liquid model are different, then only one picture is correct, so one of them must be discarded. I hope that qualitative lattice QCD calculations can help simplify the theories behind hadronic spectroscopy by ruling out the underlying pictures behind certain classes of models. The aim of simplifying the theory of hadronic physics is a stated aim of the current experimental program [239, 275].

There are a number of interesting "challenges" for hadron spectrum calculations beyond the critical task of reducing the errors in lattice QCD calculations:

- Can the resonant nature of the ρ meson be determined from lattice QCD?
- Can improved lattice calculations determine the structure of the Roper resonance (first excited state of the nucleon)? For over thirty five years, there have been many speculations on the nature of the Roper resonance. Can lattice QCD close this issue?
- Can the physical mechanism behind the hyperfine splittings in mesons and baryons be determined from lattice QCD?
- Can lattice QCD calculations be used to simplify hadronic physics by ruling out (or perhaps even validating) the bag model?

Acknowledgements

This work is supported by PPARC. I thank members of the UKQCD and MILC collaborations for discussions. I thank Chris Michael for reading the manuscript.

Appendix: Technical Details

To outsiders (physicists who live in continuum), the lattice gauge theory community must seem like a very inward looking bunch. A large fraction of their research is on improving the methods used in lattice calculations. These methods then lead to smaller and more believable error bars and hence is a good thing! Currently, the biggest improvements in the methodology of lattice QCD calculations are coming from "better" lattice representations of the continuum Dirac operator. The "new" lattice representations of the Dirac operator have either reduced lattice spacing dependence or a better chiral symmetry.

The importance of the dependence of the results on the lattice spacing has been stressed throughout this review. As the results of lattice calculations are extrapolated to the continuum, the calculations would be more precise if the lattice spacing dependence of quantities was weak. The computational cost of reducing the lattice spacing used in lattice QCD calculations from Eq. 1.46 is very large, hence it is advantageous to use coarser lattice spacings [108, 315].

A standard technique from numerical analysis is to use finite differences that are closer approximations to the continuum derivatives. For example the lattice derivative

$$\frac{4}{3}\{\frac{f(x+a)-f(x-a)}{2a} - \frac{f(x+2a)-f(x-2a)}{16a}\} = \frac{df}{dx} + O(a^4) \quad \text{(A.1)}$$

should be more accurate with a larger lattice spacing than the derivative

$$\frac{f(x+a)-f(x-a)}{2a} = \frac{df}{dx} + O(a^2). \quad \text{(A.2)}$$

However, in a quantum field theory there are additional complications, such as the operators in Eq. A.1 mixing under renormalisation.

There is a formalism due to Symanzik [316, 317] called improvement, where new terms are added to the lattice action that cancel $O(a)$ terms (irrelevant operators) in a way that is consistent with quantum field theory. The required terms in the improved Lagrangian can be simplified by the use of field redefinitions in the path integral [318]. A very elegant numerical procedure to improve the Wilson fermion action has been developed by the ALPHA collaboration (see the review [192] by Lüscher).

The Wilson fermion operator in Eq. 1.9 differs from the continuum Lagrangian by $O(a)$ terms. The improvement scheme used in most lattice

QCD calculations with Wilson fermions is called clover improvement. The clover term [318] is added to the Wilson fermion operator in Eq. 1.9 to give

$$S_f^{\text{clover}} = S_f^W + c_{SW}\frac{ia\kappa}{2}\sum_x(\overline{\psi_x}\sigma_{\nu\mu}F_{\nu\mu}\psi_x), \tag{A.3}$$

where $F_{\nu\mu}$ is the lattice field strength tensor.

If the c_{SW} coefficient computed in perturbation theory is used then the errors in the results from the lattice calculation are $O(ag^4)$. The ALPHA collaboration [319] have computed c_{SW} to all orders in g^2 using a numerical technique. The result for c_{SW} from ALPHA is

$$c_{SW} = \frac{1 - 0.656g^2 - 0.152g^4 - 0.054g^6}{1 - 0.922g^2} \tag{A.4}$$

for $0 < g < 1$, where g is the coupling. The estimate of c_{SW}, by the ALPHA collaboration, agrees with the one loop perturbation theory for $g < 1/2$.

The clover improvement program for Wilson fermions has had many practical successes. Unfortunately, it is computationally very costly to reach light quark masses in quenched or unquenched lattice calculations that use the clover fermion operator [131]. Hence, attention has focused on also improving the eigenvalue spectrum of the lattice representation of the Dirac operator.

The design of fermion operators on the lattice has a deep connection with chiral symmetry and the global chiral anomaly. The theoretical complications with transcribing the Dirac operator to the lattice are reviewed in many places [17, 20]. Our understanding of chiral symmetry on the lattice has recently increased by the rediscovery of the Ginsparg-Wilson relation [320]

$$M\gamma_5 + \gamma_5 M = aM\gamma_5 M, \tag{A.5}$$

where M is the fermion operator in Eq. 1.8 at zero mass. Equation A.5 smoothly matches onto the chiral symmetry equation in the continuum as the lattice spacing is taken to zero. Lattice fermion operators that obey this Ginsparg-Wilson relation have a form of lattice chiral symmetry [321]. Explicit solutions — such as overlap-Dirac [322] or perfect actions [323] — to Eq. A.5 are known. Those actions that obey the Ginsparg-Wilson relation are increasingly being used for quenched QCD calculations [324]. Domain Wall actions, that can loosely be thought of as being approximate solutions to the Ginsparg-Wilson relation, have been used in calculations [325, 326] of the matrix elements for ϵ'/ϵ.

The main downside of fermion operators that obey the Ginsparg-Wilson relation is that they are computationally expensive. In a review of the literature, Jansen argues [327] that overlap-Dirac type operators are roughly 100 times more expensive computationally than calculations with standard Wilson fermions. The development of new algorithms should reduce this difference in computational cost [200]. The various versions of the overlap-Dirac operator are cheap enough to use for quenched calculations and I, therefore, speculate that there will be an increasing trend to use overlap-Dirac type operators for quenched calculations. It will be some time before unquenched calculations, that are useful for phenomenology, are performed with overlap-Dirac operators. Unquenched calculations of QCD with domain wall fermions have just started [328].

A more pragmatic development in the design of light fermion actions is the development of improved staggered fermion actions [91, 329]. This class of action is being used for unquenched lattice QCD calculations with very light quarks (see Table 1.3) by the MILC collaboration. The problem with standard Kogut-Susskind quarks was that the formalism broke flavour symmetry. So numerical calculations usually had fifteen pions split by a considerable amount from the Goldstone boson pion. The new variants of fermion operators in the staggered formulations have much reduced flavour symmetry breaking. The improved staggered quark formalism is quite ugly compared to actions that are solutions of the Ginsparg-Wilson relation, but lattice QCD is a pragmatic subject and utility wins out over beauty. It is not understood why calculations using improved staggered quarks are much faster [83] than calculations using Wilson fermions [82].*

Bibliography

[1] D. Weingarten, Phys. Lett. **B109**, 57 (1982),
[2] H. Hamber and G. Parisi, Phys. Rev. Lett. **47**, 1792 (1981),
[3] F. Butler, H. Chen, J. Sexton, A. Vaccarino, and D. Weingarten, Phys. Rev. Lett. **70**, 2849 (1993), hep-lat/9212031,

*Ed. The above topic of Improvement is discussed in more detail in Chapter 4 Appendix B. Some believe [330] that this program of Improvement is "the most important theoretical advance in recent years" — a claim that is not universally accepted in the LQCD community [331].

[4] F. Butler, H. Chen, J. Sexton, A. Vaccarino, and D. Weingarten, Nucl. Phys. **B430**, 179 (1994), hep-lat/9405003,
[5] M. Beneke, (2002), hep-lat/0201011,
[6] Particle Data Group, K. Hagiwara *et al.*, Phys. Rev. **D66**, 010001 (2002),
[7] H. Wittig, (2002), hep-lat/0210025,
[8] C. Michael, Chapter 2 in this Volume , hep-lat/0302001,
[9] C. Michael, Phys. Scripta **T99**, 7 (2002), hep-lat/0111056,
[10] C. Davies, The heavy hadron spectrum, 1997, hep-ph/9710394.
[11] C. Davies, Heavy Flavour Physics, Scottish Graduate Textbook Series, Institute of Physics 2002, eds. C. T. H. Davies and S. M. Playfer, hep-ph/0205181.
[12] C. McNeile, (2002), hep-lat/0210026,
[13] CLAS, P. Rossi, (2003), hep-ex/0302032,
[14] V. D. Burkert, (2002), hep-ph/0210321,
[15] V. D. Burkert, (2001), hep-ph/0106143,
[16] I. Montvay and G. Munster, *Quantum fields on a lattice* , Cambridge, UK: Univ. Pr. (1994) (Cambridge monographs on mathematical physics).
[17] H. J. Rothe, *Lattice gauge theories: An Introduction* Vol. 59 (1997).
[18] J. Smit, *Introduction to quantum fields on a lattice: A robust mate* , Cambridge, UK: Univ. Pr. (2002).
[19] M. Creutz, *Quarks, Gluons and Lattices* , Cambridge, UK: Univ. Pr. (1983) (Cambridge monographs on mathematical physics).
[20] R. Gupta, Introduction to lattice QCD, Lectures given at the LXVIII Les Houches Summer School "Probing the Standard Model of Particle Interactions", July 28-Sept 5, 1997, hep-lat/9807028.
[21] A. S. Kronfeld, (2002), hep-lat/0205021,
[22] C. T. H. Davies *et al.*, (1998), hep-lat/9801024,
[23] T. DeGrand, C. DeTar, R. Sugar, and D. Toussaint, (1998), hep-lat/9811023,
[24] M. Mueller-Preussker *et al.*, (2002), hep-lat/0203004,
[25] G. S. Bali, (2003), nucl-th/0302039,
[26] R. G. Edwards, (2002), hep-lat/0210027,
[27] S. Gottlieb, Nucl. Phys. Proc. Suppl. **53**, 155 (1997), hep-lat/9608107,
[28] R. D. Mawhinney, Nucl. Phys. Proc. Suppl. **83**, 57 (2000), hep-lat/0001032,
[29] S. Aoki, Nucl. Phys. Proc. Suppl. **94**, 3 (2001), hep-lat/0011074,
[30] D. Toussaint, Nucl. Phys. Proc. Suppl. **106**, 111 (2002), hep-lat/0110010,
[31] T. Kaneko, (2001), hep-lat/0111005,
[32] M. Di Pierro, (2000), hep-lat/0009001,
[33] M. Di Pierro, (1998), hep-lat/9811036,
[34] M. Di Pierro, (2000), hep-lat/0004007,
[35] J. Glimm and A. Jaffe, New York, USA: Springer (1987) 535p.
[36] J. E. Mandula, G. Zweig, and J. Govaerts, Nucl. Phys. **B228**, 91 (1983),
[37] J. E. Mandula and E. Shpiz, Nucl. Phys. **B232**, 180 (1984),
[38] G. P. Lepage and P. B. Mackenzie, Phys. Rev. **D48**, 2250 (1993),

hep-lat/9209022,
[39] UKQCD, D. S. Henty and R. D. Kenway, Phys. Lett. **B289**, 109 (1992), hep-lat/9206009,
[40] S. Elitzur, Phys. Rev. **D12**, 3978 (1975),
[41] B. Velikson and D. Weingarten, Nucl. Phys. **B249**, 433 (1985),
[42] S. Gottlieb, Presented at Conf. 'Advances in Lattice Gauge Theory', Tallahassee, FL, Apr 10-13, 1985.
[43] A. Duncan, E. Eichten, and H. Thacker, Phys. Lett. **B303**, 109 (1993),
[44] M. W. Hecht *et al.*, Phys. Rev. **D47**, 285 (1993), hep-lat/9208005,
[45] M. W. Hecht and T. A. DeGrand, Phys. Rev. **D46**, 2155 (1992),
[46] UKQCD, P. Lacock, A. McKerrell, C. Michael, I. M. Stopher, and P. W. Stephenson, Phys. Rev. **D51**, 6403 (1995), hep-lat/9412079,
[47] A. Billoire, E. Marinari, and G. Parisi, Phys. Lett. **B162**, 160 (1985),
[48] UKQCD, C. R. Allton *et al.*, Phys. Rev. **D47**, 5128 (1993), hep-lat/9303009,
[49] J. W. Negele and H. Orland, Redwood City, USA: Addison-Wesley (1988) 459 P. (Frontiers in Physics, 68).
[50] A. Duncan, S. Pernice, and J. Yoo, Phys. Rev. **D65**, 094509 (2002), hep-lat/0112036,
[51] J. W. Negele, Nucl. Phys. Proc. Suppl. **73**, 92 (1999), hep-lat/9810053,
[52] UKQCD, C. McNeile, Data storage issues in lattice QCD calculations, 2000, hep-lat/0003009.
[53] UKQCD, C. T. H. Davies, A. C. Irving, R. D. Kenway, and C. M. Maynard, (2002), hep-lat/0209121,
[54] P. Maris and C. D. Roberts, (2003), nucl-th/0301049,
[55] C. D. Roberts and A. G. Williams, Prog. Part. Nucl. Phys. **33**, 477 (1994), hep-ph/9403224,
[56] A. Frommer, Nucl. Phys. Proc. Suppl. **53**, 120 (1997), hep-lat/9608074,
[57] G. M. de Divitiis, R. Frezzotti, M. Masetti, and R. Petronzio, Phys. Lett. **B382**, 393 (1996), hep-lat/9603020,
[58] UKQCD, C. Michael and J. Peisa, Phys. Rev. **D58**, 034506 (1998), hep-lat/9802015,
[59] A. Duncan and E. Eichten, Phys. Rev. **D65**, 114502 (2002), hep-lat/0112028,
[60] M. C. Chu, J. M. Grandy, S. Huang, and J. W. Negele, Phys. Rev. **D48**, 3340 (1993), hep-lat/9306002,
[61] UKQCD, S. J. Hands, P. W. Stephenson, and A. McKerrell, Phys. Rev. **D51**, 6394 (1995), hep-lat/9412065,
[62] T. DeGrand, Phys. Rev. **D64**, 094508 (2001), hep-lat/0106001,
[63] E. V. Shuryak, Rev. Mod. Phys. **65**, 1 (1993),
[64] G. P. Lepage *et al.*, Nucl. Phys. Proc. Suppl. **106**, 12 (2002), hep-lat/0110175,
[65] C. Michael, Phys. Rev. **D49**, 2616 (1994), hep-lat/9310026,
[66] C. Michael and A. McKerrell, Phys. Rev. **D51**, 3745 (1995),

hep-lat/9412087,
[67] C. Michael and M. Teper, Nucl. Phys. **B314**, 347 (1989),
[68] M. Luscher and U. Wolff, Nucl. Phys. **B339**, 222 (1990),
[69] UKQCD, C. McNeile and C. Michael, Phys. Rev. **D63**, 114503 (2001), hep-lat/0010019,
[70] T. Draper and C. McNeile, Nucl. Phys. Proc. Suppl. **34**, 453 (1994), hep-lat/9401013,
[71] G. P. Lepage, Nucl. Phys. Proc. Suppl. **26**, 45 (1992),
[72] T. A. DeGrand and M. W. Hecht, Phys. Rev. **D46**, 3937 (1992), hep-lat/9206011,
[73] UKQCD, C. R. Allton *et al.*, Phys. Rev. **D65**, 054502 (2002), hep-lat/0107021,
[74] M. A. Shifman, World Sci. Lect. Notes Phys. **62**, 1 (1999),
[75] D. B. Leinweber, Phys. Rev. **D51**, 6369 (1995), nucl-th/9405002,
[76] C. Allton and S. Capitani, Nucl. Phys. **B526**, 463 (1998), hep-lat/9712006,
[77] C. Allton, D. Blythe, and J. Clowser, Nucl. Phys. Proc. Suppl. **109**, 192 (2002), hep-lat/0202024,
[78] A. Bochkarev and P. de Forcrand, Nucl. Phys. **B477**, 489 (1996), hep-lat/9505025,
[79] M. Asakawa, T. Hatsuda, and Y. Nakahara, Prog. Part. Nucl. Phys. **46**, 459 (2001), hep-lat/0011040,
[80] Y. Nakahara, M. Asakawa, and T. Hatsuda, Phys. Rev. **D60**, 091503 (1999), hep-lat/9905034,
[81] CP-PACS, T. Yamazaki *et al.*, Phys. Rev. **D65**, 014501 (2002), hep-lat/0105030.
[82] TXL, T. Lippert, Nucl. Phys. Proc. Suppl. **106**, 193 (2002), hep-lat/0203009,
[83] S. Gottlieb, Nucl. Phys. Proc. Suppl. **106**, 189 (2002), hep-lat/0112039,
[84] C. W. Bernard *et al.*, Phys. Rev. **D48**, 4419 (1993), hep-lat/9305023,
[85] J.-W. Chen, Phys. Lett. **B543**, 183 (2002), hep-lat/0205014,
[86] W. Bardeen, A. Duncan, E. Eichten, N. Isgur, and H. Thacker, Phys. Rev. **D65**, 014509 (2002), hep-lat/0106008,
[87] W.-J. Lee and D. Weingarten, Phys. Rev. **D61**, 014015 (2000), hep-lat/9910008,
[88] P. H. Damgaard, M. C. Diamantini, P. Hernandez, and K. Jansen, Nucl. Phys. **B629**, 445 (2002), hep-lat/0112016,
[89] W. Bardeen, A. Duncan, E. Eichten, G. Hockney, and H. Thacker, Phys. Rev. **D57**, 1633 (1998), hep-lat/9705008,
[90] C. W. Bernard *et al.*, Phys. Rev. **D64**, 054506 (2001), hep-lat/0104002,
[91] MILC, C. W. Bernard *et al.*, Phys. Rev. **D61**, 111502 (2000), hep-lat/9912018,
[92] HPQCD, A. Gray *et al.*, (2002).
[93] R. Sommer, Nucl. Phys. **B411**, 839 (1994), hep-lat/9310022,
[94] G. S. Bali, Phys. Rept. **343**, 1 (2001), hep-ph/0001312,

[95] M. Creutz, L. Jacobs, and C. Rebbi, Phys. Rept. **95**, 201 (1983),
[96] C. R. Allton, (1996), hep-lat/9610016,
[97] CP-PACS, S. Aoki *et al.*, Phys. Rev. **D67**, 034503 (2003), hep-lat/0206009,
[98] C. J. Morningstar and M. J. Peardon, Phys. Rev. **D60**, 034509 (1999), hep-lat/9901004,
[99] C. Morningstar and M. J. Peardon, Nucl. Phys. Proc. Suppl. **83**, 887 (2000), hep-lat/9911003,
[100] M. Guagnelli, K. Jansen, and R. Petronzio, Phys. Lett. **B457**, 153 (1999), hep-lat/9901016,
[101] T. DeGrand and A. Hasenfratz, Phys. Rev. **D64**, 034512 (2001), hep-lat/0012021,
[102] T. Schafer and E. V. Shuryak, Rev. Mod. Phys. **70**, 323 (1998), hep-ph/9610451,
[103] M. Teper, Nucl. Phys. Proc. Suppl. **83**, 146 (2000), hep-lat/9909124,
[104] A. Hasenfratz and C. Nieter, Phys. Lett. **B439**, 366 (1998), hep-lat/9806026.
[105] A. Hasenfratz, Phys. Lett. **B476**, 188 (2000), hep-lat/9912053,
[106] I. Horvath, N. Isgur, J. McCune, and H. B. Thacker, Phys. Rev. **D65**, 014502 (2002), hep-lat/0102003,
[107] I. Horvath *et al.*, Phys. Rev. **D66**, 034501 (2002), hep-lat/0201008,
[108] M. G. Alford, W. Dimm, G. P. Lepage, G. Hockney, and P. B. Mackenzie, Phys. Lett. **B361**, 87 (1995), hep-lat/9507010,
[109] M. G. Alford, T. R. Klassen, and G. P. Lepage, Phys. Rev. **D58**, 034503 (1998), hep-lat/9712005,
[110] M. G. Alford and R. L. Jaffe, Nucl. Phys. **B578**, 367 (2000), hep-lat/0001023,
[111] MILC, T. DeGrand, Phys. Rev. **D58**, 094503 (1998), hep-lat/9802012,
[112] UKQCD, C. R. Allton *et al.*, Phys. Rev. **D60**, 034507 (1999), hep-lat/9808016,
[113] UKQCD, A. C. Irving *et al.*, Phys. Rev. **D58**, 114504 (1998), hep-lat/9807015,
[114] S. R. Sharpe, (1998), hep-lat/9811006,
[115] C. Bernard *et al.*, (2002), hep-lat/0209086,
[116] W. Detmold, W. Melnitchouk, J. W. Negele, D. B. Renner, and A. W. Thomas, Phys. Rev. Lett. **87**, 172001 (2001), hep-lat/0103006,
[117] H. Georgi, Ann. Rev. Nucl. Part. Sci. **43**, 209 (1993),
[118] G. P. Lepage, (1997), nucl-th/9706029,
[119] D. B. Kaplan, (1995), nucl-th/9506035,
[120] S. Scherer, (2002), hep-ph/0210398,
[121] S. Weinberg, Physica **A96**, 327 (1979),
[122] A. Manohar and H. Georgi, Nucl. Phys. **B234**, 189 (1984),
[123] J. Gasser and H. Leutwyler, Nucl. Phys. **B250**, 465 (1985),
[124] C. W. Bernard and M. F. L. Golterman, Phys. Rev. **D49**, 486 (1994), hep-lat/9306005,

[125] S. R. Sharpe and N. Shoresh, Phys. Rev. **D62**, 094503 (2000), hep-lat/0006017,
[126] S. R. Sharpe, Phys. Rev. **D56**, 7052 (1997), hep-lat/9707018,
[127] ALPHA, J. Heitger, R. Sommer, and H. Wittig, Nucl. Phys. **B588**, 377 (2000), hep-lat/0006026,
[128] W. Bardeen, A. Duncan, E. Eichten, and H. Thacker, Phys. Rev. **D62**, 114505 (2000), hep-lat/0007010,
[129] UKQCD, A. C. Irving, C. McNeile, C. Michael, K. J. Sharkey, and H. Wittig, Phys. Lett. **B518**, 243 (2001), hep-lat/0107023,
[130] D. R. Nelson, G. T. Fleming, and G. W. Kilcup, Phys. Rev. Lett. **90**, 021601 (2003), hep-lat/0112029,
[131] UKQCD, A. C. Irving, (2002), hep-lat/0208065.
[132] J. Bijnens, G. Ecker, and J. Gasser, (1994), hep-ph/9411232,
[133] G. Ecker, J. Gasser, A. Pich, and E. de Rafael, Nucl. Phys. **B321**, 311 (1989),
[134] D. B. Kaplan and A. V. Manohar, Phys. Rev. Lett. **56**, 2004 (1986),
[135] CP-PACS, A. Ali Khan *et al.*, Phys. Rev. **D65**, 054505 (2002), hep-lat/0105015,
[136] N. H. Fuchs, H. Sazdjian, and J. Stern, Phys. Lett. **B269**, 183 (1991),
[137] M. Knecht and J. Stern, (1994), hep-ph/9411253,
[138] L. Giusti, F. Rapuano, M. Talevi, and A. Vladikas, Nucl. Phys. **B538**, 249 (1999), hep-lat/9807014,
[139] P. Hernandez, K. Jansen, and L. Lellouch, Phys. Lett. **B469**, 198 (1999), hep-lat/9907022,
[140] G. Ecker, (1998), hep-ph/9805500,
[141] T. Morozumi, A. I. Sanda, and A. Soni, Phys. Rev. **D46**, 2240 (1992),
[142] A. Morel, J. Phys. (France) **48**, 1111 (1987),
[143] S. R. Sharpe, Phys. Rev. **D41**, 3233 (1990),
[144] C. W. Bernard and M. F. L. Golterman, Phys. Rev. **D46**, 853 (1992), hep-lat/9204007,
[145] M. Golterman, (1997), hep-ph/9710468,
[146] G. Rupak and N. Shoresh, (2002), hep-lat/0201019,
[147] S. R. Sharpe and J. Singleton, Robert, Phys. Rev. **D58**, 074501 (1998), hep-lat/9804028,
[148] qq+q, F. Farchioni, C. Gebert, I. Montvay, E. Scholz, and L. Scorzato, Phys. Lett. **B561**, 102 (2003), hep-lat/0302011,
[149] U. G. Meissner, Phys. Rept. **161**, 213 (1988),
[150] G. Ecker, J. Gasser, H. Leutwyler, A. Pich, and E. de Rafael, Phys. Lett. **B223**, 425 (1989),
[151] J. Bijnens, P. Gosdzinsky, and P. Talavera, Nucl. Phys. **B501**, 495 (1997), hep-ph/9704212,
[152] E. Jenkins, A. V. Manohar, and M. B. Wise, Phys. Rev. Lett. **75**, 2272 (1995), hep-ph/9506356,
[153] M. Booth, G. Chiladze, and A. F. Falk, Phys. Rev. **D55**, 3092 (1997),

hep-ph/9610532,
[154] C.-K. Chow and S.-J. Rey, Nucl. Phys. **B528**, 303 (1998), hep-ph/9708432,
[155] T. A. DeGrand, Phys. Rev. **D43**, 2296 (1991),
[156] D. B. Leinweber and T. D. Cohen, Phys. Rev. **D49**, 3512 (1994), hep-ph/9307261,
[157] P. Geiger and N. Isgur, Phys. Rev. **D41**, 1595 (1990),
[158] N. Isgur, Phys. Rev. **D60**, 054013 (1999), nucl-th/9901032,
[159] D. B. Leinweber, A. W. Thomas, K. Tsushima, and S. V. Wright, Phys. Rev. **D64**, 094502 (2001), hep-lat/0104013,
[160] A. W. Thomas, (2002), hep-lat/0208023,
[161] D. B. Leinweber, A. W. Thomas, K. Tsushima, and S. V. Wright, Phys. Rev. **D61**, 074502 (2000), hep-lat/9906027,
[162] E. Jenkins and A. V. Manohar, Phys. Lett. **B259**, 353 (1991),
[163] V. Bernard, N. Kaiser, J. Kambor, and U. G. Meissner, Nucl. Phys. **B388**, 315 (1992),
[164] B. Borasoy and U.-G. Meissner, Annals Phys. **254**, 192 (1997), hep-ph/9607432,
[165] J. F. Donoghue, B. R. Holstein, and B. Borasoy, Phys. Rev. **D59**, 036002 (1999), hep-ph/9804281,
[166] D. B. Kaplan, M. J. Savage, and M. B. Wise, Nucl. Phys. **B478**, 629 (1996), nucl-th/9605002,
[167] E. Jenkins, Nucl. Phys. **B368**, 190 (1992),
[168] A. W. Thomas, D. B. Leinweber, K. Tsushima, and S. V. Wright, Nucl. Phys. **A663**, 973 (2000), nucl-th/9909041,
[169] U.-G. Meissner, (1998), hep-ph/9810276,
[170] R. D. Young, D. B. Leinweber, and A. W. Thomas, (2002), hep-lat/0212031,
[171] D. B. Leinweber, A. W. Thomas and R. D. Young, hep-lat/0302020.
[172] R. Lewis and P.-P. A. Ouimet, Phys. Rev. **D64**, 034005 (2001), hep-ph/0010043,
[173] B. Borasoy, R. Lewis, and P.-P. A. Ouimet, Phys. Rev. **D65**, 114023 (2002), hep-ph/0203199,
[174] T. Becher and H. Leutwyler, Eur. Phys. J. **C9**, 643 (1999), hep-ph/9901384,
[175] A. R. Hoch and R. R. Horgan, Nucl. Phys. **B380**, 337 (1992),
[176] C. R. Allton, V. Gimenez, L. Giusti, and F. Rapuano, Nucl. Phys. **B489**, 427 (1997), hep-lat/9611021,
[177] UKQCD, P. Lacock and C. Michael, Phys. Rev. **D52**, 5213 (1995), hep-lat/9506009,
[178] L. Maiani and G. Martinelli, Phys. Lett. **B178**, 265 (1986),
[179] K. C. Bowler *et al.*, Phys. Lett. **B162**, 354 (1985),
[180] Ape, P. Bacilieri *et al.*, Nucl. Phys. **B317**, 509 (1989),
[181] M. Fukugita, H. Mino, M. Okawa, and A. Ukawa, Phys. Rev. Lett. **68**, 761 (1992),
[182] B. Sakita, World Sci. Lect. Notes Phys. **1**, 1 (1985),

[183] M. Luscher, Phys. Lett. **B118**, 391 (1982),
[184] M. Luscher, Nucl. Phys. **B219**, 233 (1983),
[185] ALPHA, J. Garden, J. Heitger, R. Sommer, and H. Wittig, Nucl. Phys. **B571**, 237 (2000), hep-lat/9906013,
[186] J. Gasser and H. Leutwyler, Phys. Lett. **B184**, 83 (1987),
[187] G. Colangelo, S. Durr, and R. Sommer, (2002), hep-lat/0209110,
[188] QCDSF, A. Ali Khan *et al.*, (2002), hep-lat/0209111,
[189] S. Antonelli *et al.*, Phys. Lett. **B345**, 49 (1995), hep-lat/9405012,
[190] I. Montvay, Rev. Mod. Phys. **59**, 263 (1987),
[191] B. Orth *et al.*, Nucl. Phys. Proc. Suppl. **106**, 269 (2002), hep-lat/0110158,
[192] M. Luscher, (1998), hep-lat/9802029,
[193] J. D. Bjorken, Presented at the SLAC Summer Institute on Particle Physics, Stanford, Calif., Jul 9-20, 1979.
[194] P. van Baal, (2000), hep-ph/0008206,
[195] UKQCD, K. C. Bowler *et al.*, Phys. Rev. **D62**, 054506 (2000), hep-lat/9910022,
[196] F. James and M. Roos, Comput. Phys. Commun. **10**, 343 (1975),
[197] UKQCD, C. R. Allton *et al.*, Phys. Lett. **B284**, 377 (1992), hep-lat/9205016,
[198] ALPHA, M. Guagnelli, R. Sommer, and H. Wittig, Nucl. Phys. **B535**, 389 (1998), hep-lat/9806005,
[199] CP-PACS, S. Aoki *et al.*, Phys. Rev. Lett. **84**, 238 (2000), hep-lat/9904012,
[200] BGR, C. Gattringer *et al.*, (2003), hep-lat/0307013,
[201] M. Della Morte, R. Frezzotti, and J. Heitger, Nucl. Phys. Proc. Suppl. **106**, 260 (2002), hep-lat/0110166,
[202] M. Gockeler *et al.*, Phys. Rev. **D57**, 5562 (1998), hep-lat/9707021,
[203] UKQCD, C. McNeile and C. Michael, Nucl. Phys. Proc. Suppl. **106**, 251 (2002), hep-lat/0110108,
[204] P. Hasenfratz, S. Hauswirth, T. Jorg, F. Niedermayer, and K. Holland, (2002), hep-lat/0205010,
[205] CP-PACS, A. Ali Khan *et al.*, Phys. Rev. Lett. **85**, 4674 (2000), hep-lat/0004010,
[206] TXL, U. Glassner *et al.*, Phys. Lett. **B383**, 98 (1996), hep-lat/9604014,
[207] TXL, G. S. Bali *et al.*, Phys. Rev. **D62**, 054503 (2000), hep-lat/0003012,
[208] ALPHA, K. Jansen and R. Sommer, Nucl. Phys. **B530**, 185 (1998), hep-lat/9803017,
[209] MILC, C. W. Bernard *et al.*, Phys. Rev. **D56**, 7039 (1997), hep-lat/9707008,
[210] V. Lubicz, Nucl. Phys. Proc. Suppl. **94**, 116 (2001), hep-lat/0012003,
[211] G. Martinelli and Y.-C. Zhang, Phys. Lett. **B125**, 77 (1983),
[212] A. Hasenfratz, P. Hasenfratz, Z. Kunszt, and C. B. Lang, Phys. Lett. **B117**, 81 (1982),
[213] A. Patel and R. Gupta, Phys. Lett. **B131**, 425 (1983),
[214] UKQCD, P. Lacock, C. Michael, P. Boyle, and P. Rowland, Phys. Rev.

D54, 6997 (1996), hep-lat/9605025,
[215] T. DeGrand and M. Hecht, Phys. Lett. **B275**, 435 (1992),
[216] H. B. Meyer and M. J. Teper, Nucl. Phys. **B658**, 113 (2003), hep-lat/0212026,
[217] J. D. Weinstein and N. Isgur, Phys. Rev. **D27**, 588 (1983),
[218] UKQCD, A. Hart, C. McNeile, and C. Michael, (2002), hep-lat/0209063,
[219] RBC, S. Prelovsek and K. Orginos, (2002), hep-lat/0209132,
[220] M. M. Brisudova, L. Burakovsky, and T. Goldman, Phys. Lett. **B460**, 1 (1999), hep-ph/9810296,
[221] A. J. G. Hey and R. L. Kelly, Phys. Rept. **96**, 71 (1983),
[222] J. R. Forshaw and D. A. Ross, Cambridge, UK: Univ. Pr. (1997) 248 p. (Cambridge lecture notes in physics 9).
[223] A. Tang and J. W. Norbury, Phys. Rev. **D62**, 016006 (2000), hep-ph/0004078,
[224] M. M. Brisudova, L. Burakovsky, and T. Goldman, Phys. Rev. **D61**, 054013 (2000), hep-ph/9906293,
[225] C. W. Bernard *et al.*, Phys. Rev. **D64**, 074509 (2001), hep-lat/0103012,
[226] G. F. Chew and S. C. Frautschi, Phys. Rev. Lett. **8**, 41 (1962),
[227] A. Donnachie, eConf **C010430**, T05 (2001), hep-ph/0106197,
[228] J. Hoek and J. Smit, Nucl. Phys. **B263**, 129 (1986),
[229] D. B. Leinweber, Phys. Rev. **D51**, 6383 (1995), nucl-th/9406001,
[230] MILC, C. W. Bernard *et al.*, Phys. Rev. Lett. **81**, 3087 (1998), hep-lat/9805004,
[231] MILC, C. Bernard, Phys. Rev. **D65**, 054031 (2002), hep-lat/0111051,
[232] S. Kim and S. Ohta, Phys. Rev. **D61**, 074506 (2000), hep-lat/9912001,
[233] R. D. Mawhinney, Nucl. Phys. Proc. Suppl. **47**, 557 (1996), hep-lat/9603019,
[234] QCDSF, M. Gockeler *et al.*, Nucl. Phys. Proc. Suppl. **83**, 203 (2000), hep-lat/9909160,
[235] JLQCD, S. Aoki *et al.*, (2002), hep-lat/0212039,
[236] MILC, C. W. Bernard *et al.*, Nucl. Phys. Proc. Suppl. **73**, 198 (1999), hep-lat/9810035,
[237] N. Isgur and G. Karl, Phys. Rev. **D20**, 1191 (1979),
[238] S. Capstick and W. Roberts, (2000), nucl-th/0008028,
[239] N. Isgur, (2000), nucl-th/0007008,
[240] J. N. Labrenz and S. R. Sharpe, Phys. Rev. **D54**, 4595 (1996), hep-lat/9605034,
[241] QCDSF, M. Gockeler *et al.*, Phys. Lett. **B532**, 63 (2002), hep-lat/0106022,
[242] S. J. Dong *et al.*, (2003), hep-ph/0306199,
[243] UKQCD, C. M. Maynard and D. G. Richards, (2002), hep-lat/0209165,
[244] S. Sasaki, T. Blum, and S. Ohta, Phys. Rev. **D65**, 074503 (2002), hep-lat/0102010,
[245] D. Broemmel *et al.*, (2003), hep-ph/0307073,
[246] Y. Nemoto, N. Nakajima, H. Matsufuru, and H. Suganuma, (2003),

hep-lat/0302013,
[247] N. Isgur and G. Karl, Phys. Rev. **D19**, 2653 (1979),
[248] L. Y. Glozman, "Reply to Isgur's 'Critique of a pion exchange model for interquark forces'.", nucl-th/9909021.
[249] P. R. Page, (2002), nucl-th/0204031,
[250] T. Barnes, (2000), nucl-th/0009011,
[251] W. Melnitchouk *et al.*, Phys. Rev. **D67**, 114506 (2003), hep-lat/0202022,
[252] S. Sasaki, (2003), nucl-th/0305014,
[253] M. Gockeler *et al.*, Nucl. Phys. **B487**, 313 (1997), hep-lat/9605035,
[254] S. Kim, J. B. Kogut, and M.-P. Lombardo, Phys. Rev. **D65**, 054015 (2002), hep-lat/0112009,
[255] M. Burkardt, D. B. Leinweber, and X. Jin, Phys. Lett. **B385**, 52 (1996), hep-ph/9604450,
[256] A. Duncan, E. Eichten, and H. Thacker, Phys. Rev. Lett. **76**, 3894 (1996), hep-lat/9602005,
[257] A. Duncan, E. Eichten, and H. Thacker, Phys. Lett. **B409**, 387 (1997), hep-lat/9607032,
[258] H. R. Fiebig, W. Wilcox, and R. M. Woloshyn, Nucl. Phys. **B324**, 47 (1989),
[259] L. Zhou, F. X. Lee, W. Wilcox, and J. Christensen, (2002), hep-lat/0209128,
[260] J. Christensen, F. X. Lee, W. Wilcox, and L. Zhou, (2002), hep-lat/0209043,
[261] B. R. Holstein, (2000), hep-ph/0010129,
[262] B. E. MacGibbon *et al.*, Phys. Rev. **C52**, 2097 (1995), nucl-ex/9507001,
[263] J. Smit and J. C. Vink, Nucl. Phys. **B286**, 485 (1987),
[264] M. Bander and H. R. Rubinstein, Phys. Lett. **B311**, 187 (1993), hep-ph/9204224,
[265] D. Grasso and H. R. Rubinstein, Phys. Rept. **348**, 163 (2001), astro-ph/0009061,
[266] H. R. Rubinstein, S. Solomon, and T. Wittlich, Nucl. Phys. **B457**, 577 (1995), hep-lat/9501001,
[267] C. W. Bernard, T. Draper, K. Olynyk, and M. Rushton, Phys. Rev. Lett. **49**, 1076 (1982),
[268] G. Martinelli, G. Parisi, R. Petronzio, and F. Rapuano, Phys. Lett. **B116**, 434 (1982),
[269] T. Draper, R. M. Woloshyn, W. Wilcox, and K.-F. Liu, Nucl. Phys. **B318**, 319 (1989),
[270] T. Draper, R. M. Woloshyn, and K.-F. Liu, Phys. Lett. **B234**, 121 (1990),
[271] V. Gadiyak, X.-d. Ji, and C.-w. Jung, Phys. Rev. **D65**, 094510 (2002), hep-lat/0112040,
[272] S. Aoki and A. Gocksch, Phys. Rev. Lett. **63**, 1125 (1989),
[273] S. Aoki, A. Gocksch, A. V. Manohar, and S. R. Sharpe, Phys. Rev. Lett. **65**, 1092 (1990),

[274] D. Guadagnoli, V. Lubicz, G. Martinelli, and S. Simula, JHEP **04**, 019 (2003), hep-lat/0210044,

[275] S. Capstick *et al.*, "Key issues in hadronic physics", 2000, hep-ph/0012238.

[276] N. Yamada, (2002), hep-lat/0210035,

[277] N. Isgur, Phys. Rev. **D61**, 118501 (2000), hep-lat/9908009,

[278] N. Isgur and H. B. Thacker, Phys. Rev. **D64**, 094507 (2001), hep-lat/0005006,

[279] E. V. Shuryak, Lect. Notes Phys. **583**, 251 (2002), hep-ph/0104249,

[280] A. W. Thomas and W. Weise, Berlin, Germany: Wiley-VCH (2001) 389 p.

[281] J. Greensite, (2003), hep-lat/0301023,

[282] R. D. Loft and T. A. DeGrand, Phys. Rev. **D39**, 2678 (1989),

[283] Y. Iwasaki and T. Yoshie, Phys. Lett. **B216**, 387 (1989),

[284] S. Ono, Phys. Rev. **D17**, 888 (1978),

[285] M. C. Chu, J. M. Grandy, S. Huang, and J. W. Negele, Phys. Rev. **D49**, 6039 (1994), hep-lat/9312071,

[286] T. G. Kovacs, Phys. Rev. **D62**, 034502 (2000), hep-lat/9912021,

[287] E. V. Shuryak and J. L. Rosner, Phys. Lett. **B218**, 72 (1989),

[288] L. Y. Glozman, Nucl. Phys. **A663**, 103 (2000), hep-ph/9908423,

[289] N. Isgur, Phys. Rev. **D62**, 054026 (2000), nucl-th/9908028,

[290] K. F. Liu *et al.*, Phys. Rev. **D59**, 112001 (1999), hep-ph/9806491,

[291] K. F. Liu *et al.*, Phys. Rev. **D61**, 118502 (2000), hep-lat/9912049,

[292] R. G. Edwards, Nucl. Phys. Proc. Suppl. **106**, 38 (2002), hep-lat/0111009,

[293] M. Anselmino, E. Predazzi, S. Ekelin, S. Fredriksson, and D. B. Lichtenberg, Rev. Mod. Phys. **65**, 1199 (1993),

[294] M. Hess, F. Karsch, E. Laermann, and I. Wetzorke, Phys. Rev. **D58**, 111502 (1998), hep-lat/9804023,

[295] D. B. Leinweber, Phys. Rev. **D47**, 5096 (1993), hep-ph/9302266,

[296] M. Lissia, M. C. Chu, J. W. Negele, and J. M. Grandy, Nucl. Phys. **A555**, 272 (1993),

[297] M. C. Chu, M. Lissia, and J. W. Negele, Nucl. Phys. **A570**, 521 (1994), hep-lat/9308012,

[298] J. F. Donoghue, E. Golowich, and B. R. Holstein, Cambridge Monogr. Part. Phys. Nucl. Phys. Cosmol. **2**, 1 (1992),

[299] B. Lucini and M. Teper, JHEP **06**, 050 (2001), hep-lat/0103027,

[300] R. Koniuk and N. Isgur, Phys. Rev. Lett. **44**, 845 (1980),

[301] J. B. Kogut, (2002), hep-lat/0208077,

[302] M. Luscher, Nucl. Phys. **B364**, 237 (1991),

[303] H. R. Fiebig and H. Markum, (2002), hep-lat/0212037,

[304] UKQCD, A. Hart and M. Teper, Phys. Rev. **D65**, 034502 (2002), hep-lat/0108022,

[305] C. Michael, Nucl. Phys. **B327**, 515 (1989),

[306] L. S. Brown, *Quantum field theory* , Cambridge, UK: Univ. Pr. (1992).

[307] T. Yamazaki and N. Ishizuka, Phys. Rev. **D67**, 077503 (2003), hep-lat/0210022,

[308] T. Bhattacharya, R. Gupta, G. Kilcup, and S. R. Sharpe, Phys. Rev. **D53**, 6486 (1996), hep-lat/9512021,

[309] UKQCD, C. McNeile and C. Michael, Phys. Lett. **B556**, 177 (2003), hep-lat/0212020,

[310] S. Gottlieb, P. B. Mackenzie, H. B. Thacker, and D. Weingarten, Phys. Lett. **B134**, 346 (1984),

[311] R. L. Altmeyer *et al.*, Z. Phys. **C68**, 443 (1995), hep-lat/9504003,

[312] R. D. Loft and T. A. DeGrand, Phys. Rev. **D39**, 2692 (1989),

[313] J. Sexton, A. Vaccarino, and D. Weingarten, Phys. Rev. Lett. **75**, 4563 (1995), hep-lat/9510022,

[314] A. Pickering, *Constructing Quarks. A Sociological History of Particle Physics* , Edinburgh, UK: Univ. Pr. (1984)

[315] P. Lepage, Nucl. Phys. Proc. Suppl. **60A**, 267 (1998), hep-lat/9707026,

[316] K. Symanzik, Nucl. Phys. **B226**, 187 (1983),

[317] K. Symanzik, Nucl. Phys. **B226**, 205 (1983),

[318] B. Sheikholeslami and R. Wohlert, Nucl. Phys. **B259**, 572 (1985),

[319] M. Luscher, S. Sint, R. Sommer, P. Weisz, and U. Wolff, Nucl. Phys. **B491**, 323 (1997), hep-lat/9609035,

[320] P. H. Ginsparg and K. G. Wilson, Phys. Rev. **D25**, 2649 (1982),

[321] M. Luscher, Phys. Lett. **B428**, 342 (1998), hep-lat/9802011,

[322] H. Neuberger, Ann. Rev. Nucl. Part. Sci. **51**, 23 (2001), hep-lat/0101006,

[323] P. Hasenfratz, Nucl. Phys. **B525**, 401 (1998), hep-lat/9802007,

[324] P. Hernandez, Nucl. Phys. Proc. Suppl. **106**, 80 (2002), hep-lat/0110218,

[325] CP-PACS, J. I. Noaki *et al.*, (2001), hep-lat/0108013,

[326] RBC, T. Blum *et al.*, (2001), hep-lat/0110075,

[327] K. Jansen, Nucl. Phys. Proc. Suppl. **106**, 191 (2002), hep-lat/0111062,

[328] RBC, T. Izubuchi, (2002), hep-lat/0210011,

[329] MILC, K. Orginos and D. Toussaint, Phys. Rev. **D59**, 014501 (1999), hep-lat/9805009,

[330] C. DeTar and S. Gottlieb, Physics Today (February 2004) p.45,

[331] H. Neuberger, "Lattice Field Theory: past, present and future", hep-ph/0402148

Chapter 2

Exotics

C. Michael
Theoretical Physics Division, Dept. of Mathematical Sciences
University of Liverpool, Liverpool L69 7ZL UK
E-mail:C.Michael@liv.ac.uk

We review lattice QCD results for glueballs (including a discussion of mixing with scalar mesons), hybrid mesons and other exotic states (such as $B_s B_s$ molecules).

2.1 Introduction

Quantum Chromodynamics has emerged as the unique theory to describe hadronic physics. It is formulated in terms of gluonic and quark fields. The only free parameters are the scale of the coupling (usually called Λ_{QCD}) and the quark masses defined at some conventional energy scale.

Where large momentum transfers occur, the effective coupling becomes weak and a perturbative treatment is valid: in this domain the theory has been tested directly by experiment. However, because the effective coupling is weak for these processes that can be described by perturbation theory, they are necessarily not the dominant hadronic processes. A typical hadronic process will involve small momentum transfers and so has to be treated non-perturbatively.

In this non-perturbative régime, the description of hadrons is quite far removed from the description of the gluonic and quark fields in the QCD Lagrangian. Because only colour-singlet states survive, the hadrons are all composites of quarks and gluons. One example emphasises this: the nucleon has a mass which is very much greater than the sum of the quark masses of the three valence quarks comprising it. This extra mass comes from the gluonic interactions of QCD. Another way to view this is that the naïve

Table 2.1 Non-exotic spin-parity combinations J^{PC} for both singlets($S = 0$) and triplets($S = 1$), where $J = L + S$.

L	$S = 0$	$S = 1$		
0	0^{-+}		1^{--}	
1	1^{+-}	0^{++}	1^{++}	2^{++}
2	2^{-+}	1^{--}	2^{--}	3^{--}

quark model is a useful phenomenological tool but has constituent quarks with masses much greater than the QCD masses (*i.e.* masses as defined in the Lagrangian). It is important to understand why this is approximately what QCD requires and to find where QCD departs from the naïve quark model.

One way to characterise the manner in which QCD goes beyond the naïve quark model is through the concept of exotic states. Here exotic is taken to mean 'not included in the naïve quark model'. In order to discuss exotic states, we need to summarise what the naïve quark model contains. Basically the degrees of freedom are the valence quarks (*i.e.* quark-antiquark for a meson and 3 quarks for a baryon) with masses and interactions given by some effective interaction. The consequences of this are that only certain J^{PC} values will exist and that the number of states with different quark flavours is specified. So, concentrating on mesons made of the three flavours of light quarks (u, d, s), one expects a nonet of mesons with the flavours ($\bar{u}d$, $\bar{d}u$, $\bar{u}u \pm \bar{d}d$, $\bar{s}s$, $\bar{u}s$, $\bar{d}s$, $\bar{s}u$, $\bar{s}d$). This is indeed what is found for vector mesons (ρ, ω, ϕ, K^*). It is also possible within the quark model for the flavour-singlet states ($\bar{u}u + \bar{d}d$, $\bar{s}s$) to mix, as found for the pseudoscalar mesons. What would be exotic is for a tenth state to exist. For mesons with orbital angular momentum L between the quark and antiquark the allowed J^{PC} values are shown in Table 2.1. Thus spin-parity combinations such as $0^{--}, 0^{+-}, 1^{-+}, 2^{+-}$ are termed spin-exotic since they cannot be made from a quark plus antiquark alone. It has been a considerable challenge to build a machinery that allows non-perturbative calculations in QCD with all systematic errors determined. The most controlled approach to non-perturbative QCD is via lattice techniques in which space-time is discretized and time is taken as Euclidean. The functional integral is then evaluated numerically using Monte Carlo techniques.

Lattice QCD needs as input the quark masses and an overall scale (conventionally given by Λ_{QCD}). Then any Green function can be evaluated by taking an average of suitable combinations of the lattice fields in the vacuum samples. This allows masses to be studied easily and matrix elements (particularly those of weak or electromagnetic currents) can be extracted straightforwardly. Unlike experiment, lattice QCD can vary the quark masses and can also explore different boundary conditions and sources. This allows a wide range of studies which can be used to diagnose the health of phenomenological models as well as casting light on experimental data.

One limitation of the lattice approach to QCD is in exploring hadronic decays because the lattice, using Euclidean time, has no concept of asymptotic states. One feasible strategy is to evaluate the mixing between states of the same energy — so giving some information on on-shell hadronic decay amplitudes.

There is an interesting theoretical world in which the quark degrees of freedom are removed from QCD, leaving pure gluo-dynamics. This is also known as pure Yang-Mills theory. It is a self-consistent theory which has the full non-perturbative gluonic interaction. It turns out that this gluonic interaction does produce the salient features of QCD: asymptotic freedom, confinement, *etc.* It is of interest to explore the spectrum in this case: the states are called glueballs.

It is also of interest to consider the propagation of quarks in this gluonic theory. The quarks are treated as in the Dirac equation and they propagate through the gluonic ground state. This approach is known as the quenched approximation. Again this turns out to be a very useful approximation: chiral symmetry breaking occurs for example. This quenched approximation is in contrast to the full quantum field theory of QCD where there would be quark loop effects in the ground state also. Thus in the quenched approximation there will be inconsistencies: the theory is not unitary. However, for heavy quarks it will be a good approximation since heavy quark loops are suppressed and it may be adequate to describe some features of lighter quarks. Moreover, many phenomenological models are appropriate to the quenched case and so can be compared with quenched QCD.

2.2 Glueballs and Scalar Mesons

2.2.1 *Glueballs in quenched QCD*

Glueballs are defined to be hadronic states made primarily from gluons. The full non-perturbative gluonic interaction is included in quenched QCD. A study of the glueball spectrum in quenched QCD is thus of great value. This will allow experimental searches to be guided as well as providing calibration for models of glueballs. A non-zero glueball mass in quenched QCD is the "mass-gap" of QCD. To prove this rigourously is one of the major challenges of our times. Here we will explore the situation using computational techniques.

In lattice studies, dimensionless ratios of quantities are obtained. To explore the glueball masses m, it is appropriate to combine them with another very accurately measured quantity to have a dimensionless observable. Since the potential between static quarks is very accurately measured from the lattice, it is now conventional [1] to use r_0 for this comparison. Here r_0 is implicitly defined by $r^2 dV(r)/dr = 1.65$ at $r = r_0$ where $V(r)$ is the potential energy between static quarks which is easy to determine accurately on the lattice. Conventionally $r_0 \approx 0.5$ fm.

Theoretical analysis indicates that for Wilson's discretisation of the gauge fields in the quenched approximation, the dimensionless ratio mr_0 will differ from the continuum limit value by corrections of order a^2. Thus in Fig. 2.1 the mass of the $J^{PC}=0^{++}$ glueball is plotted versus the lattice spacing a^2. The straight line then shows the continuum limit obtained by extrapolating to $a = 0$. As can be seen, there is essentially no need for data at even smaller a-values to further fix the continuum value. The value shown corresponds to $m(0^{++})r_0 = 4.33(5)$. Since several lattice groups [2, 3, 4, 5] have measured these quantities, it is reassuring to see that the purely lattice observables are in excellent agreement. The publicised difference of quoted $m(0^{++})$ from UKQCD [4] and GF11 [5] comes entirely from relating quenched lattice measurements to values in GeV.

In the quenched approximation, different hadronic observables differ from experiment by factors of up to 10%. Thus using one quantity or another to set the scale, gives an overall systematic error. Here the scale is set by taking the conventional value of the string tension (determined from potential models and from hadronic lattice studies), $\sqrt{\sigma} = 0.44$ GeV, which then corresponds to $r_0^{-1} = 373$ MeV (or $r_0 = 0.53$ fm). An over-

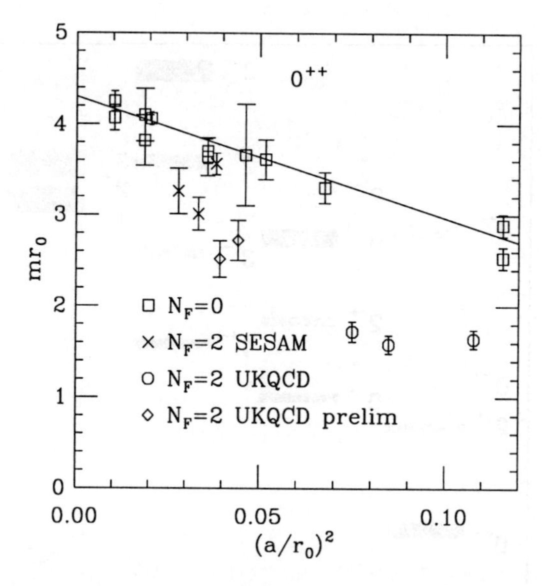

Fig. 2.1 The value of the mass of the $J^{PC} = 0^{++}$ glueball state from quenched data ($N_F = 0$) [2, 3, 4, 5] in units of r_0 where $r_0 \approx 0.5$ fm. The straight line shows a fit describing the approach to the continuum limit as $a \to 0$. Results [6, 7, 8] for the lightest scalar meson with $N_F = 2$ flavours of sea quarks are also shown.

all systematic error of 10% is then to be included to any extracted mass. This yields $m(0^{++}) = 1611(30)(160)$ MeV where the second error is the systematic scale error. Note that this is the glueball mass in the quenched approximation — in the real world significant mixing with $q\bar{q}$ states *etc.* may modify this value substantially, as we discuss below.

In the Wilson approach, the next lightest glueballs are [3, 4] the tensor $m(2^{++})r_0 = 6.0(6)$ [resulting in $m(2^{++}) = 2232(220)(220)$ MeV] and the pseudoscalar $m(0^{-+})r_0 = 6.0(1.0)$. Although the Wilson discretisation provides a definitive study of the lightest (0^{++}) glueball in the continuum limit, other methods are competitive for the determination of the mass of heavier glueballs. Namely, using an improved gauge discretisation which has even smaller discretisation errors than the a^2 dependence of the Wilson discretisation, so allowing a relatively coarse lattice spacing a to be used. To extract mass values, one has to explore the time dependence of correlators and for this reason, it is optimum to use a relatively small time lattice spacing. Thus an asymmetric lattice spacing is most appropriate. The results [9] are shown in Fig. 2.2 and for low lying states are that

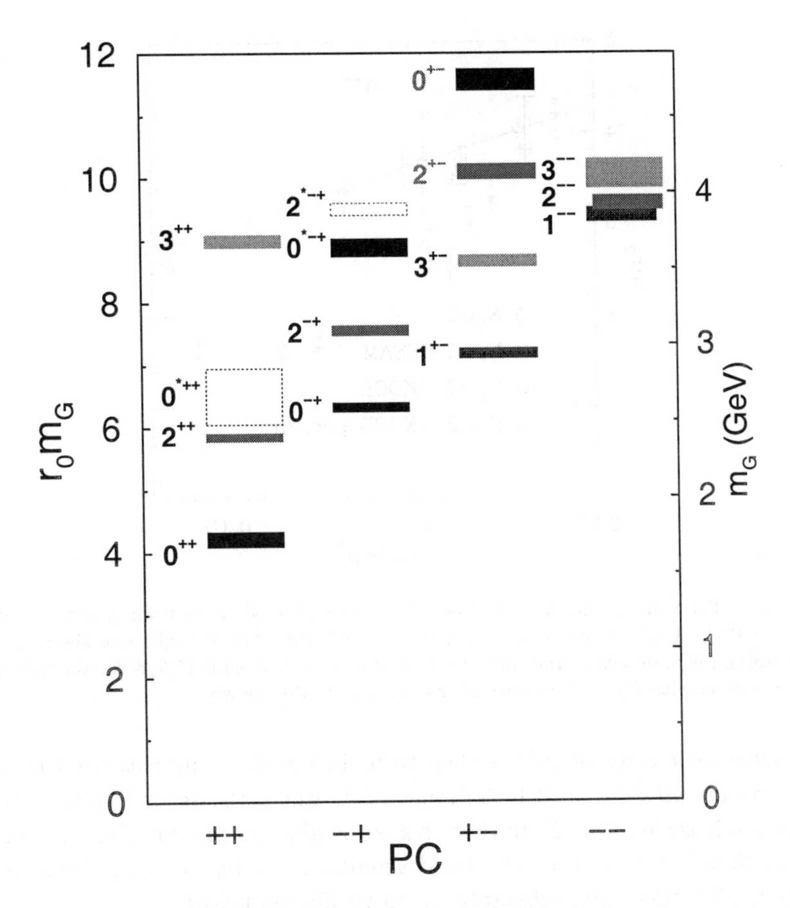

Fig. 2.2 The continuum glueball spectrum[9].

$m(0^{++})r_0 = 4.21(11)(4)$, $m(2^{++})r_0 = 5.85(2)(6)$, $m(0^{-+})r_0 = 6.33(7)(6)$ and $m(1^{+-})r_0 = 7.18(4)(7)$. It will be very difficult to identify experimentally states corresponding to these heavier glueballs since the spectrum is rich in $q\bar{q}$ states of those quantum numbers at those mass values and there will thus be considerable mixing.

One signal of great interest would be a glueball with J^{PC} not allowed for $q\bar{q}$ — a spin-exotic glueball or *oddball* — since it would not mix with $q\bar{q}$ states. These states are found [3, 4, 9] to be high lying: considerably above $2m(0^{++})$. Thus they are likely to be in a region around 4 GeV where it is very difficult to separate states unambiguously by experiment.

As well as the mass of a glueball, it is possible to study their physical size. In principle this is determined by measuring the matrix element $\langle G|J|G\rangle$ where J is some local current which couples to the glueball. Since glueballs have no flavour, the energy-momentum tensor is the most appropriate choice. A preliminary study has been made, albeit with large systematic errors [10] and finds a radius 0.9 ± 0.3 fm. This approach should be contrasted with what has come to be called the Bethe-Salpeter wavefunction which is obtained from $\langle G|L|0\rangle$ where L is the lattice operator used to create a glueball state from the vacuum. Within a lattice calculation, it is easy to measure the dependence of this overlap on the spatial extent of L [11], but difficult to interpret the result.

Related information can be obtained from a study of the gluelump: the state with one static colour source in the octet representation with a gluonic field making it a colour singlet. This would be of physical relevance should a massive gluino exist: it would be the glueballino, a gluino-gluon bound state. The spectrum [12] and spatial distribution [13] have been studied.

Another topic which is of mainly theoretical interest is the glueball mass (as a dimensionless ratio to the string tension) as the number of colours N_C is varied from 3. The SU(2) Yang-Mills theory has often been studied, especially as it is computationally simpler. Recently a study of SU(N_C) for N_C=4 and 5 has been made. The summary [14] is that $N_C = \infty$ is relatively close to N_C=3. This has theoretical implications since the $N_C = \infty$ theory is formally simpler. For a comparison of these lattice results with ADS supergravity see Ref. [15].

2.2.2 *Scalar mesons in quenched QCD*

In quenched QCD the flavour singlet (f_0) and non-singlet (a_0) scalar mesons are degenerate. In full QCD this degeneracy is split by disconnected quark diagrams but these are omitted from the quenched approximation. This same feature of the quenched approximation implies that the η meson is wrongly treated - it will be degenerate with the π. This implies that the scalar meson propagation can have the wrong sign [16] because the $\eta\pi$ intermediate state is mistreated (once quark loops are allowed in the vacuum then this anomaly is removed). For light quarks of mass corresponding to the strange quark or heavier, it is expected that this anomaly is relatively unimportant. Thus the measurement of the mass of the $q\bar{q}$ scalar meson can be particularly unreliable in the quenched approximation.

Even though the mixing of the glueball and $q\bar{q}$ states is not implemented in the quenched approximation, one can determine the mixing matrix element. This can then be used to estimate the result of the mixing by hand (by using a mass matrix for example). On a rather coarse lattice, where $a^{-1} \approx 1.2$ GeV, two groups have attempted to measure this mixing [7, 17]. Their results expressed as the mixing for two degenerate quarks of mass around the strange quark mass are similar, namely $E \approx 0.3$ GeV [17] and 0.5 GeV [7]. This is a relatively large mixing (if the glueball and scalar meson states were degenerate they would be split by $\pm E$).

An exploratory attempt to extrapolate this mixing to the continuum [17] gave a very small mixing of 61(45) MeV, while the other determination [7] uses clover improvement so order a effects in the extrapolation to the continuum are suppressed and one would not expect a significant decrease in going to the continuum limit. What this discussion shows is that precision studies of the mixing on a quenched lattice have not yet been achieved. Furthermore the problems with the scalar meson propagation in the quenched approximation discussed above also limit progress.

As well as this mixing of the glueball with $q\bar{q}$ states, there will be mixing with $q\bar{q}q\bar{q}$ states which will be responsible for the hadronic decays. A first attempt to study this [18] at a coarse lattice spacing yields an estimated width for decay to two pseudoscalar mesons from the scalar glueball of order 100 MeV. A more realistic study would involve taking account of mixing with the $n\bar{n}$ and $s\bar{s}$ scalar mesons as well.

2.2.3 *Scalar mesons in full QCD*

It is now feasible to explore the flavour-singlet scalar meson spectrum including the quark loops in the vacuum, *i.e.* in full QCD. From dynamical fermion studies with $N_f = 2$, one can determine the flavour singlet and non-singlet mass spectrum. What is found [7, 8] is that the lightest flavour-singlet scalar meson (f_0) is lighter than the lightest flavour non-singlet (a_0).

The interpretation of this study is hampered by the same issue that hampers the interpretation of experimental data, namely, the mass eigenstates are not distinguished as 'glueball' or as 'quark-antiquark'. What one can do is explore the output spectrum and deduce what mixing might have occurred. To give an example, where we restrict here to $N_f = 2$ flavours of degenerate quarks, the f_0 masses will be m_0 and $m_0{}'$ where the latter is the first excited state and the flavour non-singlet a_0 mass will be m_1.

Results for m_0 are given in Fig. 2.1. Then one would expect in a simple 2×2 mixing scenario (*i.e.* glueball and $q\bar{q}$ meson) a mass matrix

$$\begin{pmatrix} m_G & E \\ E & m_1 \end{pmatrix},$$

where m_G is the glueball mass and E the mixing matrix element. This will have two mass eigenstates which can be identified with m_0 and $m_0{}'$ so determining the two free parameters in the matrix. This approach explains what is going on but obtains two numbers with two parameters, so there is no cross-check.

One can directly address the issue of the mass of the lightest scalar singlet meson from the lattice with $N_f = 2$. It is advantageous to use as full a basis of lattice operators as possible, including Wilson loops and quark-antiquark loops. Including the latter can in practice lead to a lower value of the ground state scalar meson mass - see Refs. [20, 8]. Most studies have shown no significant change of the scalar glueball mass as dynamical quarks are included [6]. However the larger lattice spacing result [7] shows a significant reduction in the lightest scalar mass, as shown in Fig. 2.1. Before concluding that this implies a lower scalar mass in the continuum limit, one needs to check whether an enhanced order a^2 correction might be present. The origin of the large coefficient of a^2 in quenched glueball studies is usually ascribed to the presence of a critical point in the fundamental-adjoint coupling plane which is close by in the usual Wilson approach with zero adjoint coupling. The extent to which this will be enhanced/reduced when dynamical quarks are introduced is not clear. Studies using the same approach at a finer lattice spacing [8, 20] do suggest that this large order a^2 effect is significant for dynamical quarks, but studies even nearer to the continuum or with improved actions are needed to resolve this fully.

A further complication is that as the quark mass is reduced towards the physical light quark mass, the decay to $\pi\pi$ becomes energetically allowed. The study of unstable particles is a difficult problem in a Euclidean time formalism [19]. We return to this topic later.

2.2.4 *Experimental evidence for scalar mesons*

In full QCD, for the flavour-singlet states of any given J^{PC}, there will be mixing between the $s\bar{s}$ state, the $u\bar{u} + d\bar{d}$ state and the glueball as well as with multi-meson channels. It may indeed turn out that no scalar meson in the physical spectrum is primarily a glueball — all states are mixtures of glue, $q\bar{q}$, $q\bar{q}q\bar{q}$, *etc.*

To help with understanding the experimental situation [21], we first discuss the flavour non-singlet states, the a_0 with isospin 1. The observed states are at 980 and 1450 MeV. The lighter state has dynamics which appears to be closely associated with the $K\bar{K}$ threshold. The heavier state is not yet very well established but seems to be a candidate for a state mostly comprised of $q\bar{q}$, while the lighter state would be $q\bar{q}q\bar{q}$.

The flavour singlet states (f_0) are more numerous. There is a very broad enhancement in the $\pi\pi$ S-wave phase shift around 700 MeV (sometimes called the σ), there is a state near the $K\bar{K}$ threshold at 980 MeV and there are more states at 1370, 1500 and 1710 MeV. Again assuming that the state at 980 MeV is predominantly $q\bar{q}q\bar{q}$, this suggests that the three states in the 1300-1750 MeV energy range are admixtures of the glueball, $u\bar{u} + d\bar{d}$ and $s\bar{s}$. The fact that there are indeed three states in this energy region close to the quenched glueball mass of 1600 MeV is the strongest evidence for the presence of a glueball. This has led to several phenomenological attempts [17, 22] to describe these three observed states in terms of the lattice input.

As we emphasised above, in full QCD on a lattice one just obtains values for the a_0 and f_0 masses. In the simplified case of $N_f = 2$ flavours of degenerate quark, one does indeed find [8] two f_0 states, and they can be interpreted as mixtures of the $q\bar{q}$ and glueball states with the $q\bar{q}$ state having the properties found for the a_0.

One useful lattice input would be a determination of the a_0 mass as the quark mass is varied in full QCD, especially because of the problems of determining the a_0 mass in the quenched approximation. At present, the full QCD studies [7, 8] are limited to relatively coarse lattice spacing, so the continuum limit is not close. Furthermore, as the quark mass is reduced the a_0 can decay (to $\pi\eta$) and this will influence the lattice analysis [23].

2.3 Hybrid Mesons

A hybrid meson is a meson in which the gluonic degrees of freedom are excited non-trivially. The most direct sign of this would be a spin-exotic meson, since that could not be created from a $q\bar{q}$ state with unexcited

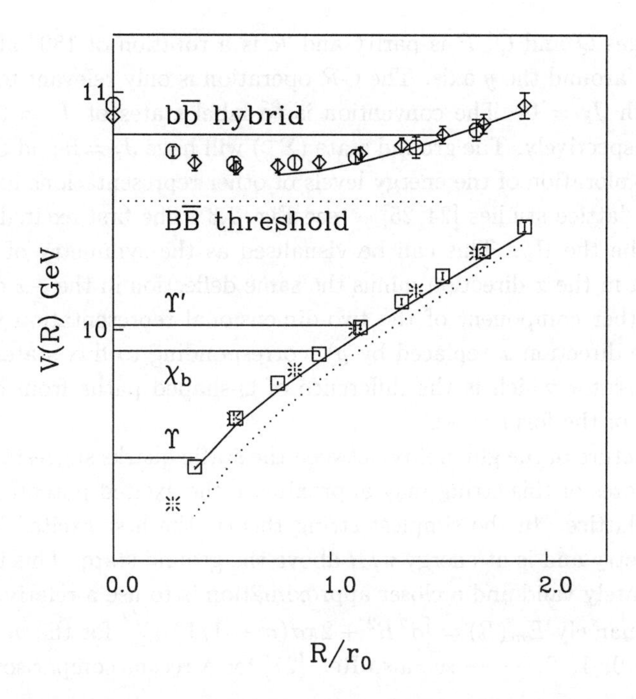

Fig. 2.3 The potential energy between static quarks at separation R (in units of $r_0 \approx 0.5$ fm) [25]. The symmetric gluonic field configuration is shown by the lower points while the Π_u excited gluonic configuration is shown above. The energy levels in these potentials for b quarks are shown using the adiabatic approximation.

glue. A spin-exotic meson could, however, be a $q\bar{q}q\bar{q}$ or meson-meson state and that possibility will be discussed. We first discuss hybrid mesons with static heavy quarks where the description can be thought of as an excited colour string. The situation concerning light quark hybrid mesons is then summarised

2.3.1 *Heavy quark hybrid mesons*

Consider $Q\bar{Q}$ states with static quarks in which the gluonic contribution may be excited. We classify the gluonic fields according to the symmetries of the system. This discussion is very similar to the description of electron wave functions in diatomic molecules. The symmetries are (i) rotation around the separation axis z with representations labelled by J_z (ii) CP with representations labelled by $g(+)$ and $u(-)$ and (iii) $C\mathcal{R}$. Here C

interchanges Q and $\bar{Q}$, P is parity and $\mathcal{R}$ is a rotation of 180^0 about the mid-point around the y axis. The $C\mathcal{R}$ operation is only relevant to classify states with $J_z = 0$. The convention is to label states of $J_z = 0, 1, 2$ by Σ, Π, Δ respectively. The ground state (Σ_g^+) will have $J_z = 0$ and $CP = +$.

The exploration of the energy levels of other representations has a long history in lattice studies [24, 25] — see Fig. 2.3. The first excited state is found to be the Π_u. This can be visualised as the symmetry of a string bowed out in the x direction minus the same deflection in the $-x$ direction (plus another component of the two-dimensional representation with the transverse direction x replaced by y), corresponding to flux states from a lattice operator which is the difference of ⊔-shaped paths from quark to antiquark of the form ⊓ − ⊔.

The picture of the gluon flux between the static quarks suggests that the excited states of this string may approximate the excited potentials found from the lattice. In the simplest string theory, the first excited level has Π_u symmetry and is at energy π/R above the ground state. This is indeed approximately valid and a closer approximation is to use a relativistic version [26], namely $E_m(R) = \left[\sigma^2 R^2 + 2\pi\sigma(m - 1/12)\right]^{1/2}$ for the m-th level, with $m = 0,\ 1,\ 2, \cdots$ — see also Ref. [27] for a recent comparison of this expression.

Recent lattice studies [27] have used an asymmetric space/time spacing which enables excited states to be determined comprehensively. These results confirm the finding that the Π_u excitation is the lowest lying and hence of most relevance to spectroscopy.

From the potential corresponding to these excited gluonic states, one can determine the spectrum of hybrid quarkonia using the Schrödinger equation in the Born-Oppenheimer approximation. This approximation will be good if the heavy quarks move very little in the time it takes for the potential between them to become established. More quantitatively, we require that the potential energy of gluonic excitation is much larger than the typical energy of orbital or radial excitation. This is indeed the case [24], especially for b quarks. Another nice feature of this approach is that the self energy of the static sources cancels in the energy difference between this hybrid state and the $Q\bar{Q}$ states. Thus the lattice approach gives directly the excitation energy of each gluonic excitation.

The Π_u symmetry state corresponds to excitations of the gluonic field in quarkonium called magnetic (with $L^{PC} = 1^{+-}$) and pseudo-electric

Table 2.2 Heavy quark hybrid meson spin-parity combinations J^{PC} for singlets($S = 0$) and triplets($S = 1$), where $J = L + S$.

L^{PC}	$S = 0$	$S = 1$		
1^{-+}	1^{--}	0^{-+}	1^{-+}	2^{-+}
1^{--}	1^{++}	0^{+-}	1^{+-}	2^{+-}

(with 1^{-+}) in contrast to the usual P-wave orbital excitation which has $L^{PC} = 1^{--}$. Thus we expect different quantum number assignments from those of the gluonic ground state. Indeed combining with the heavy quark spins, we get the degenerate set of 8 states shown in Table 2.2. Note that of these, $J^{PC} = 1^{-+}$, 0^{+-} and 2^{+-} are spin-exotic and hence will not mix with $Q\bar{Q}$ states. They thus form a very attractive goal for experimental searches for hybrid mesons.

The eightfold degeneracy of the static approach will be broken by various corrections. As an example, one of the eight degenerate hybrid states is a pseudoscalar with the heavy quarks in a spin triplet. This has the same overall quantum numbers as the S-wave $Q\bar{Q}$ state (η_b) which, however, has the heavy quarks in a spin singlet. So any mixing between these states must be mediated by spin dependent interactions. These spin dependent interactions will be smaller for heavier quarks. It is of interest to establish the strength of these effects for b and c quarks. Another topic of interest is the splitting between the spin-exotic hybrids which will come from the different energies of the magnetic and pseudo-electric gluonic excitations.

One way to go beyond the static approach is to use the NRQCD approximation which then enables the spin dependent effects to be explored. One study [27] finds that the $L^{PC} = 1^{+-}$ and 1^{-+} excitations have no statistically significant splitting although the 1^{+-} excitation does lie a little lighter. This would imply, after adding in heavy quark spin, that the $J^{PC} = 1^{-+}$ hybrid was the lightest spin-exotic. Also a relatively large spin splitting was found [28] among the triplet states considering, however, only magnetic gluonic excitations. Another study [29] explores the mixing of non spin-exotic hybrids with regular quarkonium states via a spin-flip interaction using lattice NRQCD.

Confirmation of the ordering of the spin-exotic states also comes from

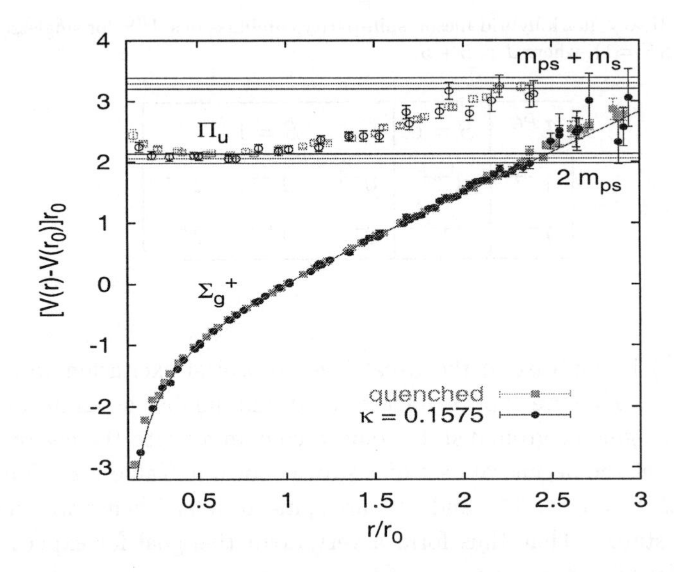

Fig. 2.4 The potential energy for quenched and 2 flavours of sea quark for the ground state and first excited gluonic state [6].

lattice studies with propagating quarks [30, 31, 32] which are able to measure masses for all 8 states. We discuss that evidence in more detail below.

Because of the similarity of the lightest hybrid wavefunction with that of the $2S$ state (which is radially excited), it is convenient to quote mass differences between these states. Within the quenched approximation, the lattice evidence for $b\bar{b}$ quarks points to a lightest hybrid spin-exotic with $J^{PC} = 1^{-+}$ at an energy given by $(m_H - m_{2S})r_0$ =1.8 (static potential [25]); 1.9 (static potential [27], NRQCD [28]); 2.0 (NRQCD [27]). These results can be summarised as $(m_H - m_{2S})r_0 = 1.9 \pm 0.1$. Using the experimental mass of the $\Upsilon(2S)$, this implies that the lightest spin-exotic hybrid is at $m_H = 10.73(7)$ GeV including a 10% scale error. Above this energy there will be many more hybrid states, many of which will be spin-exotic.

The results from a study with $N_f = 2$ flavours of sea-quarks show very little change in the static potential (see Fig. 2.4) and also relatively little change in NRQCD determinations [28] of mass ratios such as $(m_H - m_{2S})/(m_{1P} - m_{1S})$. Expressed in terms of r_0 (using $r_0 = 1.18/\sqrt{\sigma}$) this gives $(m_H - m_{2S})r_0 = 2.4(2)$, however. This is significantly larger than the quenched result and, using the $1P - 1S$ mass difference to set the scale, yields a prediction [28] for the lightest hybrid mass of 11.02(18) GeV.

2.3.2 *Hybrid meson decays*

Within this static quark framework, one can explore the decay mechanisms. One special feature is that the symmetries of the quark and colour fields about the static quarks must be preserved exactly in decay. This has the consequence that the decay from a Π_u hybrid state to the open-b mesons ($B\bar{B}$, $B^*\bar{B}$, $B\bar{B}^*$, $B^*\bar{B}^*$) will be forbidden [41, 33] if the light quarks in the B and B^* mesons are in an S-wave relative to the heavy quark (since the final state will have the light quarks in either a triplet with the wrong CP or a singlet with the wrong J_z where z is the interquark axis). The decay to B^{**}-mesons with light quarks in a P-wave is allowed by symmetry but not energetically.

The only allowed decays are when the hybrid state de-excites to a non-hybrid state with the emission of a light quark-antiquark pair. Since the Π_u hybrid state has the heavy quark-antiquark in a triplet P-wave state, the resulting non-hybrid state must also be in a triplet P-wave since the heavy quarks do not change their state in the limit of very heavy quarks. Thus the decay for b quarks will be to $\chi_b + M$ where M is a light quark-antiquark meson in a flavour singlet. This proceeds by a disconnected light quark diagram and it would be expected [34] that the scalar or pseudoscalar meson channels are the most important (*i.e.* they have the largest relative OZI-rule violating contributions). Lattice estimates [33] of these transitions have been made and the dominant mode (with a width of around 100 MeV) is found to be with M as a scalar meson, namely $H \to \chi_b + f_0$.

These estimates are in the static quark limit, in which the spin-exotic and non-spin-exotic hybrid mesons are degenerate. For the latter, however, the interpretation of any observed states is less clear cut, since they could be conventional quark-antiquark states. Moreover, the non-spin-exotic hybrid mesons can mix directly (*i.e.* without emission of any meson M) with conventional quark-antiquark states once one takes into account corrections (of order $1/M_Q$) to the static approximation applicable for heavy quarks with physical masses.

It is encouraging that the decay width comes out as relatively small, so that the spin-exotic hybrid states should show up experimentally as sufficiently narrow resonances to be detectable. This decay analysis does not take into account heavy quark motion or spin-flip and these effects will be significantly more important for charm quarks than for b-quarks.

3.3.3 Light quark hybrid mesons

I now focus on lattice results for hybrid mesons made from light quarks using fully relativistic propagating quarks. There will be no mixing with $q\bar{q}$ mesons for spin-exotic hybrid mesons and these are of special interest. The first study of this area was by the UKQCD Collaboration [30] who used operators motivated by the heavy quark studies referred to above to study all 8 J^{PC} values coming from $L^{PC} = 1^{+-}$ and 1^{-+} excitations. The resulting mass spectrum gives the $J^{PC} = 1^{-+}$ state as the lightest spin-exotic state. Taking account of the systematic scale errors in the lattice determination, a mass of 2000(200) MeV is quoted for this hybrid meson with $s\bar{s}$ light quarks. Although not directly measured, the corresponding light quark hybrid meson would be expected to be around 120 MeV lighter.

A second lattice group has also evaluated hybrid meson spectra with propagating quarks from quenched lattices. They obtain masses of the 1^{-+} state with statistical and various systematic errors of 1970(90)(300) MeV, 2170(80)(100)(100) MeV and 4390(80)(200) MeV for $n\bar{n}$, $s\bar{s}$ and $c\bar{c}$ quarks respectively [31]. For the 0^{+-} spin-exotic state they have a noisier signal but evidence that it is heavier. They also explore mixing matrix elements between spin-exotic hybrid states and 4 quark operators.

The first analysis [32] to determine the hybrid meson spectrum using full QCD used Wilson quarks. The sea quarks used had several different masses and an extrapolation was made to the limit of physical sea quark masses, yielding a mass of 1.9(2) GeV for the lightest spin-exotic hybrid meson, which again was found to be the 1^{-+}. In principle this calculation should take account of sea quark effects such as the mixing between such a hybrid meson and $q\bar{q}q\bar{q}$ states such as $\eta\pi$, although it is possible that the sea quark masses used are not light enough to explore these features.

A recent dynamical quark study from 2+1 flavours of improved staggered quarks has also produced results [36]. They also compare their results with quenched calculations and find no significant difference, except that the ambiguity in fixing the lattice energy scale is better controlled in the dynamical simulation since different reference observables are closer to experiment. Their summary result for the 1^{-+} hybrid with strange quarks is 2100 ± 120 MeV, in agreement with earlier results. They note that the energies of two-meson states [such as $\pi + b_1$ or $K + K(1^+)$] with the hybrid meson quantum numbers are close to the energies they obtain. This suggests that these two-particle states, which are allowed to mix in a dynamical quark treatment, may be influencing the masses determined. A study of hybrid meson transitions to two particle states is needed to illuminate this area, using techniques such as those used for heavy quark hybrid decay [33] and decays of light quark vector mesons [35].

The lattice calculations [30, 31, 32, 36, 37] of the light hybrid spectrum are in good agreement with each other. They imply that the natural energy range for spin-exotic hybrid mesons is around 1.9 GeV. The $J^{PC} = 1^{-+}$ state is found to be lightest. It is not easy to reconcile these lattice results with experimental indications [38] for resonances at 1.4 GeV and 1.6 GeV, especially the lower mass value. Mixing with $q\bar{q}q\bar{q}$ states such as $\eta\pi$ is not included for realistic quark masses in the lattice calculations. Such effects of pion loops (both real and virtual) have been estimated in chiral perturbation theory based models [39] and they could potentially reconcile some of the discrepancy between lattice mass estimates (with light quarks which are too heavy) and those from experiment. This can be interpreted, dependent on one's viewpoint, as either that the lattice calculations are incomplete or as an indication that the experimental states may have an important meson-meson component in them.

The light quark technique of using relativistic propagating quarks can also be extended to charm quarks, as was note above [31]. Another group has explored the charm quark hybrid states also using a fully relativistic action, albeit with an anisotropic lattice formulation [40]. Their quenched study is in agreement with the isotropic lattice result quoted above, finding a mass value of 4.428(41) GeV in the continuum limit for the 1^{-+} hybrid where the scale is set by the ${}^1P_1 - 1S$ mass splitting (458.2 MeV experimentally) in charmonium. Their result is also consistent with that from NRQCD methods [28] applied to this case. These results all have the usual caveat that in quenched evaluations the overall mass scale of the energy difference from the $1S$ state at 3.067 GeV is uncertain to 10% or so (for example the $(2S-1S)/({}^1P_1-1S)$ is found to be 15% higher than experiment) which is a major source of systematic error (approximately ± 140 MeV). They also produce estimates for other charmonium spin-exotic states: 0^{+-} at 4.70(17) GeV and 2^{+-} at 4.89(9) GeV. The 0^{--} state is not resolved.

Thus masses near 4.4 GeV are found for the charmonium 1^{-+} state using relativistic quarks. The non-relativistic approach using NRQCD is expected to have big systematic errors for quarks as light as charm, but results [28] do agree with this value. The heavy quark effective theory approach has a leading term which corresponds to a static heavy quark, resulting in an estimate [25] of the spin-exotic charm state mass of 4.0 GeV. Here again the systematic error is potentially large for charm quarks.

2.4 Hadronic Molecules

By exotic state we mean any state which is not dominantly a $q\bar{q}$ or qqq state. For example, a state made from hadrons bound in a molecule would be exotic. Examples of hadronic molecules have been known for a long time: the deuteron is a proton-neutron molecule for example. It is very weakly bound (2 MeV) and is quite extended. It is more efficiently described in terms to a neutron and a proton than as six quarks.

The residual hadronic interaction, the force between two colour-singlet hadrons, is much weaker than the colour force between quarks. Although it is called a 'strong interaction', it is relatively weak. At large distance, it will be dominated by the exchange of the lightest hadrons allowed (typically one or two pions). For example, the scale of nuclear binding is around 8 MeV whereas the gluonic forces binding three quarks to make a nucleon contribute most of its mass of 938 MeV. For this reason lattice methods need to be developed specially to tackle this problem. Basically, one is interested in binding energies, so it is the energy difference between the two hadrons and the hadronic molecule that is of interest. This difference can sometimes be determined better than the total energy itself. Even so, the detailed dynamics of such molecular states will depend on the long range forces (typically one or two pion exchange) and this will be modified considerably in lattice studies with light quark masses which are too heavy (typically down to 50% of the strange quark mass only). So only qualitative input can be obtained from the lattice, but this can still be used to validate models.

Because of the small binding energy and the dominance of pion exchange in the binding, it is not feasible to obtain the deuteron binding direct from QCD using lattice methods at present. There has been speculation that other di-baryon systems might be more strongly bound: the H dibaryon (a $\Lambda\Lambda$ state) being the best known. If it were strongly bound then it might be stable to weak decay which might have astrophysical consequences. The current status of lattice studies [42] is that finite box size effects are large but there is no convincing lattice evidence that this state is bound. At the largest volumes studied, with $L \approx 4$ fm, in quenched simulations it is found that the ratio $(m_H/2m_\Lambda) - 1$ is positive and in the range 5 to 15 %.

Other molecular states involving two hadrons have been conjectured. Several meson resonances are known which are closely connected with nearby thresholds: the $\Lambda(1405)$ which is just below the $\bar{K}N$ threshold and the $a_0(980)$ and $f_0(980)$ which are close to the $K\bar{K}$ threshold. Another state close to a threshold is the $N(1535)$ which is just above the ηN threshold. Again lattice studies are not able to shed very much light directly on these states since the quark masses used in the lattice studies are unphysical. However, if they are not produced in a lattice study which explores $q\bar{q}$ and qqq states, this may help to support the conclusion that they are primarily molecular in structure.

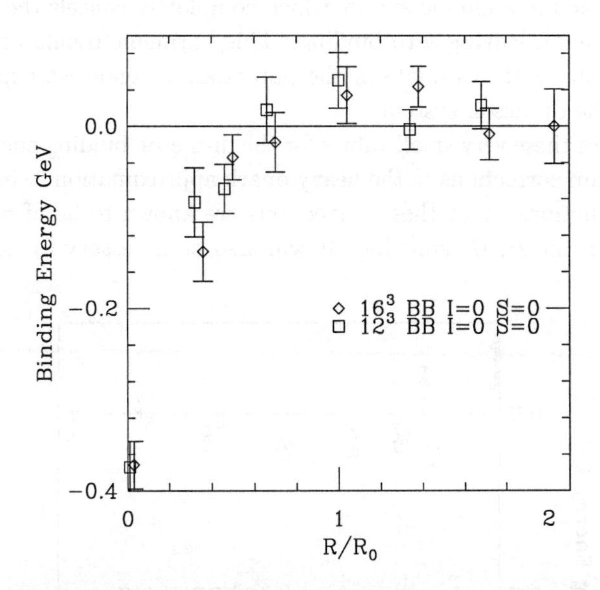

Fig. 2.5 The binding energy [43] between two heavy-light mesons (with static heavy quarks and light quarks of mass corresponding to strange) at separation R (in units of $r_0 \approx 0.5$ fm) with the two light quarks having $I = 0$ and $S = 0$.

One case which is relatively easy to study is the BB system, idealised as two static quarks and two light quarks. Then a potential as a function of the separation R between the static quarks can be determined. Because the static quark spin is irrelevant, the states can be classified by the light quark spin and isospin. Lattice results [43] (using a light quark mass close to strange) have been obtained for the potential energy for $I_q = 0, 1$ and

$S_q = 0, 1$. For very heavy quarks, a potential below $2M_B$ will imply binding of the BB molecules with these quantum numbers and $L = 0$. For the physically relevant case of b quarks of around 5 GeV, the kinetic energy will not be negligible and the binding energy of the BB molecular states is less clear cut. One way to estimate the kinetic energy for the BB case with reduced mass circa 2.5 GeV is to use analytic approximations to the potentials found. For example the I_q, S_q=(0,0) case (see Fig. 2.5) shows a deep binding at $R = 0$ which can be approximated as a Coulomb potential of $-0.1/R$ in GeV units. This will give a di-meson binding energy of only 10 MeV. For the other interesting case shown in Fig. 2.6, (I_q, S_q)=(0,1), a harmonic oscillator potential in the radial coordinate of form $-0.04[1 - (r - 3)^2/4]$ in GeV units leads to a kinetic energy which completely cancels the potential energy minimum, leaving zero binding. This harmonic oscillator approximation lies above the estimate of the potential, so again we expect weak binding of the di-meson system.

Because of these very small values for the di-meson binding energies, one needs to retain corrections to the heavy quark approximation to make more definite predictions, since these corrections are known to be of magnitude 46 MeV from the B, B^* splitting. It will also be necessary to extrapolate

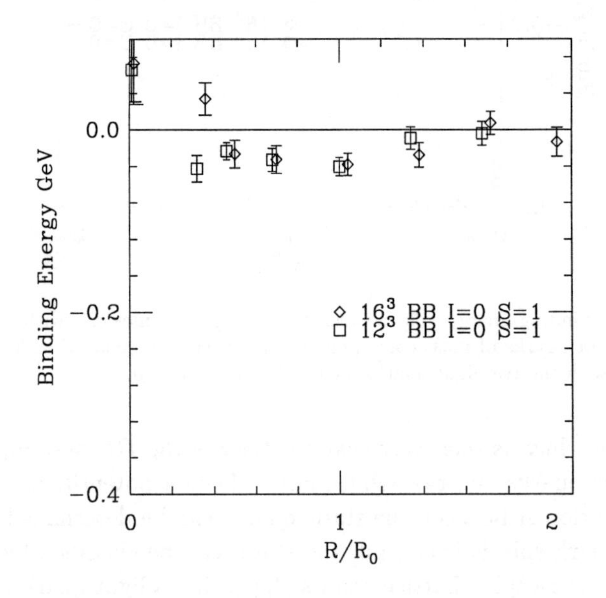

Fig. 2.6 Same as Fig. 2.5 but for $I = 0$ and $S = 1$.

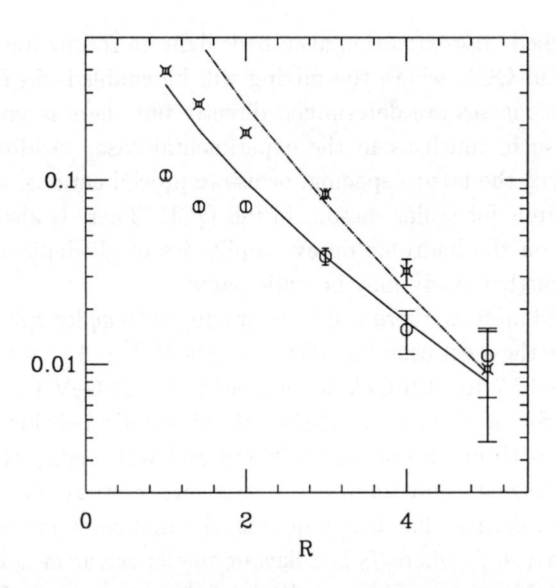

Fig. 2.7 The contribution [43] to the binding energy for the spin and isospin combinations corresponding to π exchange (octagons) and ρ exchange (fancy squares). The solid line gives the π exchange contribution which is normalised by the $B^*B\pi$ coupling. The ρ exchange prediction has a free normalisation and is shown by the dash-dotted line. The results are plotted versus interquark separation R (in units of $a \approx 0.17$ fm).

the light quark mass from strange to the lighter u, d values to make more definite predictions about the binding of BB molecules.

Models for the binding of two B mesons involve, as in the case of the deuteron, pion exchange. The lattice study [43] is able to make a quantitative comparison of lattice pion exchange with the data described above using lattice determinations of the $B^*B\pi$ coupling [44] and excellent agreement is obtained at larger R values as shown in Fig. 2.7, as expected.

2.5 Conclusions and Outlook

Quenched lattice QCD is well understood and accurate predictions in the continuum limit are increasingly becoming available. The lightest glueball is scalar with mass $m(0^{++}) = 1611(30)(160)$ MeV where the second error is an overall scale error. The excited glueball spectrum is known too. The quenched approximation also gives information on quark-antiquark scalar mesons and their mixing with glueballs. This determination of the mixing in the quenched approximation also sheds light on results for the spectrum directly in full QCD where the mixing will be enabled. In full QCD, the scalar meson masses are determined directly but there is no concept of a glueball as such, much as in the experimental case. Additional work is need to reduce the lattice spacing, or use improved actions, to explore the continuum limit for scalar mesons in full QCD. There is also some lattice information on the hadronic decay amplitudes of glueballs and this is an area where further study may be anticipated.

For hybrid mesons, there will be no mixing with $q\bar{q}$ for spin-exotic states and these are the most useful predictions. The $J^{PC} = 1^{-+}$ state is expected in the range 10.7 to 11.0 GeV for b quarks, 2.0(2) GeV for s quarks and 1.9(2) GeV for u, d quarks. Mixing of spin-exotic hybrids with $q\bar{q}q\bar{q}$ or equivalently with meson-meson is allowed and will modify the predictions from the quenched approximation. A first lattice study has been made of hybrid meson decays. For heavy quarks, the dominant mode is string de-excitation to $\chi + f_0$ where f_0 is a flavour singlet scalar meson (or possibly two pions in this state). The magnitude of the decay rate is found to be of order 100 MeV, so this decay mode should still leave a detectably narrow resonance to be observed.

The topic of possible multi-quark bound states is difficult because the scale of the expected binding energies is a few MeV and this small value is a challenge for lattice studies. As an example, some evidence was presented for a possible B_sB_s molecular state.

Acknowledgements

The author thanks Craig McNeile for helpful discussions.

Bibliography

[1] R. Sommer, *Nucl. Phys.* **B411**, 839 (1994).
[2] P. De Forcrand *et al.*, *Phys. Lett.* **B152**, 107 (1985).
[3] C. Michael and M. Teper, *Nucl. Phys.* **B314**, 347 (1989).
[4] UKQCD collaboration, G. Bali, *et al.*, *Phys. Lett.* **B309**, 378 (1993).
[5] H. Chen *et al.*, *Nucl. Phys. B (Proc. Suppl.)* **34**, 357 (1994); A. Vaccarino and D. Weingarten, Phys. Rev. *D* **60**, 114501 (1999).
[6] SESAM and TχL Collaboration, G. Bali *et al.*, *Nucl. Phys. B (Proc. Suppl.)* **63** (1998) 209; *Phys. Rev.* **D62**, 054503 (2000).

[7] C. McNeile and C. Michael, *Phys. Rev.* **D63**, 114503 (2001).
[8] UKQCD collaboration; A. Hart, C. McNeile and C. Michael, hep-lat/0209063.
[9] C. Morningstar and M. Peardon, *Phys. Rev.* **D56**, 4043 (1997); *ibid.*, **D60**, 034509 (1999).
[10] C. Michael and G. A. Tickle, *Nucl. Phys.* **B333**, 593 (1990).
[11] P. de Forcrand and K.-F. Liu *Phys. Rev. Lett.* **69**, 245 (1992).
[12] UKQCD Collaboration, M Foster and C. Michael, *Phys. Rev.* **D59** 094509 (1999).
[13] I.H. Jorysz and C. Michael, *Nucl. Phys.* **B302**, 448 (1988).
[14] B. Lucini and M. Teper, *JHEP* **06**, 050 (2001).
[15] R. C. Brower, S. D. Mathur and C.-I. Tan, *Nucl. Phys.* **B587**, 249 (2000).
[16] W. Bardeen *et al.*, *Phys. Rev.* **D65**, 014509 (2002).
[17] W. Lee and D. Weingarten, *Nucl. Phys. B (Proc. Suppl)* **53**, 236(1997); *Nucl. Phys. B (Proc. Suppl)* **63**, 194 (1998); hep-lat/9805029; *Nucl. Phys. B (Proc. Suppl)* **73**, 249 (1999) Phys. Rev. *D* **61**, 014015 (2000).
[18] J. Sexton, A. Vaccarino and D. Weingarten, *Nucl. Phys. B (Proc. Suppl.)* **42**, 279 (1995); *Phys. Rev. Lett.* **75** 4563 (1995).
[19] C. Michael, *Nucl. Phys.* **B327**, 517 (1989).
[20] UKQCD Collaboration, A. Hart and M. Teper, *Phys. Rev.* **D65**, 034502 (2002).
[21] K. Hagiwara *et al.*, *Phys. Rev.* **D 66**, 010001 (2002).
[22] F. Close and A. Kirk, *Eur. Phys. J.* **C21**, 531 (2001).
[23] C. W. Bernard *et al.*, (MILC Collaboration), *Phys. Rev.* **D64**, 054506 (2001).
[24] L.A. Griffiths, C. Michael and P.E.L. Rakow, *Phys. Lett.* **B129**, 351 (1983).
[25] S. Perantonis and C. Michael, *Nucl. Phys.* **B347**, 854 (1990).
[26] S. Perantonis, A. Huntley and C. Michael, *Nucl. Phys.* **B326**, 544 (1989).
[27] K. Juge , J. Kuti and C. Morningstar, *Phys. Rev. Lett.* **82**, 1999 (4400); *Nucl. Phys. B (Proc. Suppl)* **83**, 304 (2000); hep-lat/0207004.
[28] CP-PACS Collaboration, T. Manke *et al.*, *Phys. Rev. Lett.* **82**, 4396 (1999); *Phys. Rev.* **D64**, 097505 (2001).
[29] T. Burch, K. Originos and D. Toussaint, *Phys. Rev.* **D64**, 074505 (2001).
[30] UKQCD Collaboration, P. Lacock, C. Michael, P. Boyle and P. Rowland, *Phys. Rev.* **D54**, 6997 (1996); *Phys. Lett.* **B401**, 308 (1997).
[31] C. Bernard *et al.*, *Phys. Rev.* **D56**, 7039 (1997); *Nucl. Phys. B (Proc. Suppl.)* **73**, 264 (1999).
[32] P. Lacock and K. Schilling, *Nucl. Phys. B (Proc. Suppl.)* **73**, 261 (1999).
[33] UKQCD Collaboration, C. McNeile, C. Michael and P. Pennanen, *Phys. Rev.* **D65**, 094505 (2002).
[34] UKQCD Collaboration, C. McNeile, C. Michael and K.J. Sharkey, *Phys. Rev.* **D65**, 014508 (2002).
[35] UKQCD Collaboration, C. McNeile and C. Michael, hep-lat/0212020.
[36] C. Bernard *et al.*, *Nucl. Phys. B (Proc. Suppl.)* (in press), hep-lat/0209097;

C. Bernard *et al.*, hep-lat/0301024.
[37] Z.H. Mei and X.Q. Luo, hep-lat/0206012.
[38] D. Thompson *et al.*, *Phys. Rev. Lett.* **79**, 1630 (1997);
S. U. Chung *et al.*, *Phys. Rev.* **D60**, 092001 (1999);
D. Adams *et al.*, *Phys. Rev. Lett.* **81**, 5760 (1998);
E.I.Ivanov *et al.*, *Phys. Rev. Lett.* **86**, 3977 (2001).
[39] A. W. Thomas and A. P. Szczepaniak, *Phys. Lett.* **B526**, 72 (2002).
[40] X. Liao and T. Manke, hep-lat/0210030.
[41] C. Michael, Proc. Heavy Flavours 8, Southampton, (ed. P. Dauncey and C. Sachrajda), JHEP, PRHEP-hf8/001, 1-10 (2000); hep-ph/9911219.
[42] A. Pochinsky, J.W. Negele and B. Scarlet, *Nucl. Phys.B (Proc. Suppl)* **73**, 255 (1999);
I. Wetzorke, F. Karsch and E. Laermann, *Nucl. Phys. (Proc. Suppl.)* **83** 218 (2000);
I. Wetzorke and F. Karsch, hep-lat/0208029.
[43] C. Michael and P. Pennanen, *Phys. Rev.* **D60**, 054012 (1999).
[44] UKQCD Collaboration, G. M. de Divitiis *et al.*, *JHEP* **10**, 010 (1998).

Chapter 3
Two Quark Potentials

G. Bali
Department of Physics & Astronomy, University of Glasgow,
Glasgow G12 8QQ, Scotland
E-mail: g.bali@physics.gla.ac.uk

In this Chapter QCD interactions between a quark and an anti-quark are discussed. In the heavy quark limit these potentials can be related to quarkonia and $1/m$ corrections can be systematically determined. Excitations of the ground state potential provide an entry point into the phenomenology of quark-gluon hybrids. The short-distance behaviour of non-perturbative potentials can serve as a test of resummation and convergence of perturbative expansions. Torelons and potentials between non-fundamental colour charges offer a window into the origin of the confinement mechanism and relate to effective string descriptions of low energy aspects of QCD.

3.1 Motivation

In order to avoid excessive overlap with the many introductions that already exist in different places of this Volume, we shall elaborate on a very subjective motivation of studying interquark potentials in QCD, centred around the general theme of *building bridges* between models and QCD.

QCD contains two related, non-perturbative features: the breaking of (approximate) chiral symmetry and the (effective) confinement of coloured objects such as quarks and gluons. Similarly, models of the QCD vacuum can roughly be divided into two classes: those that are primarily based on chiral symmetry and those that have confinement, for instance in the form of a confining potential, as their starting point. Often it is difficult to *microscopically* relate a particular model to QCD. For example, should the instantons that are evidenced in lattice simulations within a given prescription at a finite cut-off, be the same that are supposed to appear within instanton liquid models? The answer to this question is not known.

There exist, however, two ways of systematically ***building bridges***. Perhaps the most obvious one is the comparison of *global* properties like form factors, charge distributions in position space, potential energies or particle masses. Such a comparison circumvents the problem of identifying a one-to-one mapping of the degrees of freedom within a given model, which often might not qualify as a quantum field theory, onto objects that appear in the QCD vacuum. The model would then very much resemble the pragmatic use of analogies in the popular science literature: it is not that important to get things 100 % right if we cannot understand them 100 % anyway.

A lot of purpose-engineered QCD information that is not directly accessible to experiment can be *manufactured* in lattice simulations. Moreover, in experiment there is only one world. On the lattice one has the freedom to vary the number of sea quark flavours n_f, the quark masses, the number of gauge group colours, N_C, the volume *etc. etc.*, away from the values that happen to be realized in nature for one reason or another. Sometimes (like in the case of quark masses) this is at present a necessity, in other cases it is pure virtue. In having not only one ***physical world*** but many ***virtual worlds*** at ones disposal, the applicability range and precision to be expected from any model can, in principle, be tested very stringently. Unfortunately, in practice, such interaction between lattice practitioners and model builders is still rather under-developed for various psychological, communication-related and dogmatic reasons and, in some cases, even out of fear, mistrust or over-confidence, one might speculate.

The other way of ***building bridges*** is when a separation of scales occurs, in which the symmetries of QCD constrain the number of possible terms and allow for the systematic construction of an effective field theory. The most prominent examples are the chiral effective field theory (χEFT) that governs the low energy interactions of QCD as well as heavy quark effective theory (HQET) and non-relativistic QCD (NRQCD) for heavy quark physics. While in χEFT the scale separation occurs through the spontaneous breaking of the chiral symmetry by some collective gluonic effect, in HQET and NRQCD the scale separation is provided by the heavy quark mass m. Ideally, one would calculate the "high" energy Wilson coefficients of χEFT, such as the pion decay constant F_π, in lattice simulations (or determine them from fits to experimental data). The low energy expansion, as a function of m_π, can be determined analytically (chiral perturbation theory). On the other hand, in HQET and NRQCD the low energy matrix elements can be provided by non-perturbative lattice simulations while the Wilson coefficients are calculable in perturbation theory, as long as $m \gg \Lambda$, where Λ denotes a typical non-perturbative scale of order 400 MeV.

It is this latter heavy quark limit, in which the static QCD potentials that are discussed here — as well as in Chapter 4 — can be related to mesonic bound states and, in the case of three body potentials, baryonic bound states. In doing this, QCD can be reduced to non-relativistic quantum mechanics within this particular sector, with the help of reasonable assumptions which themselves can be tested in lattice simulations. For example, in the case of quarkonia, QCD itself tells us what "potential model" we have to choose.

It is also possible to fit the spectra of light mesons and baryons, assuming phenomenological potentials. These potentials cannot be related in a systematic way either to QCD or to static potentials as calculated from Wilson loops. However, in spite of some rather dubious assumptions that are implicitly folded into such non-relativistic or relativized quark-potential models, it can still be instructive to compare the parametrizations that are commonly used in this context with those that are relevant in quarkonium physics. After all, most *physics* appears to interpolate smoothly between hadrons made out of light and heavy quarks. There also appears to be an intimate connection between, on one hand, the broken chiral symmetry that is most relevant in the light hadronic sector — but plays little direct rôle for quarkonia with masses much larger than the chiral symmetry breaking scale — and, on the other hand, the confining potential and flux tubes that seem to be the vacuum excitations that are responsible for interactions between the slowly moving heavy quark degrees of freedom.

3.2 The Static QCD Potential

The very definition of a "potential" requires the concept of "instantaneous" interactions: a test particle has to interact with the field induced by a source on a time scale short enough to guarantee that the relative distance remains unaffected. Whenever the relative speed of the two particles becomes relativistic, the underlying assumption of a constant time difference between cause and effect is obviously violated: only in non-relativistic systems, *i.e.* as long as the typical interaction energies (E) within bound states are small compared to the particle masses (M), can we define a potential as a function of coordinates such as the distance r, the spin S, angular momentum L and relative momentum p. While for interactions between elementary charges in QED this is always the case as $E/M = O(\alpha_F) \ll 1$, QCD implies typical binding energies of order 400 MeV and only the bottom, and eventually the charm quark, can be regarded as non-relativistic.

Another example in which the non-relativistic approximation is justified are nucleon-nucleon potentials, $V_{NN}(r)$. Although there exist attempts to extract this information also from QCD, by employing lattice simulations (cf. Chapter 4 Subsec. 4.4.2), it is fair to predict that quantitatively reliable information, using the presently available methodology, might not become available within this decade. However, in this case a wealth of phenomenological information exists from experiment. For the quark-antiquark potential it is exactly the opposite. All experimental information is model dependent and rather indirect as quarks never appear as free particles in nature. However, this potential is among the most precisely determined quantities that have been calculated so far on the lattice

As a starting point one can make the test charges infinitely heavy, prohibiting any change in their relative speed and study the "static" limit. To this end we shall introduce the Wegner-Wilson loop and derive its relation to the static potential. Subsequently, expectations of this potential from exact considerations, strong coupling and string arguments as well as from perturbation theory are presented. Lattice results are then reviewed. Finally, the model is extended to non-static quarks and the form of the resultant potential in coordinate space is compared with its counterpart (ω-meson exchange) in the nucleon-nucleon interaction.

3.2.1 *Wilson loops*

We will derive the relationship between the expectation values of Wegner-Wilson loops and the potential energy $V(r) = E_0(r)$ between two colour charges, separated by a distance r. This is a technical but instructive exercise. The final result is displayed in Eq. 3.17.

The Wegner-Wilson loop was originally introduced by Wegner [1] as an order parameter in Z_2 gauge theory. It is defined as the trace of the product of gauge variables $U_{x,\mu}$ along a closed oriented contour δC, enclosing an area C,

$$W(C) = \mathrm{Tr}\left\{\mathcal{P}\left[\exp\left(i\int_{\delta C} dx_\mu\, A_\mu(x)\right)\right]\right\} = \mathrm{Tr}\left(\prod_{(x,\mu)\in\delta C} U_{x,\mu}\right). \quad (3.1)$$

While the loop, determined on a gauge configuration* $\{U_{x,\mu}\}$, is in general complex, its expectation value is real, due to charge invariance: in Euclidean space we have $\langle W(C)\rangle = \langle W^*(C)\rangle = \langle W(C)\rangle^*$. It is straight forward to generalise the above Wilson loop to any non-fundamental representation D of the gauge field, just by replacing the variables $U_{x,\mu}$ with the corresponding links $U^D_{x,\mu}$. The arguments below, relating the Wilson loop to the potential energy of static sources, go through independent of the representation according to which the sources transform under local gauge transformations. In what follows, we will denote a Wilson loop, enclosing a rectangular contour with one purely spatial distance, $\mathbf{r}$, and one temporal separation, t, by $W(\mathbf{r}, t)$. Examples of Wilson loops on a lattice for two different choices of contours δC are displayed in Figure 3.1.

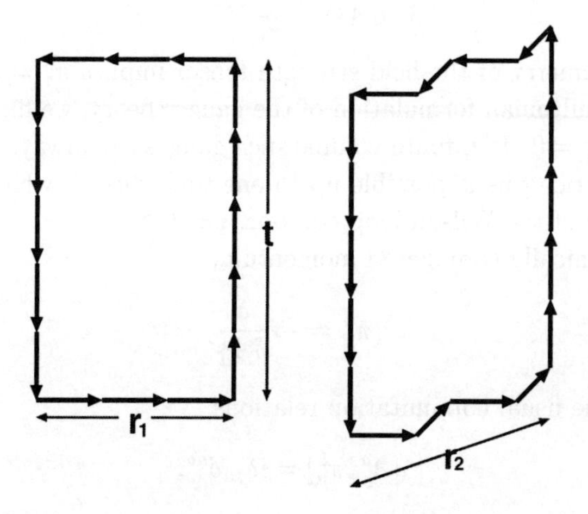

Fig. 3.1 Examples of rectangular on- and off-axis Wilson loops with temporal extent $t = 5a$ and spatial extents $r_1 = 3a$ and $r_2 = 2\sqrt{2}\,a$, respectively.

*$U_\mu \in SU(3)$ a gauge group element, pointing into direction $\mu \in \{1, 2, 3, 4\}$, located between positions x and $x + a\hat{\mu}$, where a is the lattice spacing and x is a lattice point within the (finite) 4-volume.

In Wilson's original work [2], the Wilson loop had been related to the potential energy of a pair of static colour sources by using transfer matrix arguments. However, it took a few years until Brown and Weisberger attempted to derive the connection between the Wilson loop and the effective potential between heavy, not necessarily static, quarks in a mesonic bound state [3]. Later on, mass dependent corrections to the static potential have been derived along similar lines [4, 5] and the approach has been made systematic within the framework of EFTs [6, 7, 8, 9]. In Sec. 3.3, we will discuss these developments in more detail. Here, we derive the connection between a Wilson loop and the static potential between colour sources which highlights similarities with the situation in classical electrodynamics.

For this purpose we start from the Euclidean Yang-Mills action,

$$S = \frac{1}{4g^2}\int d^4x\, F^a_{\mu\nu}F^a_{\mu\nu}. \tag{3.2}$$

The canonically conjugated momentum to the field, A^a_i, is given by the functional derivative,

$$\pi^a_i = \frac{\delta S}{\delta(\partial_4 A^a_i)} = \frac{1}{g^2}F^a_{4i} = -\frac{1}{g}E^a_i. \tag{3.3}$$

The anti-symmetry of the field strength tensor implies $\pi^a_4 = 0$. In order to obtain a Hamiltonian formulation of the gauge theory, we fix the temporal gauge *i.e.* $A^a_4 = 0$. In infinite volume such gauges can always be found. On a toroidal lattice this is possible up to one time slice t', which we demand to be outside of the Wilson loop contour, *i.e.* $t' > t$.

The canonically conjugated momentum,

$$\pi^a_\mu = -i\frac{\delta}{\delta A^a_\mu}, \tag{3.4}$$

now fulfils the usual commutation relations,

$$[A^a_j, \pi^b_\mu] = i\delta_{j\mu}\delta^{ab}, \tag{3.5}$$

and we can construct the Hamiltonian,

$$H = \int d^3x \left(\pi^a_\mu \partial_4 A^a_\mu - \frac{1}{4g^2}F^a_{\mu\nu}F^a_{\mu\nu}\right) = \frac{1}{2}\int d^3x\, (E^a_i E^a_i - B^a_i B^a_i) \tag{3.6}$$

that acts onto states $\Psi[A_\mu]$. In the Euclidean metric the magnetic contribution to the total energy is negative. Note that we can also add a fermionic term $\sum_f \bar{q}_f[\gamma_\mu D_\mu + m_f]q_f$ to the action above. In this case the momentum conjugate to the Dirac spinor field q_f^α is given by $-i\delta/\delta q_f^\alpha = \bar{q}_f^\alpha \gamma_4$, where $\alpha = 1, \cdots, N_C$ runs over the colour in the fundamental representation. Here f denotes the quark flavour and m_f the respective mass.

A gauge transformation Ω can be represented as a bundle of $SU(N_C)$ matrices in some representation R, $\Omega_R(\mathbf{x}) = e^{i\omega^a(\mathbf{x})T_R^a}$. We wish to derive the operator representation of the group generators T_R^a, that acts on the Hilbert space of wave functionals. For this purpose we start from the definition

$$R(\Omega)\Psi = \left[1 + i\int d^3x\, \omega^a(\mathbf{x})T_R^a(\mathbf{x}) + \cdots\right]\Psi = \Psi + \delta\Psi. \tag{3.7}$$

One easily verifies that $\delta A_i = A_i^\Omega - A_i = -(\partial_i\omega + i[A_i, \omega]) \equiv D_i\omega(\mathbf{x})$. We then obtain

$$\begin{aligned} \delta\Psi &= \int d^3x\, \delta A_i(\mathbf{x})\frac{\delta\Psi}{\delta A_i(\mathbf{x})} = \int d^3x\, D_i\omega(\mathbf{x})\frac{\delta\Psi}{\delta A_i(\mathbf{x})} \\ &= -\frac{i}{g}\int d^3x\, \omega^a(\mathbf{x})(D_iE_i)^a(\mathbf{x})\Psi, \end{aligned} \tag{3.8}$$

where we have performed a partial integration and have made use of the equivalence

$$\frac{\delta}{\delta A_i} = -\frac{i}{g}E_i \tag{3.9}$$

of Eqs. 3.3 and 3.4. Hence we obtain the representation

$$T_R^a = -\frac{1}{g}(D_iE_i)^a, \tag{3.10}$$

i.e. the covariant divergence of the electric field operator is the generator of gauge transformations! Again note that had we included sea quarks into the action, we would have encountered an additional term $-\frac{1}{g}\sum_f \bar{q}_f\gamma_4 T^a q_f$ on the right hand side of this equation, where the generator is to be taken in the fundamental representation. It is trivial to generalize the equations below accordingly and the physical meaning is clear too: the vacuum has an intrinsic charge density distribution due to sea quarks. This then in turn allows for *string breaking* of the static potential.

Let us assume that the wave functional is a singlet under gauge transformations $R(\Omega)\Psi[A_\mu] = \Psi[A_\mu]$. This implies that

$$(D_i E_i)^a \Psi = 0, \tag{3.11}$$

which is Gauss' law in the absence of sources: Ψ lies in the eigenspace of $D_i E_i$ that corresponds to the eigenvalue zero. Let us next place an external source in the fundamental representation of the colour group at position $\mathbf{r}$. In this case, the associated wave functional $\Psi_\alpha, \alpha = 1, \cdots, N_C$ transforms in a non-trivial way, namely

$$[R(\Omega)\Psi]_\alpha = \Omega_{\alpha\beta}\Psi_\beta. \tag{3.12}$$

This implies that

$$(D_i E_i)^a \Psi = -g\delta^3(\mathbf{r}) T^a \Psi, \tag{3.13}$$

which again resembles Gauss' law, this time for a point-like colour charge at position† $\mathbf{r}$. For non-fundamental representations D, Eq. 3.13 remains valid under the replacement $T^a \to T^a_D$.

Let us now place a fundamental source at position $\mathbf{0}$ and an anti-source at position $\mathbf{r}$. The wave functional $\Psi_\mathbf{r}$, which is an $N_C \times N_C$ matrix in colour space will transform according to

$$\Psi^\Omega_{\mathbf{r},\alpha\beta} = \Omega_{\alpha\gamma}(\mathbf{0})\Omega^*_{\beta\delta}(\mathbf{r})\Psi_{\mathbf{r},\gamma\delta}. \tag{3.14}$$

One object with the correct transformation property is a gauge transporter (Schwinger line) from $\mathbf{0}$ to $\mathbf{r}$,

$$\Psi_\mathbf{r} = \frac{1}{\sqrt{N_C}} U^\dagger(\mathbf{r}, t) = \frac{1}{\sqrt{N_C}} \mathcal{P}\left[\exp\left(i\int_\mathbf{0}^\mathbf{r} d\mathbf{x}\, \mathbf{A}(\mathbf{x}, t)\right)\right], \tag{3.15}$$

which on the lattice corresponds to the ordered product of link variables along a spatial connection between the two points. Since we are in the temporal gauge, $A_4(x) = 0$, the correlation function between two such lines at time-like separation t is the Wilson loop

$$\langle W(\mathbf{r}, t)\rangle = \frac{1}{N_C}\langle U_{\alpha\beta}(\mathbf{r}, t) U^\dagger_{\beta\alpha}(\mathbf{r}, 0)\rangle, \tag{3.16}$$

†Of course, on a torus, such a state cannot be constructed. Note that in our Euclidean space-time conventions Gauss' law reads $[\mathbf{DE}]^a(\mathbf{x}) = -\rho^a(\mathbf{x})$, where ρ denotes the charge density. Again note that, in general, ρ^a will automatically contain a contribution $g\bar{q}_f\gamma_4 T_a q_f$ for each sea quark flavour f.

which, being a gauge invariant object, will give the same result in any gauge. Other choices of $\Psi_{\mathbf{r}}$, *e.g.* linear combinations of spatial gauge transporters connecting $\mathbf{0}$ with $\mathbf{r}$, define generalised (or smeared) Wilson loops, $W_\Psi(\mathbf{r},t)$.

We insert a complete set of transfer matrix eigenstates, $|\Phi_{\mathbf{r},n}\rangle$, within the sector of the Hilbert space that corresponds to a charge and anti-charge in the fundamental representation at distance $\mathbf{r}$, and expect the Wilson loop in the limit of large temporal lattice extent, $L_\tau a \gg t$, to behave like

$$\langle W_\Psi(\mathbf{r},t)\rangle = \sum_n |\langle \Phi_{\mathbf{r},n}|\, \Psi_{\mathbf{r}}\,|0\rangle|^2\, e^{-E_n(\mathbf{r})t}, \tag{3.17}$$

where the normalisation convention is such that $\langle\Phi_n|\Phi_n\rangle = \langle\Psi^\dagger\Psi\rangle = 1$ and the completeness of eigenstates implies $\sum_n |\langle\Phi_n|\Psi|0\rangle|^2 = 1$. Note that no disconnected part has to be subtracted from the correlation function since $\Psi_{\mathbf{r}}$ is distinguished from the vacuum state by its colour indices. $E_n(\mathbf{r})$ denote the energy levels. The ground state contribution $E_0(\mathbf{r})$ — that will dominate in the limit of large t — can now be identified with the static potential $V(\mathbf{r})$, which we have been aiming to calculate.

The gauge transformation properties of the colour state discussed above, which determine the colour group representation of the static sources and their separation $\mathbf{r}$, do not yet completely determine the state in question: the sources will be connected by an elongated chromo-electric flux tube. This vortex can, for instance, be in a rotational state with spin $\Lambda \neq 0$ about the inter-source axis. Moreover, under interchange of the ends the state can transform evenly ($\eta = \mathrm{g}$) or oddly ($\eta = \mathrm{u}$), where η denotes the combined CP parity. Finally, in the case of the one-dimensional $\Lambda = 0$ representations, it can transform symmetrically or anti-symmetrically under reflections with respect to a plane containing the sources ($\sigma_v = \pm$). It is possible to single out sectors within a given irreducible representation of the relevant cylindrical symmetry group [10], $D_{\infty h}$, with an adequate choice of Ψ. A straight line connection between the sources corresponds to the $D_{\infty h}$ quantum numbers Σ_g^+, where $\Lambda = 0, 1, 2, \cdots$ is replaced by capital Greek letters, $\Sigma, \Pi, \Delta, \cdots$. Any static potential that is different from the Σ_g^+ ground state will be referred to as a "hybrid" potential. ‡ Since these potentials are gluonic excitations they can be thought of as being hybrids between pure "glueballs" and a pure static-static state; indeed, high hybrid excitations are unstable and will decay into lower lying potentials via the

‡Compare the discussion in Sec. 2.3.1 of Chapter 2

radiation of glueballs.

3.2.2 *Exact results*

We identify the static potential $V(\mathbf{r})$ with the ground state energy $E_0(\mathbf{r})$ of Eq. 3.17 that can be extracted from the Wilson loop of Eq. 3.1. By exploiting the symmetry of a Wilson loop under an interchange of the space and time directions, it can be proven that the static potential cannot rise faster than linearly as a function of the distance r in the limit $r \to \infty$ [11]. Moreover, reflection positivity of Euclidean n-point functions [12, 13] implies convexity of the static potential [14], *i.e.*

$$V''(r) \leq 0. \tag{3.18}$$

The proof also applies to ground state potentials between sources in non-fundamental representations. However, it does not apply to hybrid excitations, since in this case the required creation operator extends into spatial directions, orthogonal to the direction of $\mathbf{r}$. Due to positivity, the potential is bound from below.[§] Therefore, convexity implies that $V(r)$ is a monotonically rising function of r, *i.e.*

$$V'(r) \geq 0. \tag{3.19}$$

In Ref. [15], which in fact preceded Ref. [14], somewhat more strict upper and lower limits on Wilson loops, calculated on a lattice, have been derived: Let a_σ and a_τ be temporal and spatial lattice resolutions. The main result for rectangular Wilson loops in representation D and d space-time dimensions then is

$$\langle W(a_\sigma, a_\tau)\rangle^{rt/(a_\sigma a_\tau)} \leq \langle W(r,t)\rangle \leq (1-c)^{r/a_\sigma + t/a_\tau - 2}, \tag{3.20}$$

with $c = \exp[-4(d-1)D\beta]$. The resulting bounds on $V(r)$ for $r > a_\sigma$ read

$$-\ln(1-c) \leq a_\tau V(r) \leq -\frac{r}{a_\sigma} \ln\langle W(a_\sigma, a_\tau)\rangle; \tag{3.21}$$

[§] The potential that is determined from Wilson loops depends on the lattice cut-off, a, and can be factorised into a finite potential $\hat{V}(r)$ and a (positive) self energy contribution: $V(r;a) = \hat{V}(r) + V_{\text{self}}(a)$. The latter diverges in the continuum limit (see Sec. 3.2.5), whereas the potential $\hat{V}(r)$ will become negative at small distances. Thus $V(r;a)$ is indeed bounded from below by $V(0)$=0.

in consistency with Ref. [11], the potential (measured in lattice units a_τ) is bound from above by a linear function of r and it takes positive values everywhere.

3.2.3 *Strong coupling expansions*

Expectation values can be approximated by expanding the exponential of the lattice action, in terms of the inverse coupling $\beta = 2N_C/g^2$, giving $\exp(-\beta S) = 1 - \beta S + \cdots$. This strong coupling expansion is similar to a high temperature expansion in statistical mechanics. When the Wilson action is used each factor β is accompanied by a plaquette and certain diagrammatic rules can be derived [2, 16, 17, 18, 19]. Let us consider a strong coupling expansion of the Wilson loop, Eq. 3.1. Since the integral over a single group element vanishes,

$$\int dU\, U = 0, \tag{3.22}$$

to zeroth order, we have $\langle W \rangle = 0$. To the next order in β, it becomes possible to cancel the link variables on the contour δC of the Wilson loop by tiling the whole minimal enclosed (lattice) surface C with plaquettes. Hence one obtains the expectation value [18, 20, 21, 22]

$$\langle W(C) \rangle = \left\{ \begin{array}{ll} [\beta/4]^{-\mathrm{area}(\delta C)} + \cdots, & N_C = 2 \\ \left[\beta/2N_C^2\right]^{-\mathrm{area}(\delta C)} + \cdots, & N_C > 2 \end{array} \right\} \tag{3.23}$$

for $SU(N_C)$ gauge theory. If we now consider the case of a rectangular Wilson loop that extends r/a lattice points into a spatial and t/a points into the temporal direction, we find the area law,

$$\langle W(\mathbf{r}, t) \rangle = \exp\left[-\sigma_d r t\right] + \cdots, \tag{3.24}$$

with a string tension

$$\sigma_d a^2 = -d_r \ln \frac{\beta}{18}. \tag{3.25}$$

The numerical value of the denominator applies to $SU(3)$ gauge theory; the potential is linear with slope σ_d, and so colour sources are confined at strong coupling. Here $d_r = (|r_1| + |r_2| + |r_3|)/r \geq 1$ denotes the ratio between lattice and continuum norms and deviates from $d_r = 1$ for source separations $\mathbf{r}$ that are not parallel to a lattice axis. The string tension of Eq. 3.25 depends on d_r and, therefore, on the lattice direction; $O(3)$ rotational symmetry is explicitly broken down to the cubic subgroup O_h. The extent of violation will eventually be reduced as one increases β and considers higher orders of the expansion. Such high order strong coupling expansions have indeed been performed for Wilson loops [23] and glueball masses [24]. Unlike standard perturbation theory, whose convergence is known to be at best asymptotic [25, 26], the strong coupling expansion is analytic around $\beta = 0$ [27] and, therefore, has a finite radius of convergence.

Strong coupling $SU(3)$ gauge theory results seem to converge for $\beta < 5$ — see, for example, Ref. [20]. One would have hoped to eventually identify a crossover region of finite extent between the validity regions of the strong and weak coupling expansions [28], or at least a transition point between the leading order strong coupling behaviour, $a^2 \propto -\ln(\beta/18)$, of Eq. 3.25 and the weak coupling limit, $a^2 \propto \exp[-2\pi\beta/(3\beta_0)]$, set by the asymptotic freedom of the QCD β-function. However, even after re-summing the strong coupling series in terms of improved expansion parameters and applying sophisticated Padé approximation techniques [29], nowadays such a direct crossover region does not appear to exist, necessitating one to employ Monte Carlo simulation techniques. In Fig. 3.2 we compare the strong coupling expansion for the string tension, calculated to $O(g^{-24})$ *i.e.* $O(\beta^{12})$ [23] with results from lattice simulations. While at large β the lattice results approach the weak coupling limit, there appears to be no overlap between weak and strong coupling and neither between strong coupling and lattice results in the region of interest ($\beta \approx 6$ corresponds to a lattice spacing $a^{-1} \approx 2$ GeV). We have taken the $n_F = 0$ value of the QCD Λ parameter as determined non-perturbatively in Ref. [30] as normalization for the weak coupling expansion. The error band of the $O(g^6)$ expectation is due to the corresponding statistical uncertainty. The $O(g^4)$ central value lies within this band. There is no normalization ambiguity in the strong coupling results. Also at $\beta > 5$, the quality of convergence of the strong coupling expansion diminishes. This break-down might be related to a roughening transition that is discussed, for example, in Refs. [31, 32].

We would like to remark that the area law of Eq. 3.24 is a rather general result for strong coupling expansions in the fundamental representation of compact gauge groups. In particular, it applies also to $U(1)$ gauge theory which we do not expect to confine in the continuum. In fact, based on duality arguments, Banks, Myerson and Kogut [33] have succeeded in proving the existence of a confining phase in the four-dimensional theory and suggested the existence of a phase transition while Guth [34] has proven that, at least in the non-compact formulation of $U(1)$, a Coulomb phase exists. Indeed, in numerical simulations of (compact) $U(1)$ lattice gauge theory two such distinct phases were found [35, 36], a Coulomb phase at weak coupling and a confining phase at strong coupling. The question whether the confinement one finds in $SU(N_C)$ gauge theories in the strong coupling limit survives the continuum limit, $\beta \to \infty$, can at present only be answered by means of numerical simulation (and has been answered positively).

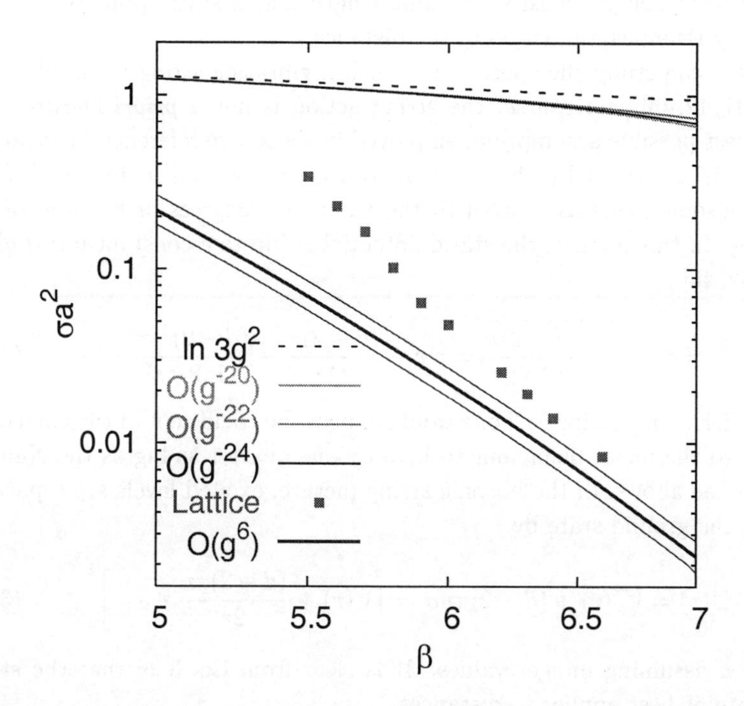

Fig. 3.2 Comparison of strong and weak coupling expansions with non-perturbative results for the string tension. The upper lines correspond to the strong coupling limit beginning with $\ln 3g^2$ (dashed line) — the lowest order term from Eq. 3.25 for $SU(3)$. The other upper lines show the additional approximations up to $O(g^{-24})$. The lower lines correspond to the weak coupling limit up to $O(g^6)$

3.2.4 *String picture*

The infra-red properties of QCD might be reproduced by effective theories of interacting strings. String models share many aspects with the strong coupling expansion. Originally, the string picture of confinement had been discussed by Kogut and Susskind [37] as the strong coupling limit of the Hamiltonian formulation of lattice QCD. The strong coupling expansion of a Wilson loop can be cast into a sum of weighted random deformations of the minimal area world sheet. This sum can then be interpreted to represent a vibrating string. The physical picture behind such an effective string description is that of the electric flux between two colour sources being squeezed into a thin, effectively one-dimensional, flux tube or Abrikosov-Nielsen-Olesen (ANO) vortex [38, 39, 40, 41]. As a consequence, this yields a constant energy density per unit length and a static potential that is linearly rising as a function of the distance.

One can study the spectrum of such a vibrating string in simple models [31, 42, 43]. Of course, the string action is not *a priori* known. The simplest possible assumption, employed in the above references, is that the string is described by the Nambu-Goto action [44, 45] in terms of $d-2$ free bosonic fields associated to the transverse degrees of freedom of the string. In this picture, the static potential is (up to a constant term) given by [42, 46]

$$V(r)=\sigma r\sqrt{1-\frac{(d-2)\pi}{12\,\sigma\,r^2}}=\sigma\,r-\frac{(d-2)\pi}{24\,r}-\frac{(d-2)^2\pi^2}{1152\,\sigma\,r^3}-\cdots. \tag{3.26}$$

For a fermionic string [47] one would expect the coefficient of the correction term to the linear behaviour to be only one quarter as big as the Nambu-Goto one above. In the bosonic string picture, excited levels are separated from the ground state by

$$V_n^2(r)=V^2(r)+(d-2)\pi n\sigma=\left[V(r)+\frac{(d-2)\pi n}{2r}-\cdots\right]^2, \tag{3.27}$$

with n assuming integer values. It is clear from Eq. 3.26 that the string picture at best applies to distances

$$r\gg r_c=\sqrt{\frac{(d-2)\pi}{12\,\sigma}}. \tag{3.28}$$

For $d=4$ one obtains $r_c\approx 0.33$ fm, when using the value $\sqrt{\sigma}\approx 430$ MeV from the $\rho, a_2, \cdots$ Regge trajectory.

The expectation of Eq. 3.26 has been very accurately reproduced in numerical simulations of Z_2 gauge theory in $d = 3$ space-time dimensions [48, 49]. In contrast, for $d = 4$ $SU(3)$ gauge theory the spectrum of hybrid potentials still differs significantly from the expectation of Eq. 3.27 for distances as large as 2 fm [50, 51]. However, qualitatively the string picture is supported by the $SU(3)$ data too, since the hybrid potentials at large r are found to group themselves into various bands that are separated by approximately equidistant gaps. These will eventually converge to values π/r, at even larger distances than accessible at present. Such an observation would support the existence of a bosonic string description of confining gauge theories in the very low energy regime [52, 53, 54, 55, 56]. Of course, in $d < 26$, the string Lagrangian is not renormalisable — but only effective — and higher order correction terms like torsion and rigidity will in general have to be added [57].

It is hard to disentangle in $d = 4$ the (large distance) $1/r$ term, expected from string vibrations, from the perturbative Coulomb term at short distances. However, a high precision attempt has been made recently, with promising results [58]. As an alternative, three-dimensional investigations (where perturbation theory yields a logarithmic contribution) have been suggested [59, 60]. Another way out is to determine the mass of a closed string, encircling a boundary of the lattice with a spatial extent $l = L_\sigma a$ (a torelon), which is not polluted by a perturbative tail. The bosonic string expectation in this case would be [59]

$$E_n(l) = \sigma\, l - \frac{(d-2)\pi}{6\, l} + \cdots . \tag{3.29}$$

The naïve range of validity of the picture is $l \gg l_c = 2\, r_c \approx 0.66$ fm. The numerical value applies to $d = 4$ from Eq. 3.28. An investigation of the finite size dependence of the torelon mass in $d = 4$ $SU(2)$ gauge theory has been performed some time ago by Michael and Stephenson [61] who found excellent agreement — on the 3 % level — with the bosonic string picture already for distances $1\,\text{fm} \leq l \leq 2.4$ fm, quite close to l_c. Qualitative agreement has also been reported by Teper [62] from simulations of $SU(2)$, $SU(3)$, $SU(4)$ and $SU(5)$ gauge theories in three dimensions as well as in a recent study of four-dimensional $SU(2N)$ gauge theories by Lucini *et al.* [63].

The bosonic string picture prediction of the free energy, calculated from

Polyakov line correlators, at finite temperatures T is similar to Eq. 3.29

$$-\frac{1}{\beta}\ln\langle P^*(r)P(0)\rangle = \sigma(\beta)r + \cdots, \quad \sigma(T) = \sigma - \frac{(d-2)\pi}{6}T^2 + \cdots, \quad (3.30)$$

with validity for $r \gg T = aL_\tau$ [64]. The Polyakov line is defined as

$$P(\mathbf{x}) = \mathrm{Tr}\left\{\mathcal{T}\left[\exp\left(i\int_0^{aL_\tau} dx_4\, A_4(x)\right)\right]\right\} = \mathrm{Tr}\left(\prod_{x_4=0}^{aL_\tau} U_{x,4}\right), \quad (3.31)$$

where $\mathcal{T}$ denotes time ordering of the argument. The dependence of the effective string tension on the temperature has been checked for rather low $T^{-1} < 1.24\, T_c^{-1} \approx 0.93\,\mathrm{fm}$ in studies of $SU(3)$ gauge theory [65, 66]. Although the sign of the leading correction term to the zero temperature limit is correct, the difference comes out to be larger than predicted. It would be interesting to check whether the result will converge towards the string expectation at lower temperatures.

3.2.5 *The potential in perturbation theory*

The strong coupling expansion is specific to the lattice regularisation.¶ However, the expectation value of a Wilson loop can also be approximated using standard perturbative techniques.

We will discuss the leading order weak coupling result that corresponds to single gluon exchange between two static colour sources which, although we neglect the spin structure, we will call "quarks" for convenience. From the Lagrangian, $\mathcal{L}_{YM} = \frac{1}{2g^2}\mathrm{Tr}F_{\mu\nu}F_{\mu\nu}$, one can easily derive the propagator of a gluon with four-momentum q,

$$G^{ab}_{\mu\nu}(q) = g^2\frac{\delta^{ab}\delta_{\mu\nu}}{q^2}, \quad (3.32)$$

where μ, ν are Lorentz indices and $a, b = 1, \cdots N_A$ label the colour generators with $N_A^2 = N_C^2 - 1$ for $SU(N_C)$. The same calculation can be done starting from a lattice discretised action. The Wilson action yields the result of Eq. 3.32 up to the replacement

$$q_\mu \to \hat{q}_\mu = \frac{2}{a}\sin\left(\frac{aq_\mu}{2}\right). \quad (3.33)$$

¶However, new strong coupling methods have been developed in the large N_C limit, based on the Maldacena conjecture of QFT/AdS correspondence.

Other lattice actions yield slightly different results but they all approach Eq. 3.32 in the continuum limit, $a \to 0$. Momentum space potentials can be obtained from the on-shell static quark anti-quark scattering amplitude: the gluon interacts with two static external currents pointing into the positive and negative time directions, $A^a_{\mu,\alpha\beta} = \delta_{\mu,4}T^a_{\alpha\beta}$ and $A'^b_{\nu,\gamma\delta} = -\delta_{\nu,4}T^b_{\gamma\delta}$. Hence, we obtain the tree level interaction kernel

$$K_{\alpha\beta\gamma\delta}(q) = -\frac{g^2}{q^2}T^a_{\alpha\beta}T^a_{\gamma\delta}. \tag{3.34}$$

For sources in the fundamental representation, the Greek indices denote the colour indices of the external currents running from 1 to N_C and the quark anti-quark state can be decomposed into two irreducible representations of $SU(N_C)$,

$$\mathbf{N}_C \otimes \mathbf{N}_C^* = \mathbf{1} \oplus \mathbf{N}_A. \tag{3.35}$$

We can now either start from a singlet or an octet[||] initial $\Phi_{\beta\gamma} = Q_\beta Q^*_\gamma$ state,

$$\Phi^{\mathbf{1}}_{\beta\gamma} = \delta_{\beta\gamma}, \tag{3.36}$$

$$\Phi^{\mathbf{N}_A}_{\beta\gamma} = \Phi_{\beta\gamma} - \frac{1}{N_C}\delta_{\beta\gamma}, \tag{3.37}$$

where the normalisation is such that $\Phi^i_{\alpha\beta}\Phi^j_{\beta\alpha} = \delta^{ij}$. A contraction with the group generators of Eq. 3.34 yields

$$\Phi^{\mathbf{1}}_{\beta\gamma}T^a_{\alpha\beta}T^a_{\gamma\delta} = C_F\Phi^{\mathbf{1}}_{\alpha\delta}, \tag{3.38}$$

$$\Phi^{\mathbf{N}_A}_{\beta\gamma}T^a_{\alpha\beta}T^a_{\gamma\delta} = -\frac{1}{2N_C}\Phi^{\mathbf{N}_A}_{\alpha\delta}, \tag{3.39}$$

where $C_F = N_A/(2N_C)$ is the quadratic Casimir charge of the fundamental representation.

We end up with the potentials in momentum space,

$$V_s(q) = -C_F g^2\frac{1}{q^2}, \quad V_o(q) = \frac{g^2}{2N_C}\frac{1}{q^2} = -\frac{1}{N_A}V_s(q), \tag{3.40}$$

governing interactions between fundamental charges coupled to a singlet and to an octet, respectively: the force in the singlet channel is attractive while that in the octet channel is repulsive and smaller in size.

[||] We call the state $\mathbf{N}_A$ an "octet" state, having the group $SU(3)$ in mind.

How are these potentials related to the static position space inter-quark potential, defined non-perturbatively through the Wilson loop,

$$V(\mathbf{r}) = -\lim_{t\to\infty}\frac{d}{dt}\ln\langle W(\mathbf{r},t)\rangle? \tag{3.41}$$

The quark anti-quark state creation operator, $\Psi_{\mathbf{r}}$, within the Wilson loop contains a gauge transporter and couples to the gluonic degrees of freedom. Thus, in general, it will have overlap with both, QQ^* singlet and octet channels**. Since the singlet channel is energetically preferred, *i.e.* $V_s < V_o$, we might expect the static potential to correspond to the singlet potential. Up to order g^6 this is indeed the case: to lowest order, the Wilson loop — defined by the closed contour δC — is given by the Gaussian integral

$$\langle W(\mathbf{r},t)\rangle = \exp\left\{-\frac{1}{2}\int d^4x\,d^4y J^a_\mu(x)G^{ab}_{\mu\nu}(x-y)J^b_\nu(y)\right\}, \tag{3.42}$$

where $J^a_\mu = \pm T^a$ if $(x,\mu)\in\delta C$ and $J^a_\mu = 0$ elsewhere††. Eq. 3.42 implies for $t \gg r$

$$\langle W(\mathbf{r},t)\rangle = \exp\left(C_F g^2 t\int_{-t/2}^{t/2}dt'\;[G(\mathbf{r},t') - G(\mathbf{0},t')]\right). \tag{3.43}$$

We have omitted gluon exchanges between the spatial closures of the Wilson loop from the above formula. Up to order g^6 (two loops), such contributions result in terms whose exponents are proportional to r and r/t and, therefore, do not affect the potential of Eq. 3.41. The propagator $G^{ab}_{\mu\nu}(x)$, the Fourier transform of $G^{ab}_{\mu\nu}(q)$ in Eq. 3.32, contains the function

$$G(x) = \int\frac{d^4q}{(2\pi)^4}\frac{e^{iqx}}{q^2}, \quad \int_{-\infty}^{\infty}dx_4\,G(x) = \frac{1}{4\pi}\frac{1}{r}. \tag{3.44}$$

After performing the t-integration, we obtain

$$V(\mathbf{r},\mu) = -C_F\frac{\alpha_s}{r} + V_{\text{self}}(\mu), \tag{3.45}$$

**Of course, for a quark and anti-quark being at different spatial positions, the singlet-octet classification should be consumed with caution in a non-perturbative context.

††Note that this formula, which automatically accounts for multi-photon exchanges, is exact in non-compact QED (excluding fermion loops) to any order of perturbation theory. However, in theories containing more complicated vertices, like non-Abelian gauge theories or compact lattice $U(1)$ gauge theory, correction terms have to be added at higher orders in g^2.

where $\alpha_s = g^2/(4\pi)$. The piece

$$V_{\text{self}}(\mu) = C_F g^2 \int_{q\leq\mu} \frac{d^3q}{(2\pi)^3}\frac{1}{q^2} = C_F\alpha_s\frac{2}{\pi}\mu, \tag{3.46}$$

that linearly diverges with the ultra-violet cut-off μ, results from self-interactions of the static (infinitely heavy) sources. Comparing Eqs. 3.40 and 3.45 we indeed find

$$V(q) = V_s(q), \tag{3.47}$$

where

$$V(\mathbf{q},0) = \int d^3r\, e^{i\mathbf{q}\cdot\mathbf{r}}\hat{V}(\mathbf{r}), \quad \hat{V}(\mathbf{r}) = V(\mathbf{r},\mu) - V_{\text{self}}(\mu). \tag{3.48}$$

This self-energy "problem" is well known on the lattice and has also received attention in continuum QCD, in the context of renormalon ambiguities in quark mass definitions [67, 68].

Note that while V_s corresponds to the static potential, the perturbation theory relevant for hybrid excitations of the ground state potential corresponds to V_o [69].

At order α_s^4 a class of diagrams appears in a perturbative calculation of the Wilson loop that results in contributions to the static potential that diverge logarithmically with the interaction time [70]. In Ref. [71], within the framework of effective field theories, this effect has been related to ultra-soft gluons due to which an extra scale, $V_o - V_s$, is generated. Moreover, a systematic procedure has been suggested to isolate and subtract such terms to obtain finite singlet and octet interaction potentials between heavy quarks.

The logarithmic divergence is related to the fact that Eq. 3.43 contains an integration over the interaction time. For large times and any fixed distance r, Wilson loops will decay exponentially with t. However, the tree level propagator in position space is proportional to $(r^2+t^2)^{-1}$, *i.e.* asymptotically decays like t^{-2} only. We notice that the integral receives significant contributions from the region of large t as demonstrated by the finite $t \gg r$ tree level result

$$-\ln\langle W(r,t)\rangle = -\frac{C_F\alpha_s}{r}t\frac{2}{\pi}\left\{\arctan\frac{t}{r} - \frac{r^2}{2t}\left[\ln\left(1+\frac{t^2}{r^2}\right)\right]\right\} + (r+t)V_{\text{self}}. \tag{3.49}$$

The tree level lattice potential can easily be obtained by replacing q_μ by $\hat{q}_\mu$ from Eq. 3.33 and (in the case of finite lattice volumes) the integrals by discrete sums over lattice momenta,

$$q_i = \frac{2\pi}{L_\sigma}\frac{n_i}{a}, \quad n_i = -\frac{L_\sigma}{2}+1, \cdots, \frac{L_\sigma}{2}. \tag{3.50}$$

The lattice potential reads

$$V(\mathbf{r}) = V_{\rm self}(a) - C_F\alpha_s\left[\frac{1}{\mathbf{r}}\right], \tag{3.51}$$

where

$$\left[\frac{1}{\mathbf{r}}\right] = \frac{4\pi}{L_\sigma^3 a^3}\sum_{\mathbf{q}\neq 0}\frac{e^{i\mathbf{q}R}}{\sum_i \hat{q}_i\hat{q}_i} \tag{3.52}$$

and $V_{\rm self}(a) = C_F\alpha_s\,[1/\mathbf{0}]$. We have neglected the zero mode contribution that is suppressed by the inverse volume $(aL_\sigma)^{-3}$. In the continuum limit, $[1/\mathbf{r}]$ approaches $1/r$ up to quadratic lattice artefacts whose coefficients depend on the direction of $\mathbf{r}$ while $V_{\rm self}(a)$ diverges like

$$V_{\rm self}(a) = C_F\alpha_s a^{-1}\times 3.1759115\cdots. \tag{3.53}$$

The numerical value applies to the limit, $L_\sigma \to \infty$. Note that under the substitution $\mu \approx 1.5879557\,\pi/a$, Eq. 3.53 is identical to Eq. 3.46. One loop computations of on-axis lattice Wilson loops can be found in Refs. [72, 73, 74], while off-axis separations in QCD with and without sea quarks have been realized in Ref. [75]. The tree level form, Eq. 3.51, is often employed to parameterise lattice artefacts — see Subsec. 5.9.3.1 of Chapter 5.

3.3 Quark-Antiquark Potentials between Non-Static Quarks

‡‡ So far in this chapter the rôle of the interquark potential has been to test various limits and models for QCD. For example, in Fig. 3.2 a comparison was made between the weak and strong coupling limits of QCD and also with the corresponding lattice results. Likewise, in Subsec. 3.2.4 string models were compared with the infra-red properties of QCD. This rôle

‡‡At this point the author "ran-out-of-steam" and so, because of the pressure of time, the editor felt that this chapter should be rapidly concluded — a task carried out by the editor himself with the semi-approval of the author.

is in stark contrast to that of the Nucleon-Nucleon potential [$V(NN)$] in nuclear physics, where $V(NN)$ is mainly used as a stepping stone to the understanding of multi-nucleon systems. To this end, the many parameters needed to define $V(NN)$ are first adjusted, more or less freely, to fit two nucleon experimental data, with — in some cases — values being imposed from meson-nucleon data or theories. Unlike $V(Q\bar{Q})$, the NN-potential is not able to predict reliable quantitative information about the input parameters to its theory — the only exception perhaps being an estimate of the πNN coupling constant [76]. The reason why the form of $V(NN)$ is so complicated compared with that so far discussed for $V(Q\bar{Q})$, *i.e.* essentially $V_{Q\bar{Q}}(r) = -e/r + cr$, is because the latter is the interaction between two *static* quarks. If, however, we go away from this limit, then immediately the spins of the quarks begin to play a rôle resulting in forms similar to those encountered in $V(NN)$. These potentials should be reliable for describing $b\bar{b}$, $b\bar{c}$ and $c\bar{c}$ states, since the b and c quarks are still sufficiently heavy ($\approx$ 4.5 and 1.2 GeV respectively) to not require a relativistic treatment. However, as soon as s, d, u quarks are involved, relativistic effects become important and even the whole concept of an interquark potential should be questioned. Furthermore,, the hope that this potential between *two* quarks can account for multiquark systems has yet to be justified. Even so, this has not deterred its use as an effective interaction — a topic discussed in Chapter 5.

3.3.1 *Radial form of* $V(Q\bar{Q})$

What is the form of $V(Q\bar{Q})$? To this question there is no unique answer, since — as with meson-exchange models of $V(NN)$ — forms depend on the theory from which the potential is derived and in which it should be utilized. A good example of this ambiguity is the momentum dependence of the potential. Even though the correct relativistic scattering equation for two quarks is the Bethe-Salpeter equation, for practical reasons, this needs to be simplified to, say, the semi-relativistic Blankenbecler–Sugar or non-relativistic Schrödinger equations — see Subsec. 5.1.2.2 of Chapter 5 for a very brief discussion of this. If the basic Blankenbecler-Sugar equation is used directly then the appropriate potential contains relativistic factors of the form $E_Q = \sqrt{M^2 + p^2}$, where p is the momentum of the quark with mass M. However, it is often convenient to expand these momentum factors in powers of p/M resulting in a potential appropriate for a Lippmann–

Schwinger or Schrödinger approach.

Just as the interaction between two static quarks contained two distinct parts — *i.e.* the gluon exchange term $V_G = -\frac{4}{3}\alpha_s/r$ and the confining term $V_C = cr$ — so can the interaction between a heavy quark of mass M and an anti-quark of mass m be expressed as

$$V(Mm) = V_G(Mm) + V_C(Mm), \tag{3.54}$$

where

$$V_G(Mm)(-\frac{4}{3}\alpha_s)^{-1} = \frac{1}{r} - \frac{2\pi}{3Mm}\delta^{(3)}(r)\boldsymbol{\sigma_M}.\boldsymbol{\sigma_m} - \frac{1}{4Mmr^3}S_{12}$$

$$-\frac{1}{2r^3}\left[\frac{M^2+m^2}{2M^2m^2} + \frac{2}{Mm}\right]\boldsymbol{L}.\boldsymbol{S} + \frac{1}{8r^3}\frac{M^2-m^2}{M^2m^2}(\boldsymbol{\sigma}_M - \boldsymbol{\sigma}_m).\boldsymbol{L} + \cdots, \tag{3.55}$$

obtained by a non-relativistic reduction of the one-gluon-exchange mechanism and

$$V_C(Mm) = cr - \frac{c}{r}\frac{M^2+m^2}{4M^2m^2}\boldsymbol{L}.\boldsymbol{S} + \frac{c}{r}\frac{M^2-m^2}{8M^2m^2}(\boldsymbol{\sigma}_M - \boldsymbol{\sigma}_m).\boldsymbol{L} + \cdots. \tag{3.56}$$

These expressions for $V_G(Mm)$ and $V_C(Mm)$ do not include effects from expanding the $E = \sqrt{M^2+p^2}$ terms mentioned above — see, for example, Ref. [80]. Also here, for simplicity, they do not include spin-independent terms proportional to p^2 and $\delta^{(3)}(r)$, since their form becomes ambiguous when going from a momentum space to a coordinate space representation. It should be added that it is convenient to have $V_G(Mm)$ in coordinate space, since it is hard to deal with the linearly rising term cr from $V_C(Mm)$ in momentum space. In addition to this ambiguity there are three others that enter in practice:

(1) A crucial term in Eq.3.55 is the hyperfine interaction proportional to $\boldsymbol{\sigma_M}.\boldsymbol{\sigma_m}$, which splits the energies of pseudoscalar and vector mesons. However, it is seen that it is proportional to $\delta(\mathbf{r})$ — a radial form that needs to be regulated before it can be used in a wave-equation. This can be accomplished in a variety of ways , of which the most simple is to make the replacement

$$\delta(\mathbf{r}) \to \frac{a^3}{\pi^{3/2}}\exp(-a^2r^2) \tag{3.57}$$

as was done in the original work of Godfrey and Isgur [77]. Another approach is to consider this term as an effective interaction that is added in first order perturbation theory using those wavefunctions generated by the rest of $V(Q\bar{Q})$. In this case the overall constant is considered to be essentially a free parameter although reasonable estimates can be made from models involving instantons [78].

(2) In Eq. 3.45, α_s is the quark-gluon running coupling "constant" *i.e.* it depends on momentum k. This can be parameterised in a variety of ways *e.g.* from Ref. [79]

$$\alpha_s(k^2) = \frac{12\pi}{27} \frac{1}{\ln[(k^2 + 4m_g^2)/\Lambda^2]}, \tag{3.58}$$

where there are two parameters
i) the gluon effective mass parameter $m_g \approx 240$ MeV and
ii) the QCD scale parameter $\Lambda \approx 280$ MeV.
In Ref. [80] this is combined with the regulator of $\delta(\mathbf{r})$ by the replacement

$$\alpha_s\delta(\mathbf{r}) \to \frac{1}{2\pi^2 r} \int_0^\infty dkk \sin(kr) \left(\frac{M+m}{E_M+E_m}\right)\left(\frac{Mm}{E_M E_m}\right) \alpha_s(k^2), \tag{3.59}$$

where $E_M = \sqrt{M^2 + (k^2/4)}$ and $E_m = \sqrt{m^2 + (k^2/4)}$.

(3) In the above it is implicitly assumed that $V_C(Mm)$ is purely *scalar*. However, there are reasons — both theoretical and phenomenological — suggesting that there could be a sizeable *vector* component. For example, the spin-orbit splitting from $V_G(Mm)$ is $\propto 1/r^3$, whereas that in $V_C(Mm)$ is $\propto -1/r$. This indicates that the natural spin-orbit ordering of a coulomb-like potential should be inverted for high partial waves — a feature not seen experimentally or in lattice calculations. This is discussed in Subsec. 5.9.3.2 in Chapter 5.

Lattice calculations are able to isolate several of the components in the potential $V(Mm)$ in Eq. 3.54. For example, the leading confinement term $V_C(Mm) = cr$ of Eq. 3.56 is shown in Fig. 3.3. Another point of interest in this figure is that there is no sign of the expected flattening at $r/r_0 \approx 2.4$, where it becomes energetically favourable to create two mesons — denoted by the horizontal band [81]. For other components of the potential the lattice data is less precise, but still there is reasonable agreement with the expectations from Eqs. 3.55 and 3.56. Probably the most extensive lattice study of the various forms contained in the potential appear in Ref. [8].

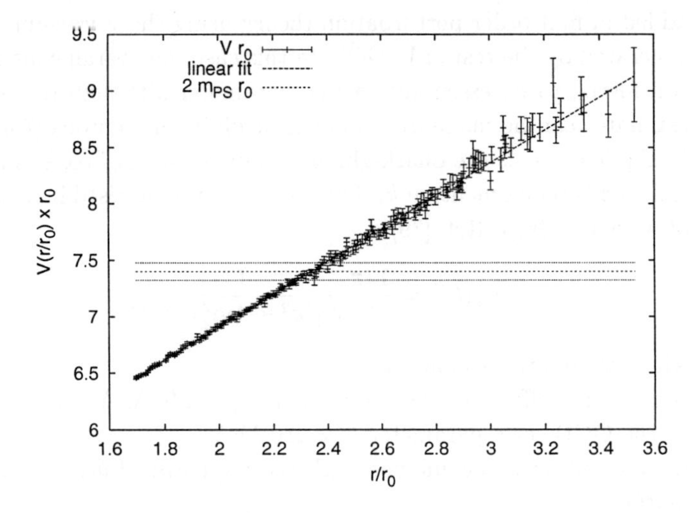

Fig. 3.3 The linearly rising confinement as calculated on a lattice [81] compared with $V_C(Mm) = cr$ in Eq. 3.56.

3.3.2 *Comparison with the form of $V(NN)$*

For those more familiar with internucleon potentials it may be of interest to compare the forms that appear in $V(NN)$ with those in $V(Q\bar{Q})$. This is, probably, best done by writing down explicitly the ω-meson contribution to the familiar One-Boson-Exchange-Potential, since this meson is the one that resembles most closely a gluon both being vector particles independent of isospin/flavour. In $V(NN)$, the ω-meson contributes much of the short range repulsion and also has a strong Spin-orbit potential. The form given in Ref. [82] is

$$\frac{V_\omega}{g^2} = \frac{\exp(-\bar{m}r)}{r} - \frac{1}{2M^2}\left[\boldsymbol{\nabla}^2 \frac{\exp(-\bar{m}r)}{r} + \frac{\exp(-\bar{m}r)}{r}\boldsymbol{\nabla}^2\right]$$

$$+\frac{1}{2M^2}\left[\bar{m}^2\frac{\exp(-\bar{m}r)}{r} - 4\pi\delta^{(3)}(r)\right](1 + \frac{\boldsymbol{\sigma_1}.\boldsymbol{\sigma_2}}{3})$$

$$-\frac{\bar{m}^2}{4M^2}\left[\frac{1}{3} + \frac{1}{\bar{m}r} + \frac{1}{(\bar{m}r)^2}\right]\frac{\exp(-\bar{m}r)}{r}S_{12} + \frac{3}{2}\frac{1}{M^2}\frac{1}{r}\frac{d}{dr}\left[\frac{\exp(-\bar{m}r)}{r}\right]\boldsymbol{L}.\boldsymbol{S}, \tag{3.60}$$

where M is the nucleon mass and $\bar{m} = 783$ MeV $\approx 4\text{fm}^{-1}$ is the mass of the ω. When $\bar{m} \to 0$ to compare with one-gluon-exchange, this reduces to

$$\frac{V(m\to 0)}{g^2} = \frac{1}{r} - \frac{1}{2M^2}\left[\nabla^2\frac{1}{r} + \frac{1}{r}\nabla^2\right] - \frac{2\pi}{M^2}\delta^{(3)}(r)\left[1 + \frac{\sigma_1.\sigma_2}{3}\right]$$

$$-\frac{1}{4M^2}\frac{1}{r^3}S_{12} - \frac{3}{2}\frac{1}{M^2}\frac{1}{r^3}\boldsymbol{L}.\boldsymbol{S}. \quad (3.61)$$

This form is now quite similar to the one-gluon-exchange potential V_G in Eq. 3.55 — the only differences being the appearance of the spin-independent terms containing ∇^2 and $\delta^{(3)}(r)$, which are not unique in going from momentum space to coordinate space. Of course, it is possible to stay throughout in momentum space, since — unlike $V_C(Mm)$ — here there is no linearly rising potential cr. However, there are also differences in how Eqs. 3.55 and 3.60 are treated. Normally in Eq. 3.60 the $\delta^{(3)}(r)$ terms do not appear, since there are physically motivated models of form factors at the $NN\omega$-vertices, which smooth out such singularities. Also the results are not crucially dependent on these form factors since the main rôle of the ω-meson is to generate a strong spin-isospin independent *repulsion* that essentially excludes the NN-wavefunction from the region of the origin. In contrast, the one-gluon-exchange potential in Eq. 3.55 has an *attractive* spin-isospin independent term $(-\frac{4}{3}\alpha_s/r)$. This means that uncertainties in the parametrizations in Eqs. 3.57 and 3.58 are more important.

3.4 Conclusions

This chapter has attempted to describe a few of the many topics that could be covered by the title. It has not aimed at being in anyway comprehensive — the selection being rather subjective with the following ideas in mind.

1) The chapter began and ended with the theme that the interquark potential $V(Q\bar{Q})$ could be viewed as a bridge for modelling QCD with somewhat the same rôle that the nucleon-nucleon potential $V(NN)$ plays in nuclear physics. To further bring out the analogy between $V(Q\bar{Q})$ and $V(NN)$ the radial forms were compared and contrasted in Sec. 3.3. However, it should be added that this possible "interdisciplinary bridge" is not well understood and its study is rather neglected.

2) The potential $V(Q\bar{Q})$ can be used for testing approximations to QCD and its lattice formulation. This is illustrated in Fig. 3.2, where it is shown that the weak and strong coupling limits *do not* overlap for the energy range of interest *i.e.* between 100 MeV and a few GeV. Furthermore, over this range neither limit agrees with the corresponding (non-perturbative) lattice calculation. This feature shows that we are "forced" into performing lattice calculations, since there seems to be no other way of treating QCD with the couplings of most interest.

Here no mention has been made about deriving a discretized form of $V(Q\bar{Q})$. This topic has been the scene of much activity and comes under the heading of Non-Relativistic QCD (NRQCD). Here an expansion is made in terms of Λ/M_Q with $\Lambda \sim 1$ GeV being a characteristic energy scale for non-perturbative effects. A few comments can be found in Subsec. 5.1.2.1 of Chapter 5.

Acknowledgements

I express sincere gratitude to Tony Green, for his initiative in suggesting this book and in particular for his patience in waiting for a ridiculously long time for my contribution to arrive. Discussions with Nora Brambilla, Antonio Pineda, Joan Soto and Antonio Vairo are acknowledged. I received support from a PPARC Advanced Fellowship (grant PPA/A/S/2000/00271) as well as by PPARC grant PPA/G/0/2002/0463.

Bibliography

[1] F. J. Wegner, *J. Math. Phys.*, 12:2259, 1971.
[2] K. G. Wilson, *Phys. Rev.*, D10:2445, 1974.
[3] L. S. Brown and W. I. Weisberger, *Phys. Rev.*, D20:3239, 1979.
[4] E. Eichten and F. L. Feinberg, *Phys. Rev. Lett.*, 43:1205, 1979.
[5] E. Eichten and F. L. Feinberg, *Phys. Rev.*, D23:2724, 1981.
[6] A. Pineda and A. Vairo *Phys. Rev.*, D63:054007, 2001.
[7] Yu-Qi Chen, Yu-Ping Kuang, and R. J. Oakes, *Phys. Rev.*, D52:264, 1995.
[8] G. S. Bali, K. Schilling, and A. Wachter, *Phys. Rev.*, D56:2566, 1997.
[9] N. Brambilla, A. Pineda, J. Soto, and A. Vairo, *Phys. Rev.*, D63:014023, 2001
[10] L. D. Landau and E. M. Lifschitz. *Lehrbuch der theoretischen Physik, Band 3: Quantenmechanik*, Akademie Verlag, Berlin, DDR, 1979.

[11] E. Seiler, *Phys. Rev.*, D18:482, 1978.
[12] K. Osterwalder and R. Schrader, *Commun. Math. Phys.*, 31:83, 1973.
[13] K. Osterwalder and R. Schrader, *Commun. Math. Phys.*, 42:281, 1975.
[14] C. Bachas, *Phys. Rev.*, D33:2723, 1986.
[15] B. Simon and L. G. Yaffe, *Phys. Lett.*, 115B:145, 1982.
[16] K. G. Wilson, *Phys. Rept.*, 23:331, 1976.
[17] R. Balian, J. M. Drouffe, and C. Itzykson, *Phys. Rev.*, D11:2104, 1975. *erratum, ibid.* D19:2514, 1979.
[18] M. Creutz, *Rev. Mod. Phys.*, 50:561, 1978.
[19] J. M. Drouffe and J. B. Zuber, *Phys. Rept.*, 102:1, 1983.
[20] M. Creutz, *Quarks, Gluons and Lattices.* Cambridge University Press, Cambridge, UK, 1983.
[21] I. Montvay and G. Münster, *Quantum fields on a lattice.* Cambridge University Press, Cambridge, UK, 1994.
[22] J. Smit, Introduction to quantum fields on a lattice: A robust mate, *Cambridge Lect. Notes Phys.*, 15:1, 2002.
[23] G. Münster and P. Weisz, *Phys. Lett.*, 96B:119, 1980. erratum, *ibid.* 100B:519, 1981.
[24] G. Münster, *Nucl. Phys.*, B190:439, 1981. errata, *ibid.* B200:536, 1982 and B205:648, 1982.
[25] F. J. Dyson, *Phys. Rev.*, 85:631, 1952.
[26] J. Zinn-Justin, *Phys. Rept.*, 70:109, 1981.
[27] K. Osterwalder and E. Seiler, *Ann. Phys.*, 110:440, 1978.
[28] J. B. Kogut and J. Shigemitsu, *Phys. Rev. Lett.*, 45:410, 1980.
[29] J. Smit, *Nucl. Phys.*, B206:309, 1982.
[30] S. Capitani, M. Lüscher, R. Sommer, and H. Wittig, *Nucl. Phys.*, B544:669, 1999.
[31] M. Lüscher, *Nucl. Phys.*, B180:317, 1981.
[32] J. M. Drouffe and J. B. Zuber, *Nucl. Phys.*, B180:264, 1981.
[33] T. Banks, R. Myerson, and J. Kogut, *Nucl. Phys.*, B129:493, 1977.
[34] A. H. Guth, *Phys. Rev.*, D21:2291, 1980.
[35] M. Creutz, L. Jacobs, and C. Rebbi, *Phys. Rev.*, D20:1915, 1979.
[36] B. Lautrup and M. Nauenberg, *Phys. Lett.*, 95B:63, 1980.
[37] J. Kogut and L. Susskind, *Phys. Rev.*, D11:395, 1975.
[38] A. A. Abrikosov, *Sov. Phys. JETP*, 5:1174, 1957.
[39] H. B. Nielsen and P. Olesen, *Nucl. Phys.*, B61:45, 1973.
[40] G. 't Hooft, *Nucl. Phys.*, B72:461, 1974.
[41] A. A. Migdal, *Phys. Rept.*, 102:199, 1983.
[42] M. Lüscher, K. Symanzik, and P. Weisz, *Nucl. Phys.*, B173:365, 1980.
[43] M. Lüscher, G. Münster, and P. Weisz, *Nucl. Phys.*, B180:1, 1981.
[44] T. Goto, *Prog. Theor. Phys.*, 46:1560, 1971.
[45] Y. Nambu, *Phys. Rev.*, D10:4262, 1974.
[46] J. F. Arvis, *Phys. Lett.*, 127B:106, 1983.
[47] M. Caselle, R. Fiore, and F. Gliozzi, *Phys. Lett.*, B200:525, 1988.

[48] M. Caselle, R. Fiore, F. Gliozzi, M. Hasenbusch, and P. Provero, *Nucl. Phys.*, B486:245, 1997.
[49] M. Caselle, M. Hasenbusch, and M. Panero, *JHEP*, 01:057, 2003.
[50] C. J. Morningstar, K. J. Juge, and J. Kuti, *Nucl. Phys. Proc. Suppl.*, 73:590, 1999.
[51] K. Jimmy Juge, J. Kuti, and C. Morningstar, QCD string formation and the Casimir energy, hep-lat/0401032
[52] E. T. Akhmedov, M. N. Chernodub, M. I. Polikarpov, and M. A. Zubkov, *Phys. Rev.*, D53:2087, 1996.
[53] A. M. Polyakov, *Nucl. Phys.*, B486:23, 1997.
[54] M. N. Chernodub and D. A. Komarov, *JETP Lett.*, 68:117, 1998.
[55] D. Antonov and D. Ebert, *Phys. Lett.*, B444:208, 1998.
[56] M. Baker and R. Steinke, *Phys. Lett.*, B474:67, 2000.
[57] J. Polchinski and A. Strominger, *Phys. Rev. Lett.*, 67:1681, 1991.
[58] M. Luscher and P. Weisz, *JHEP*, 07:049, 2002.
[59] J. Ambjørn, P. Olesen, and C. Peterson, *Nucl. Phys.*, B244:262, 1984.
[60] P. Majumdar, *Nucl. Phys.*, B664:213, 2003.
[61] C. Michael and P. W. Stephenson, *Phys. Rev.*, D50:4634, 1994.
[62] M. J. Teper, *Phys. Rev.*, D59:014512, 1999.
[63] B. Lucini, M. Teper, and U. Wenger, Glueballs and k-strings in SU(N) gauge theories : calculations with improved operators, hep-lat/0404008 .
[64] P. de Forcrand, G. Schierholz, H. Schneider, and M. Teper, *Phys. Lett.*, 160B:137, 1985.
[65] O. Kaczmarek, F. Karsch, E. Laermann, and M. Lütgemeier, *Phys. Rev.*, D62:034021, 2000.
[66] O. Kaczmarek, S. Ejiri, F. Karsch, E. Laermann, and F. Zantow, Heavy quark free energies and the renormalized Polyakov loop in full QCD, hep-lat/0312015.
[67] I. I. Bigi, M. A. Shifman, N. G. Uraltsev, and A. I. Vainshtein, *Phys. Rev.*, D50:2234, 1994.
[68] M. Beneke, *Phys. Lett.*, B434:115, 1998.
[69] G. S. Bali and A. Pineda, *Phys. Rev.*, D69:094001, 2004.
[70] T. Appelquist, M. Dine, and I. J. Muzinich, *Phys. Lett.*, 69B:231, 1977.
[71] N. Brambilla, A. Pineda, J. Soto, and A. Vairo, *Phys. Rev.*, D60:091502, 1999.
[72] G. Curci, G. Paffuti, and R. Tripiccione, *Nucl. Phys.*, B240:91, 1984.
[73] R. Wohlert, P. Weisz, and Werner Wetzel, *Nucl. Phys.*, B259:85, 1985.
[74] U. Heller and F. Karsch, *Nucl. Phys.*, B251:254, 1985.
[75] G. S. Bali and P. Boyle, Perturbative Wilson loops with massive sea quarks on the lattice, hep-lat/0210033.
[76] V. Stoks, R. Timmermans and J. J. de Swart, *Phys. Rev.*, C47:512, 1993, nucl-th/9211007
[77] S. Godfrey and N. Isgur, *Phys. Rev.*, D32:189, 1985.
[78] S. Chernyshev, M. A. Nowak, and I. Zahed, *Phys. Rev.*, D53:5176, 1996.

[79] A. C. Mattingly and P. M. Stevenson, *Phys. Rev.*, D49:437, 1994.
[80] T. A. Lähde, C. J. Nyfält and D .O. Riska, *Nucl. Phys.*, A674:141, 2000, hep-ph/9908485
[81] B. Bolder *et al.*, *Phys. Rev.*, D63:074504, 2001.
[82] R. Bryan and B.L. Scott, *Phys. Rev.*, 177:1435, 1969.

Chapter 4

Interactions between Lattice Hadrons

H.R. Fiebig

Physics Department, FIU – University Park, 11200 SW 8th Street, Miami, Florida 33199, USA
E-mail: fiebig@fiu.edu

H. Markum

Atominstitut, Technische Universität Wien, A-1040 Vienna, Austria
E-mail: markum@tuwien.ac.at

The effective residual interaction for a system of hadrons has a long tradition in theoretical physics. It has been mostly addressed in terms of boson exchange models. The aim of this review is to describe approaches based on lattice field theory and numerical simulation. At the present time this subject matter is in an exploratory stage. A large array of problems waits to be tackled, so that known features of hadron-hadron interactions will eventually be understood in a model-independent way. The lattice formulation, being capable of dealing with the nonperturbative regime, describes strong-interaction physics from first principles, *i.e.* quantum chromodynamics (QCD). Although the physics of hadron-hadron interactions may be intrinsically complicated, the methods used in lattice simulations are simple: For the most part they are based on standard mass calculations. This chapter addresses commonly used techniques, within QCD_{3+1} and also simpler lattice models, describes important results, and also gives some insight into numerical methods for multi-quark systems.

4.1 Introductory Overview on Goals, Strategies, Methods

4.1.1 *Modeling nuclear forces*

The theory of nuclear forces was pioneered by Yukawa [1] in 1935. A particle of intermediate mass, a meson, accounts for the interaction energy of the proton and neutron. In the framework of a quantum field theory the meson appears as the quantum of an effective field that describes the strong interaction [2]. The approximate range of the interaction

$$R = \frac{\hbar c}{mc^2} \tag{4.1}$$

is roughly consistent with $R \approx 2$ fm, for a meson mass in the region of $m \approx 100\,\mathrm{MeV}$. Those numbers are crude estimates based on the uncertainty relation (for energy and time) and the contemporary experimental evidence.

An effective neutron-proton interaction can be obtained from solutions of the Klein-Gordon equation for the meson field φ

$$(\Box + m^2)\varphi(x) = g\bar{\psi}(x)\psi(x)\,, \tag{4.2}$$

with the pion-nucleon coupling g. In the static limit, where the product of nucleon fields $\bar{\psi}\psi$ is replaced with a generic point-like source,

$$(-\Delta + m^2)\varphi(\vec{r}) = g\delta(\vec{r})\,, \tag{4.3}$$

the solution, *i.e.* the static Green function,

$$\varphi(x) = \frac{g}{4\pi}\frac{e^{-mr}}{r} \tag{4.4}$$

is interpreted as a potential energy function, the 'Coulomb potential' of the meson theory. (Units are such that $\hbar = c = 1$, and the conversion factor $\hbar c = 0.197$ GeV fm.) This line of thought is in close analogy to electrodynamics where the Coulomb potential emerges as the solution of a Poisson equation with a static point-like charge (source).

Since the early work, the meson theory of nuclear forces has received much attention and refinements. One of the more significant insights [3] has been that the range of the nucleon-nucleon force can be subdivided into three regions:

- The long-range region, inter-nucleon distances are above $r \approx 1.5$ fm. Here we have attraction. This is the region described by the Yukawa potential.

- Intermediate-range attraction, around $r \approx 1.0$ fm.
- Short-range repulsion, below $r \approx 0.5$ fm.

Other salient features are:

- A tensor force, important for the quadrupole moment of the deuteron.
- The presence of a spin-orbit force.
- Charge independence of the strong interaction.

In order to describe the short-distance region, scalar and vector mesons $(\eta, \rho, \omega, \cdots)$ are needed in the boson-exchange model. Remarkably, a π–π correlated state (known as the σ meson) is relied upon to produce intermediate range attraction. The pseudoscalar meson (π), which also gives rise to the long-range (though finite) Yukawa potential, contributes significantly to the tensor force.

After approximately half a century the meson theory of nuclear forces is at a mature, sophisticated stage. Exhaustive review articles exist, for example Refs. [4, 5, 6, 7] may serve the interested reader. Boson-exchange models of this type presently constitute the state-of-the-art for effective theories of nuclear forces.

Typical for this class of models is an effective Lagrangian, say for the nucleon-nucleon interaction, such as

$$\begin{aligned}\mathcal{L}_{\overline{\psi}\varphi\psi} &= \sum_{\varphi}\Big(- g_{\mathrm{P}}\overline{\psi} i\gamma_5\psi\varphi^{\mathrm{P}} + g_{\mathrm{S}}\overline{\psi}\psi\varphi^{\mathrm{S}} \\ &\qquad - g_{\mathrm{V}}\overline{\psi}\gamma_\mu\psi\varphi^{\mathrm{V}}_\mu - \frac{f_{\mathrm{V}}}{2M}\overline{\psi}\sigma^{\mu\nu}\psi\partial_\mu\varphi^{\mathrm{V}}_\nu\Big) , \end{aligned} \tag{4.5}$$

with pseudoscalar g_{P}, scalar g_{S}, vector g_{V} and tensor f_{V} couplings for the various bosons φ, and M is the nucleon mass [7]. Potentials are obtained from tree-level contributions to the nucleon-nucleon propagator. Those are most naturally written in momentum space. The simplest contribution derived from (4.5) is

$$V(\vec{k}) = -\frac{g_{\mathrm{P}}^2}{12M}\left[\vec{\sigma}_1\cdot\vec{\sigma}_2 - \vec{\sigma}_1\cdot\vec{\sigma}_2\frac{m_{\mathrm{P}}^2}{\vec{k}^2+m_{\mathrm{P}}^2} + \frac{S_{12}(\vec{k})}{\vec{k}^2+m_{\mathrm{P}}^2}\right]\vec{\tau}_1\cdot\vec{\tau}_2\cdots , \tag{4.6}$$

where m_{P} is the pseudoscalar mass. This already is the non-relativistic reduction where the three terms correspond to a (coordinate-space) δ-function, a Yukawa, and a tensor potential, the latter containing

$$S_{12} = 3(\vec{\sigma}_1 \cdot \vec{k})(\vec{\sigma}_2 \cdot \vec{k}) - \vec{\sigma}_1 \cdot \vec{\sigma}_2 \vec{k}^2 \,. \tag{4.7}$$

The momentum transfer is $\vec{k} = \vec{q}\,' - \vec{q}$ and $\vec{\sigma}_1$, $\vec{\sigma}_2$ are the nucleon spins. In addition, for each type of coupling, momentum cutoff form factors are included with the coupling constants $g_{\mathrm{P}}, g_{\mathrm{S}}, g_{\mathrm{V}}$ and f_{V}, for example

$$g_{\mathrm{P}}^2 \rightarrow g_{\mathrm{P}}^2 \left(\frac{\Lambda_{\mathrm{P}}^2 - m_{\mathrm{P}}^2}{\Lambda_{\mathrm{P}}^2 + \vec{k}^2} \right)^{n_{\mathrm{P}}} \,. \tag{4.8}$$

The inverse of a momentum cutoff parameter Λ_{P} is a length, and as such indicates the region of validity of the model. Values of ≈ 1 GeV, or equivalently ≈ 0.2 fm are typical. Other refinements of the model include isobar (Δ) degrees of freedom, and the extension to the strange hadron sector [8, 9]. Typical potentials are illustrated in Fig. 4.1.

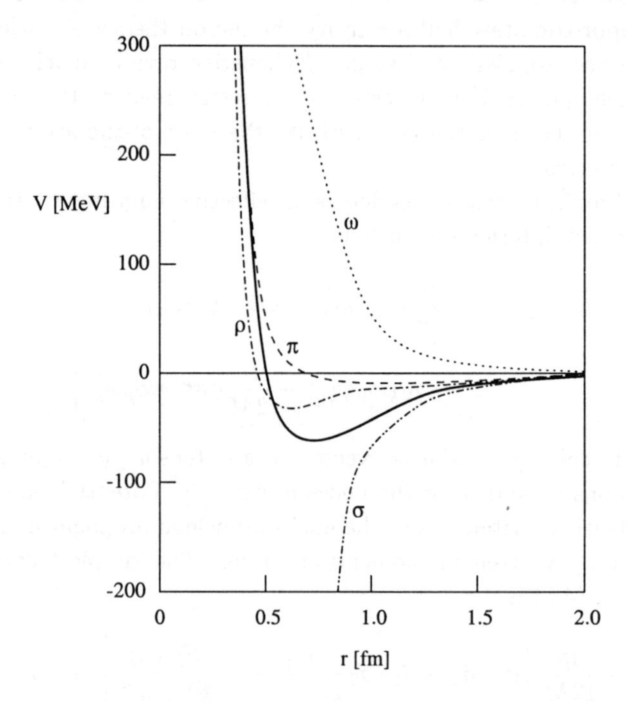

Fig. 4.1 Generic form of a central nucleon-nucleon potential according to boson exchange models(solid line) and its contributions from various bosons. Schematic drawing after Ref. [7].

The philosophy of boson exchange models is to take the phenomenology of low-energy excitations of the hadronic vacuum as a starting point. The experimentally known spectrum of mesons and hadrons is used to reconstruct the forces between those particles. The result is an effective theory with adjustable parameters and predictive power in a reasonably well-defined realm of validity. An example is shown in Fig. 4.2. Naturally, the question of how an hadronic interaction results from the underlying microscopic structure of hadrons and the, as we now know, complicated structure of the hadronic vacuum, is beyond the scope, and aim, of the model.

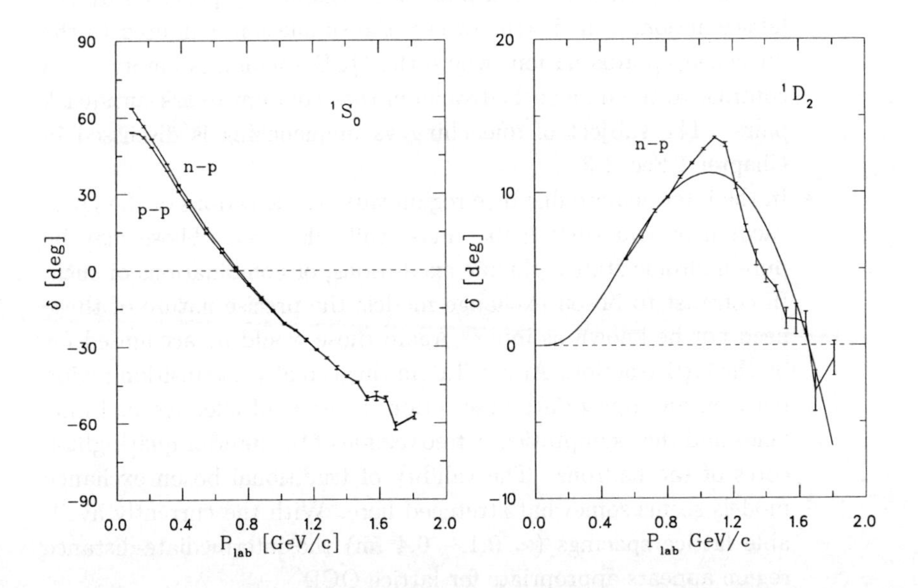

Fig. 4.2 Nucleon-nucleon scattering phase shift for channels $^1S_0$ and $^1D_2$ compiled from Ref. [10]. In the $^1D_2$ channel the smooth curve is a 'Bonn-potential' result.

In the present context we wish to look at hadronic interactions from the perspective of lattice QCD. Since the physics of hadronic interactions is contained in the low-energy, non-perturbative, regime of QCD the lattice formulation is well suited for this task.

4.1.2 *The lattice QCD perspective*

From the lattice QCD point of view the three-fold subdivision of the inter-nucleon interaction region, see Fig. 4.1, appears in a new light:

- The long-range region is expected to be dominated by excitations of the QCD vacuum that are recovered only in the chiral limit of the lattice simulation. If Wilson fermions are used this region can be monitored by extrapolating to the zero-mass limit of the pseudoscalar meson, which is the Goldstone boson of QCD. The mechanism for pion exchange is available on the lattice because correlated quark-antiquark pairs are dynamically provided by the lattice action. This is true even if the simulation is limited to the quenched approximation, where the QCD vacuum is empty — in contrast to the unquenched vacuum that contains quark-antiquark pairs. The subject of quenching vs unquenching is discussed in Chapter 1 Sec. 1.3.1.
- In the intermediate-distance region various excitations of the QCD vacuum of somewhat higher mass will take over. Those may be pure hadronic states, gluonic excitations, or combinations of such. In contrast to boson exchange models the precise nature of those need not be known as input. Again those would be accounted for by the lattice action. At $r \approx 1.0$ fm this is really a transition region between an appropriate description in terms of effective hadronic fields and the asymptotically free regime of the interior quark-gluon cores of the hadrons. The validity of traditional boson exchange models seems somewhat stretched here. With the currently available lattice spacings ($\approx 0.1 - 0.4$ fm) the intermediate-distance region appears appropriate for lattice QCD.
- For small distances the compositeness of the hadrons, as finite-size quark-gluon systems, will play the dominant role. Not surprisingly, quark models have been used to study the nucleon-nucleon interaction. The energies involved are naturally larger, for example ≈ 2 GeV, and thus point to lattice spacings of about ≈ 0.1 fm, values not out of reach of contemporary lattice simulations. Lattice QCD provides dynamically for the compositeness of the hadrons. Internal hadronic structure is, in turn, crucial for the features of the residual interaction.

Thus lattice QCD is in a position to combine the classic nucleon-nucleon interaction regions in a single unified approach. Moreover, the strange hadron sector, or effects of strangeness in the light-quark sector are straightforwardly incorporated.

4.1.3 *Short and long term goals for lattice QCD*

While boson exchange models have a long tradition, attempts to study residual interactions between systems of hadrons within lattice QCD are only in an experimental stage. The physics of hadronic interactions in terms of first principles is rather involved, see Fig. 4.3. Conceptual, technical, and computational questions are very much in the foreground at this time. The tentative nature of this situation will be apparent throughout this chapter.

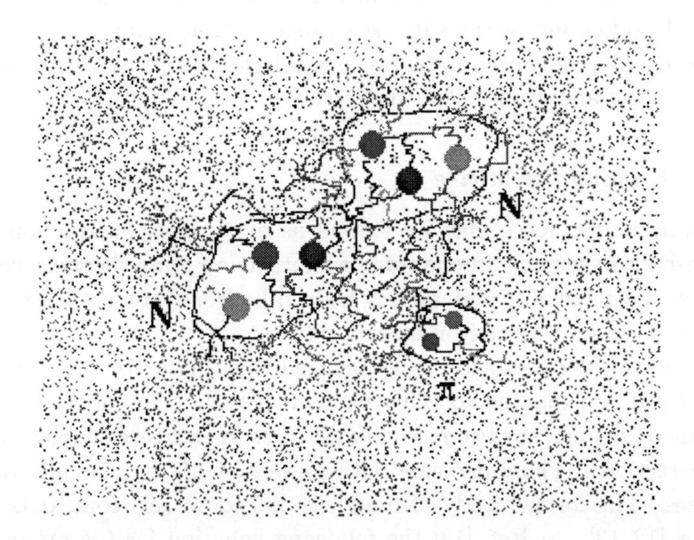

Fig. 4.3 Cartoon of a strongly interacting hadron-hadron system from the QCD perspective.

Nevertheless, there is potential for new physics, or at least for understanding known phenomena in a new light. In particular, probing hadronic matter at somewhat higher energies, say in the 4 – 8 GeV region with electron beams, has become a new focus in experimental nuclear physics. The interest here lies in the boundary between particle (high energy) and nuclear (low energy) physics which we will refer to as hadronic physics. Our short-term working list is as follows:

- Discuss various basic concepts and numerical methods using simple lattice models.
- Study simple systems, mostly meson-meson, and extract non-relativistic potentials.
- Explore alternative approaches to the problem, like momentum space and coordinate space formulations.

As a long-term perspective we have in mind:

- Study realistic meson-meson, meson-baryon, and baryon-baryon systems. For example, the pion-nucleon system features the prominent delta resonance, a hallmark of strong-interaction physics.
- Residual interactions in the strange-hadron sector.
- Develop advanced methods, possibly the direct extraction of scattering amplitudes from the lattice.

4.1.4 *Probing the lattice*

A hadron-hadron force may be viewed as a hyperfine interaction in the hadron-hadron system. In one way or another, we must therefore compare the mass spectrum of the interacting hadron-hadron system to the spectrum of the non-interacting hadron-hadron system.

About two decades ago it was realized that finite volume effects may in principle be exploited to achieve this end [11]. The energy levels of a two-hadron system, say two mesons with mass m each, are subject to finite-size corrections. Those decrease with powers of $1/L$ as the lattice size L increases. The formal study of finite-size effects in this context is due to Lüscher [12, 13]. In Ref. [13] the following equation for the ground-state two-body energy shift

$$W_0 - 2m = -\frac{4\pi a_0}{mL^3}\left\{1 + c_1\frac{a_0}{L} + c_2\frac{a_0^2}{L^2}\right\} + o(L^{-6}) \tag{4.9}$$

is derived. It is valid for s-wave elastic scattering, below the particle production threshold. Above $c_1 = -2.837297$, $c_2 = 6.375183$, and a_0 is the s-wave scattering length. This result has been the basis for a number of studies [14, 15, 16, 17, 18, 19, 20, 21, 22, 23, 24, 25, 26]. The extraction of scattering information beyond the static limit, like phase shifts at nonzero relative energy, is formally more involved [27, 28].

In general it will be necessary to probe the lattice (vacuum) with one- and two-hadron operators. Let $\phi_{\vec{p}}(t)$ be a composite field, say, constructed from quark–antiquark (meson) or from three–quark (baryon) fields. Let $\vec{p}$ be the hadron momentum, for simplicity we disregard additional quantum numbers. The time correlation matrix built from one-hadron operators

$$C^{(2)}_{\vec{p}\,\vec{q}}(t,t_0) = \langle \phi^\dagger_{\vec{p}}(t)\,\phi_{\vec{q}}(t_0)\rangle - \langle \phi^\dagger_{\vec{p}}(t)\rangle\langle \phi_{\vec{q}}(t_0)\rangle \tag{4.10}$$

possesses eigenvalues $\lambda^{(2)}_{m_1}(t,t_0)$ which behave exponentially,

$$\lambda^{(2)}_{m_1}(t,t_0) \propto e^{-w_{m_1}(t-t_0)}\,. \tag{4.11}$$

The time correlation matrix is just the euclidean version of a propagator, in our example a 2-point function $C^{(2)}$ on the hadronic level (a 4- or 6-point function on the quark level). Thus from (4.11) the energies w_{m_1} of the one-meson system can be extracted. Typically, this is done numerically. A practical example is shown in Fig. 4.4. The interpretation of the w_{m_1} is that those are energies of the hadron moving through the lattice. In principle, the hadron could be in an internally excited state.

The fields $\Phi_{\vec{p}}(t) = \phi_{-\vec{p}}(t)\,\phi_{+\vec{p}}(t)$ can be used for probing two-hadron states. These are operators with total momentum $\vec{P}=0$. Note that $\vec{p}$ in this example means the relative momentum of the two-hadron system. We construct a 4-point correlation matrix

$$C^{(4)}_{\vec{p}\,\vec{q}}(t,t_0) = \langle \Phi^\dagger_{\vec{p}}(t)\,\Phi_{\vec{q}}(t_0)\rangle - \langle \Phi^\dagger_{\vec{p}}(t)\rangle\langle \Phi_{\vec{q}}(t_0)\rangle\,, \tag{4.12}$$

and, correspondingly, the eigenvalues

$$\lambda^{(4)}_{n}(t,t_0) \propto e^{-W_n(t-t_0)} \tag{4.13}$$

yield the energy levels W_n of the two-hadron system. Most importantly, these levels somehow contain the desired residual interaction.

A sensible definition of the latter, of course, also involves the eigenvectors of $C^{(2)}$ and $C^{(4)}$, however, comparison of the level schemes

$$\{\overline{W}_{m=(m_1,m_2)} \stackrel{\text{def}}{=} w_{m_1} + w_{m_2}\} \quad \text{to} \quad \{W_n\} \tag{4.14}$$

already gives important insight into the nature of the hadron-hadron force.

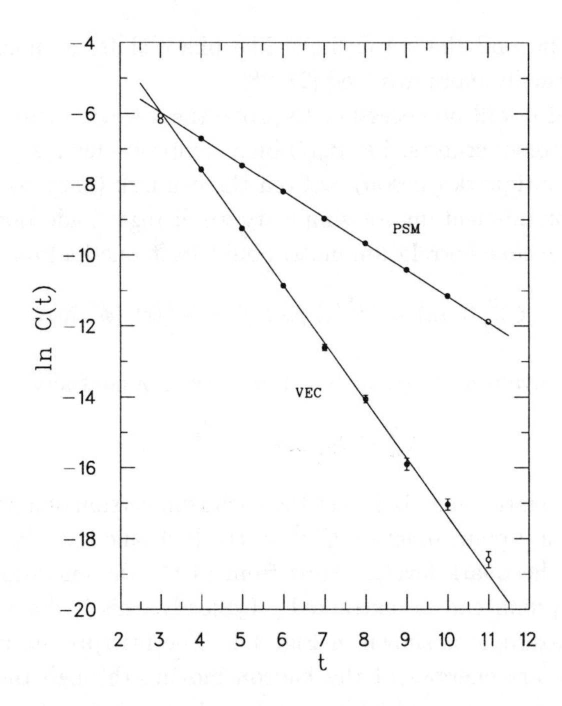

Fig. 4.4 Typical time correlation functions (eigenvalues) extracted from a lattice simulation. The examples are for a pseudoscalar meson (PSM) and a vector meson (VEC) using QCD_{3+1} and Wilson fermions.

Probing the lattice with momentum-type operators is only one of several alternatives. The use of coordinate space operators is another popular choice [29, 30, 31, 32, 33, 34, 35].

4.1.5 *Finite-size methods*

The spectrum $\{W_n\}$ of the interacting two-body system may be utilized in a variety of ways. Lüscher has shown how to extract scattering phase shifts directly from the spectrum [28]. The idea is to match the poles of (interacting) Green functions that live on the finite-sized periodic lattice to the poles of (free) Green functions in infinite volume. Since the lattice Green functions comprise the interaction, comparison in the asymptotic (free) region yields the phase shifts. Using methods similar to the 'Jost' formalism of scattering theory [36] an equation of the following form

$$\det\left[e^{2i\delta} - (M+i)^{-1}(M-i)\right] = 0 \tag{4.15}$$

is derived. Here $M = M(q^2)$ is an analytically constructed energy-dependent matrix ($W^2 = m^2 + q^2$). It reflects features of the 'natural' (free) spectrum of the periodic lattice. The other matrix $e^{2i\delta}$ in (4.15) contains a set of scattering phase shifts $\delta_\ell(q^2)$. The lattice simulation gives a discrete spectrum $\{W_n\}$. A certain energy W_n, for a fixed n, is then used to calculate $M(q_n^2)$ and (4.15) is solved for the corresponding $\delta_\ell(q_n^2)$. This yields a set of phase shifts, as many as the size of M, at each energy W_n. The physical $\delta_\ell(q_n^2)$, at a discrete q_n^2, is contained in this set.

A simple illustration of the idea behind this formalism [27] is the following: Consider a solution $\psi(x)$ of a Schrödinger equation in the noninteracting case with a phase $kL = 2\pi n$. It satisfies $\psi(x) = \psi(x+L)$, see Fig. 4.5(a). If the interaction is turned on, say an attractive potential with range $|x| \leq 0.2L$, the periodic boundary condition would be violated, see Fig. 4.5(b), unless compensated for by a simultaneous change in the kinematical phase kL. In other words, the change in kL necessary to restore the original periodic boundary conditions, see Fig. 4.5(c), is a measure for the phase shift.

The kinematical phase kL is changed by varying k or L. Varying L is reminiscent of a finite-size effect, see (4.9). Finding the change in k at fixed L is equivalent to finding all (discrete) two-body levels $\{W_n\}$ for the interacting case. This method, which already appears in Ref. [13], has been used by Lüscher and Wolff [27] to obtain scattering phase shifts in an O(3) nonlinear sigma model on a 1+1 dimensional lattice.

4.1.6 *Residual interaction*

Scattering phase shifts are very close to observable quantities (cross sections). This is an appealing feature of Lüscher's method, as the above procedure came to be known. On the other hand an effective residual interaction between hadrons is more in line with the historic development and has a heuristic value which should not be underestimated. We will therefore place considerable emphasis on this aspect. In this approach the information contained in the 2- and 4-point correlation matrices, see (4.10) and (4.12), will be utilized in a different way.

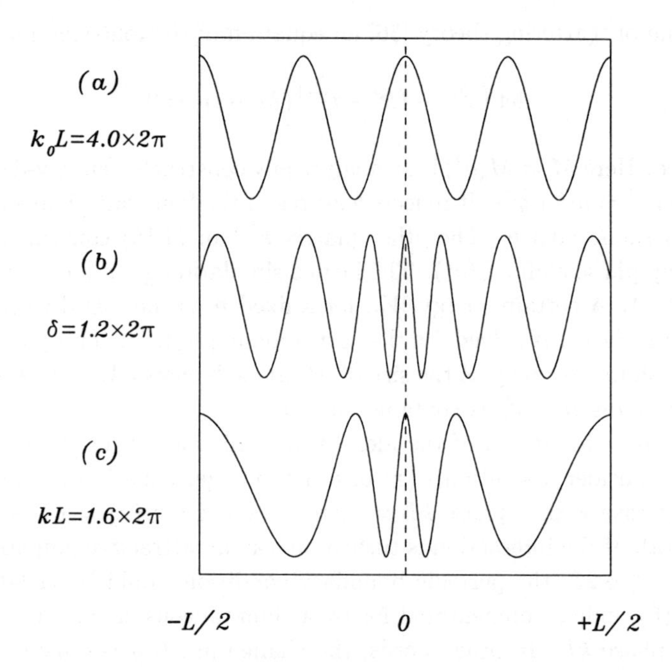

Fig. 4.5 Generic solutions $\psi(x)$ of a Schrödinger equation: (a) no interaction, (b) an attractive potential around $x \approx 0$, and (c) with decreased wave number k to restore the original phase conditions.

In principle an effective residual interaction is subject to definition. Towards this end we will first split a free, or non-interacting, part off the 4-point correlation matrix (4.12)

$$C^{(4)}(t,t_0) = \overline{C}(t,t_0) + C_I(t,t_0)\,, \tag{4.16}$$

where $\overline{C}$ is built from 2-point correlators $C^{(2)}$ and describes two non-interacting hadrons on the lattice. This can be accomplished by means of diagrammatic considerations. In terms of (4.16) we then define an effective residual interaction as

$$H_I = -\frac{\partial}{\partial t}\ln\left(\overline{C}^{-1/2}C_I\overline{C}^{-1/2}\right). \tag{4.17}$$

In principle the interaction Hamiltonian (potential) is non-local. Our focus is to compute H_I from a lattice simulation and eventually calculate the corresponding phase shifts.

The above method relies on lattice sizes large enough to accommodate two hadrons. Assuming a manageable lattice $L^3 \times T$ with, say, $L = 10$ and physical extent of about 4.0 fm this translates into a lattice spacing of $a = 0.4$ fm. As experience with hadron spectroscopy has shown the standard Wilson plaquette action [37] requires $a \approx 0.1 - 0.2$ fm in order to work reasonably close to the continuum limit. This does not sound very encouraging for putting multi-hadron systems on the lattice.

4.1.7 *Use of improved actions*

Fortunately, considerable progress in recent years has provided us with so-called 'improved' lattice actions [38]. Here, in the simplest form, one supplements the Wilson action, built from elementary 1×1 plaquettes (pl), with larger 1×2 rectangles (rt)

$$S_G[U] = \beta \left(S_{\text{pl}}[U] + c_{\text{rt}} S_{\text{rt}}[U] \right) . \tag{4.18}$$

The Wilson part of the action (pl) approaches the classical (tree-level) limit like $o(a^4)$ as $a \to 0$. If $c_{\text{rt}} = -1/20$ the (rt) term in (4.18) improves this behavior to order $o(a^5)$. We refer to such an action as $o(a)$ tree-level improved. Better yet, one can also account for a large part of the quantum fluctuations of the gluon field through renormalizing the link variables $U_\mu(x)$. Of course, this would destroy the tree-level cancellations between the (pl) and (rt) terms in (4.18). Within mean-field theory it can be 'repaired' by changing the factor c_{rt} to

$$c_{\text{rt}} = -\frac{1}{20u^2} , \tag{4.19}$$

where u, known as the tadpole factor, is the fourth root of the average (4-link) plaquette

$$u = \left\langle \frac{1}{3} \text{Re}\,\text{Tr}\, U_{\text{pl}} \right\rangle^{1/4} . \tag{4.20}$$

The tadpole factor can easily be computed self-consistently within a running lattice simulation. There are also radiative corrections which can be incorporated into the lattice action to some extent [39, 40, 41].

The benefit of improvement is that the physical continuum limit, $a \to 0$, is approached 'faster' compared to the Wilson action. The 'same physics' is described with improved actions at larger lattice constants a. For example, a Wilson action at $a \approx 0.1$ fm may give a similar mass spectrum as an improved action at $a \approx 0.3$ fm. For our present purpose of putting two hadrons on the lattice, this means that, stretching things a bit and choosing $a \approx 0.4$ fm, we would only need an $L^3 \times T$ lattice with about $L = 10$ for a lattice of 4.0 fm across. This should give us a first window on two-hadron systems on the lattice.

Improved actions have been a big innovation for lattice field theory [42].* There are versions of improved actions including fermions [45], with active work pushing the improvement to higher order, for example $o(a^2)$ improved actions [46, 47, 48]. In Appendix B we have collected some topics that are relevant to this chapter. Strong-interaction physics, in particular the boundary between particle and nuclear physics, is now open to studies within quantum chromodynamics.

4.2 A Simple U(1) Lattice Model in 2+1 Dimensions

In order to discuss some practical work we look at a simple but non-trivial model. We consider U(1) gauge theory on an $L^d \times T$ lattice, in $d+1$ dimensions, with $d = 2$. The intent is to study the interaction between two mesons living in a plane. For the sake of simplicity we work with the (unimproved) Wilson action and use staggered fermions.

4.2.1 *Lattice action*

For a U(1) gauge group the link variables are

$$U_\mu(x) = e^{i\Theta_\mu(x)} \in \mathrm{U}(1)\,, \tag{4.21}$$

with $\mu = 1 \cdots d+1$, $x = (\vec{x}, t)$, and $\Theta_\mu(x) \in [-\pi, \pi)$. The plaquette (pl) variables are

$$\begin{aligned} U_{pl}(\mu,\nu;x) &= U_\mu(x) U_\nu(x+\hat{\mu}) U_\mu^\dagger(x+\hat{\nu}) U_\nu^\dagger(x) \\ &= \exp[i\Theta_{pl}(\mu,\nu;x)]\,, \end{aligned} \tag{4.22}$$

where

$$\Theta_{pl}(\mu,\nu;x) = \Theta_\mu(x) + \Theta_\nu(x+\hat{\mu}) - \Theta_\mu(x+\hat{\nu}) - \Theta_\nu(x) \tag{4.23}$$

*Ed. Some even say [43] that this program of Improvement is "the most important theoretical advance in recent years" — a claim that is not universally accepted in the LQCD community [44].

is the oriented plaquette angle. We write $\hat{\mu}$ for a vector of length a in direction μ. The (compact form of the) Wilson gauge field action is

$$S_G = \beta \sum_x \sum_{\mu<\nu} \mathrm{Re}[1 - U_{pl}(\mu,\nu;x)] = \beta \sum_x \sum_{\mu<\nu} [1 - \cos\Theta_{pl}(\mu,\nu;x)] \,. \quad (4.24)$$

For simplicity, fermions are treated in the staggered, or Kogut-Susskind, scheme where Dirac indices have been 'spin-diagonalized' away [49, 50, 51] — see Appendix A. Thus we work with one-component Grassmann fields $\chi_f(x)$ defined on the lattice sites $x = (\vec{x}, t)$, with (external) flavor index $f = u, d, s$. There is just one color index, $C = 1$, which we omit. The fermionic action can be written as

$$S_F = \sum_{x,y} \sum_f \bar{\chi}_f(x)\, G^{-1}(x,y)[U]\, \chi_f(y) \,, \quad (4.25)$$

with the flavor-independent fermion matrix

$$G^{-1}(x,y) = \frac{1}{2}\sum_\mu \eta_\mu(x) \left[U_\mu(x)\delta_{x+\hat{\mu},y} - U_\mu^\dagger(y)\delta_{x,y+\hat{\mu}}\right] + m_F\, \delta_{x,y} \,, \quad (4.26)$$

including a mass term. Here $\eta_\mu(x) = (-1)^{x_1+\cdots x_{\mu-1}}, \mu = 2, \cdots, d+1$ (with $\eta_1(x) = 1$) is the staggered phase.

This lattice field model is confining in finite volume [52], and exhibits chiral symmetry breaking [53] and monopole condensation [54], for appropriately chosen inverse gauge coupling β. Its basis is the euclidean generating functional integral

$$Z[\bar{J}, J] = \int [dU][d\chi][d\bar{\chi}] e^{-S_G[U] - S_F[U,\chi,\bar{\chi}] + \bar{J}\chi + \bar{\chi}J} \,, \quad (4.27)$$

with the fermionic sources $\bar{J}\chi = \sum_x \bar{J}(x)\chi(x)$ and $\bar{\chi}J = \sum_x \bar{\chi}(x)J(x)$. We will use gauge configurations in the quenched approximation.

4.2.2 *Meson fields*

The simplest hadrons we can work with are pseudoscalar mesons. In the staggered scheme suitable one-meson operators are

$$\phi_{\vec{p}}(t) = L^{-2} \sum_{\vec{x}} e^{i\vec{p}\cdot\vec{x}} \bar{\chi}_d(\vec{x},t)\, \chi_u(\vec{x},t) \,. \quad (4.28)$$

The $\vec{x}$ sum extends over the spatial sites of the lattice. The fixed flavor assignment u,d leading to a π^+ will simplify computations. For $d = 2$, a planar lattice, the momentum parameters are

$$\vec{p} = \frac{2\pi}{L}(k_1, k_2)\,, \quad \text{where} \quad k_{1,2} = -(\frac{L}{2} - 1), \cdots, \frac{L}{2}, \quad \text{even } L\,, \tag{4.29}$$

if L is odd we have $k_i = -(L-1)/2, \cdots, (L-1)/2$. The operators (4.28) probe the lattice (vacuum) for composite meson states with momentum $\vec{p}$. Products of those operators probe for multi-meson excitations. For example

$$\Phi_{\vec{p}}(t) = \phi_{-\vec{p}}(t)\, \phi_{+\vec{p}}(t) \tag{4.30}$$

describes two-meson fields with total momentum $\vec{P} = 0$ and relative momentum $\vec{p}$. The reduced mass is $m/2$.

4.2.3 *Correlation matrices*

The time-correlation matrix for the one-meson system is

$$C^{(2)}_{\vec{p}\,\vec{q}}(t, t_0) = \langle \phi^\dagger_{\vec{p}}(t)\, \phi_{\vec{q}}(t_0) \rangle - \langle \phi^\dagger_{\vec{p}}(t) \rangle \langle \phi_{\vec{q}}(t_0) \rangle\,. \tag{4.31}$$

It can be worked out with Wick's theorem in terms of contractions between the Grassmann fields. In our simple example we use, see Appendix A,

$$\cdots \overset{n}{\chi}_f\,(x)\, \overset{n}{\bar{\chi}}_{f'}\,(x') \cdots \;=\; \cdots \delta_{ff'} G(x, x') \cdots \tag{4.32}$$

$$\cdots \overset{n}{\bar{\chi}}_f\,(x)\, \overset{n}{\chi}_{f'}\,(x') \cdots \;=\; \cdots \delta_{ff'} G^*(x, x') \cdots\,, \tag{4.33}$$

where n indicates the partners of the contraction, and G is the inverse fermion matrix, cf. (4.26). This may be used to work out the correlation matrices. For example, consider

$$\langle \phi^\dagger_{\vec{p}}(t)\, \phi_{\vec{q}}(t_0) \rangle = L^{-4} \sum_{\vec{x}} \sum_{\vec{y}} e^{-i\vec{p}\cdot\vec{x}} e^{i\vec{q}\cdot\vec{y}} \langle \bar{\chi}_u(\vec{x}, t) \chi_d(\vec{x}, t)\, \bar{\chi}_d(\vec{y}, t_0) \chi_u(\vec{y}, t_0) \rangle\,. \tag{4.34}$$

Because of the flavor assignment in (4.28) the separable term in (4.31) is zero, and there is only one group of contractions in the first term

$$C^{(2)} \sim \overset{2}{\bar{\chi}}_u\, \overset{1}{\chi}_d\, \overset{1}{\bar{\chi}}_d\, \overset{2}{\chi}_u\,, \tag{4.35}$$

with

$$\cdots \overset{1}{\chi}_d\,(\vec{x}, t)\, \overset{1}{\bar{\chi}}_d\,(\vec{y}, t_0) \cdots \;=\; \cdots G(\vec{x}t, \vec{y}t_0) \cdots \tag{4.36}$$

$$\cdots \overset{2}{\bar{\chi}}_u\,(\vec{x}, t)\, \overset{2}{\chi}_u\,(\vec{y}, t_0) \cdots \;=\; \cdots G^*(\vec{x}t, \vec{y}t_0) \cdots\,. \tag{4.37}$$

Thus we obtain

$$C^{(2)}_{\vec{p}\,\vec{q}}(t,t_0) = L^{-4}\sum_{\vec{x}}\sum_{\vec{y}} e^{-i\vec{p}\cdot\vec{x}+i\vec{q}\cdot\vec{y}}\langle|G(\vec{x}t,\vec{y}t_0)|^2\rangle\,. \tag{4.38}$$

Assuming translational invariance

$$\langle|G(\vec{x},t;\vec{y},t_0)|^2\rangle = \langle|G(\vec{x}+\vec{a},t;\vec{y}+\vec{a},t_0)|^2\rangle \tag{4.39}$$

renders the 2-point correlator diagonal

$$C^{(2)}_{\vec{p}\,\vec{q}}(t,t_0) = \delta_{\vec{p}\,\vec{q}}\,L^{-2}\sum_{\vec{x}} e^{-i\vec{p}\cdot\vec{x}}\langle|G(\vec{x}t,\vec{x}_0t_0)|^2\rangle\, e^{i\vec{p}\cdot\vec{x}_0}\,. \tag{4.40}$$

The point $\vec{x}_0$ is arbitrary, $C^{(2)}$ is independent of $\vec{x}_0$, and the phase factor is irrelevant.

The 4-point correlator describes the propagation of two interacting mesons on the lattice

$$C^{(4)}_{\vec{p}\,\vec{q}}(t,t_0) = \langle\Phi^\dagger_{\vec{p}}(t)\,\Phi_{\vec{q}}(t_0)\rangle - \langle\Phi^\dagger_{\vec{p}}(t)\rangle\langle\Phi_{\vec{q}}(t_0)\rangle\,. \tag{4.41}$$

Here $\vec{p}$ and $\vec{q}$ are the relative momenta in the meson-meson system. In terms of the fermion propagator this correlator reads

$$\begin{aligned} C^{(4)}_{\vec{p}\,\vec{q}}(t,t_0) \quad &= L^{-8}\sum_{\vec{x}_1}\sum_{\vec{x}_2}\sum_{\vec{y}_1}\sum_{\vec{y}_2} \mathrm{e}^{i\vec{p}\cdot(\vec{x}_2-\vec{x}_1)+i\vec{q}\cdot(\vec{y}_2-\vec{y}_1)} \\ &\langle\big(\, G^*(\vec{x}_2t,\vec{y}_2t_0)\;\; G(\vec{x}_2t,\vec{y}_2t_0)\;\; G^*(\vec{x}_1t,\vec{y}_1t_0)\;\; G(\vec{x}_1t,\vec{y}_1t_0) \\ &+ G^*(\vec{x}_1t,\vec{y}_2t_0)\;\; G(\vec{x}_1t,\vec{y}_2t_0)\;\; G^*(\vec{x}_2t,\vec{y}_1t_0)\;\; G(\vec{x}_2t,\vec{y}_1t_0) \\ &- G^*(\vec{x}_2t,\vec{y}_1t_0)\;\; G(\vec{x}_2t,\vec{y}_2t_0)\;\; G^*(\vec{x}_1t,\vec{y}_2t_0)\;\; G(\vec{x}_1t,\vec{y}_1t_0) \\ &- \;G(\vec{x}_2t,\vec{y}_1t_0)\;\; G^*(\vec{x}_2t,\vec{y}_2t_0)\;\; G(\vec{x}_1t,\vec{y}_2t_0)\;\; G^*(\vec{x}_1t,\vec{y}_1t_0)\big)\rangle\,. \end{aligned} \tag{4.42}$$

For an SU(N) gauge group, $G \to G_{CC'}$, there would be sums over four sets of color indices corresponding to the $\vec{x}_1,\vec{x}_2,\vec{y}_1,\vec{y}_2$ assignment in (4.42). A diagrammatic classification of the various terms which contribute to $C^{(4)}$ proves useful. Such a classification arises naturally from working out the contractions between the quark fields in (4.41) and (4.31). Let us write

$$C^{(4)} \;=\; C^{(4A)} + C^{(4B)} - C^{(4C)} - C^{(4D)} \tag{4.43}$$

$$= \;\|\; + \;\times\; - \;\bowtie\; - \;\bowtie \tag{4.44}$$

for the four terms as they occur in (4.42). Each diagram line corresponds to a fermion propagator, see Fig. 4.6. Specifically, using the notation intro-

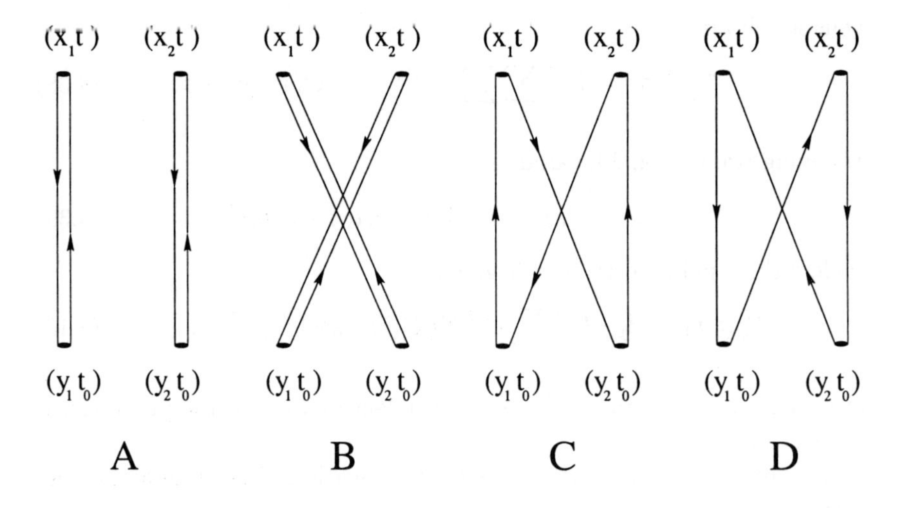

Fig. 4.6 Diagrammatic classification of the correlation matrix $C^{(4)}$ according to (4.42).

duced in (4.32) and (4.33), we have

$$C^{(4A)} = \langle \overset{43}{\phi^\dagger_{+\vec{p}}} \overset{21}{\phi^\dagger_{-\vec{p}}} \overset{12}{\phi_{-\vec{q}}} \overset{34}{\phi_{+\vec{q}}} \rangle = \langle \overset{43}{\phi^\dagger_{+\vec{p}}} \overset{34}{\phi_{+\vec{q}}} \overset{21}{\phi^\dagger_{-\vec{p}}} \overset{12}{\phi_{-\vec{q}}} \rangle \tag{4.45}$$

$$C^{(4B)} = \langle \overset{21}{\phi^\dagger_{+\vec{p}}} \overset{43}{\phi^\dagger_{-\vec{p}}} \overset{12}{\phi_{-\vec{q}}} \overset{34}{\phi_{+\vec{q}}} \rangle = \langle \overset{43}{\phi^\dagger_{-\vec{p}}} \overset{34}{\phi_{+\vec{q}}} \overset{21}{\phi^\dagger_{+\vec{p}}} \overset{12}{\phi_{-\vec{q}}} \rangle\,, \tag{4.46}$$

where the $n = 1 \cdots 4$ identify the partners χ and $\bar{\chi}$ of a contraction. In diagrams A and B no quark exchange between the mesons occurs. Diagram B describes the exchange of (composite) mesons as a whole. On the other hand diagrams C and D both contain quark exchange between the mesons

$$C^{(4C)} = \langle \overset{23}{\phi^\dagger_{+\vec{p}}} \overset{41}{\phi^\dagger_{-\vec{p}}} \overset{12}{\phi_{-\vec{q}}} \overset{34}{\phi_{+\vec{q}}} \rangle \tag{4.47}$$

$$C^{(4D)} = \langle \overset{41}{\phi^\dagger_{+\vec{p}}} \overset{23}{\phi^\dagger_{-\vec{p}}} \overset{12}{\phi_{-\vec{q}}} \overset{34}{\phi_{+\vec{q}}} \rangle\,. \tag{4.48}$$

Thus these diagrams must be considered as sources of effective interaction.

The propagation of free, noninteracting, mesons is contained in the sum $C^{(4A)} + C^{(4B)}$. However, gluonic correlations between the mesons are also present because the gauge configuration average $\langle \cdots \rangle$ is taken over the product of all four fields. (This is also true for $C^{(4C)}$ and $C^{(4D)}$.) Gluonic correlations of course contribute to the effective residual interaction. Hence it is essential to isolate the uncorrelated part contained in $C^{(4A)} + C^{(4B)}$.

4.2.4 *Correlation matrix for noninteracting mesons*

Let us examine $C^{(4A)}$. Expressing the contractions in (4.45) in terms of quark propagators with the notation introduced in (4.32) and (4.33) we have, in generic notation

$$C^{(4A)} \sim \langle \overset{4}{G^*} \overset{3}{G}\, \overset{2}{G^*} \overset{1}{G} \,\rangle\,, \tag{4.49}$$

where the $\sim$ indicates the Fourier sums, *etc.*, which carry over from (4.28), see (4.42). The gauge configuration average in (4.49) may be analyzed by means of the cumulant (or cluster) expansion theorem†. Taking advantage of $\langle G \rangle = 0$ we have

$$\begin{aligned} &\langle \overset{4}{G^*} \overset{3}{G}\, \overset{2}{G^*} \overset{1}{G} \,\rangle = \\ &\quad \langle \overset{4}{G^*} \overset{3}{G} \,\rangle \langle \overset{2}{G^*} \overset{1}{G} \,\rangle + \langle \overset{4}{G^*} \overset{1}{G} \,\rangle \langle \overset{2}{G^*} \overset{3}{G} \,\rangle + \langle \overset{4}{G^*} \overset{2}{G^*} \rangle \langle \overset{3}{G}\, \overset{1}{G} \,\rangle + \langle\!\langle \overset{4}{G^*} \overset{3}{G}\, \overset{2}{G^*} \overset{1}{G} \,\rangle\!\rangle\,. \end{aligned} \tag{4.50}$$

The last term defines the cumulant. The first three terms on the right-hand side of (4.50) are illustrated in Fig. 4.7. The dashed lines indicate that the propagators are correlated through gluons. Evidently only the first one of the three separable terms in (4.50) represents free, uncorrelated, mesons. All other terms in (4.50) are sources of residual effective interaction between the mesons. We therefore define

$$\overline{C}^{(4A)} = \langle \overset{43}{\phi^\dagger_{+\vec p}} \overset{34}{\phi_{+\vec q}} \rangle \langle \overset{21}{\phi^\dagger_{-\vec p}} \overset{12}{\phi_{-\vec q}} \rangle \sim \langle \overset{4}{G^*} \overset{3}{G} \,\rangle \langle \overset{2}{G^*} \overset{1}{G} \,\rangle\,. \tag{4.51}$$

A similar analysis of $C^{(4B)}$ leads to

$$\overline{C}^{(4B)} = \langle \overset{43}{\phi^\dagger_{-\vec p}} \overset{34}{\phi_{+\vec q}} \rangle \langle \overset{21}{\phi^\dagger_{+\vec p}} \overset{12}{\phi_{-\vec q}} \rangle\,. \tag{4.52}$$

The sum of those is the free meson-meson time correlation matrix

$$\overline{C}^{(4)}_{\vec p \vec q}(t, t_0) = \overline{C}^{(4A)}_{\vec p \vec q}(t, t_0) + \overline{C}^{(4B)}_{\vec p \vec q}(t, t_0)\,. \tag{4.53}$$

†The cluster expansion theorem involves the definition of the cumulant as a generalization of the standard deviation. For two random variables X_1 and X_2, for example, we have $\langle X_1 X_2 \rangle = \langle X_1 \rangle \langle X_2 \rangle + \langle\!\langle X_1 X_2 \rangle\!\rangle$. If $X_1 = X_2$ the square of the standard deviation is $\sigma^2 = \langle\!\langle X_1 X_2 \rangle\!\rangle$ [55].

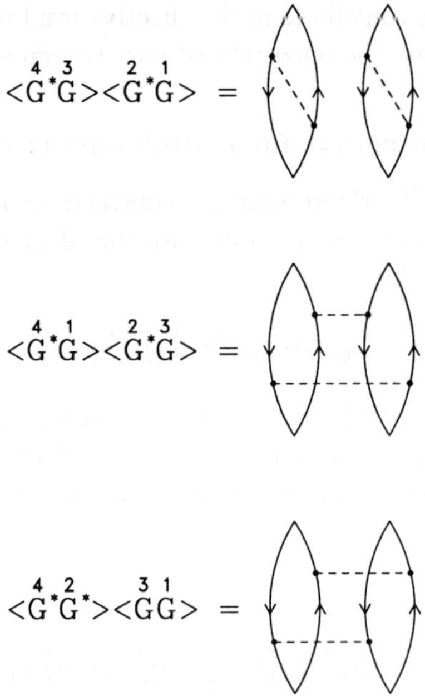

$$\langle \overset{4}{G}{}^{*}\overset{3}{G}\rangle\langle \overset{2}{G}{}^{*}\overset{1}{G}\rangle =$$

$$\langle \overset{4}{G}{}^{*}\overset{1}{G}\rangle\langle \overset{2}{G}{}^{*}\overset{3}{G}\rangle =$$

$$\langle \overset{4}{G}{}^{*}\overset{2}{G}{}^{*}\rangle\langle \overset{3}{G}\overset{1}{G}\rangle =$$

Fig. 4.7 Illustration of the cumulant expansion of $C^{(4)}$. The dashed lines represent gluonic correlations.

$\overline{C}^{(4)}$ is an additive part of the full 4-point correlator,

$$C^{(4)} = \overline{C}^{(4)} + C_I^{(4)} . \tag{4.54}$$

The remainder

$$C_I^{(4)} = \left(C^{(4A)} + C^{(4B)} - \overline{C}^{(4)}\right) - C^{(4C)} - C^{(4D)} \tag{4.55}$$

contains all sources of the residual interaction, be it from gluonic correlations, or quark exchange (or quark-antiquark loops, if the simulation is unquenched) or any other effects. The free correlator $\overline{C}^{(4)}$ describes two noninteracting identical lattice mesons. Their internal structure and masses consistently arise from the dynamics determined by the lattice field model and its numerical implementation. As already mentioned the separable term in (4.31) is identically zero because of the u, d flavor assignment to the Grassmann fields of the one-meson operators (4.28).

The correlator $\overline{C}^{(4)}$ may be expressed in terms of the 2-point correlator $C^{(2)}$. Using (4.31) and (4.51), (4.52) gives

$$\overline{C}^{(4)}_{\vec{p}\,\vec{q}} = C^{(2)}_{\vec{p},\vec{q}}\,C^{(2)}_{-\vec{p},-\vec{q}} + C^{(2)}_{-\vec{p},\vec{q}}\,C^{(2)}_{\vec{p},-\vec{q}}\,. \tag{4.56}$$

$C^{(2)}$ is diagonal in the momentum indices. Writing

$$C^{(2)}_{\vec{p}\,\vec{q}}(t,t_0) = \delta_{\vec{p}\,\vec{q}}\,c_{\vec{p}}(t,t_0)\,e^{i\vec{p}\cdot\vec{x}_0}\,, \tag{4.57}$$

where $c_{\vec{p}}(t,t_0)$ is given by (4.40) and has the obvious property

$$c_{-\vec{p}}(t,t_0) = c^*_{\vec{p}}(t,t_0)\,, \tag{4.58}$$

we obtain

$$\overline{C}^{(4)}_{\vec{p}\,\vec{q}}(t,t_0) = (\delta_{\vec{p},\vec{q}} + \delta_{-\vec{p},\vec{q}})\,|c_{\vec{p}}(t,t_0)|^2\,. \tag{4.59}$$

To be consistent with the computation of $C^{(4)}_{\vec{p}\,\vec{q}}(t,t_0)$, using (4.42), $c_{\vec{p}}(t,t_0)$ should be computed from (4.38). It is interesting to note that the property

$$\overline{C}^{(4)}_{\vec{p},\vec{q}} = \overline{C}^{(4)}_{-\vec{p},\vec{q}} = \overline{C}^{(4)}_{\vec{p},-\vec{q}} = \overline{C}^{(4)}_{-\vec{p},-\vec{q}}\,, \tag{4.60}$$

which is evident from (4.56) and is explicit in (4.42), also holds for $C^{(4)}$. This property reflects Bose symmetry with respect to the *composite* mesons. Permutation of the mesons, one with momentum $+\vec{p}$ the other with momentum $-\vec{p}$, results in the substitution $\vec{p} \to -\vec{p}$. Clearly, combining diagrams $\overline{C}^{(4A)}$ and $\overline{C}^{(4B)}$ into $\overline{C}^{(4)}$ is crucial for the symmetry (4.60) to hold.

4.2.5 *Computation with random sources*

The computation of $C^{(4)}$ from the expression (4.42) is a formidable numerical task. First, propagator matrix elements are needed for arbitrary $\vec{y}$, as opposed to only one column of G like in (4.40). Typically G is computed by solving a linear equation like $G^{-1}X^{(R)} = R$ where R is a given source vector; the solution vector $X^{(R)}$ then represents 'some column' of G. For arbitrary $\vec{y}$ such a strategy translates into L^d linear equations, all for a fixed time slice t_0 and not even counting color indices (or Dirac indices in case of Wilson fermions).

Furthermore, the sum over lattice sites in (4.42) contains L^{4d} terms. Even for $d = 2$ and our simple U(1) gauge model there are $\approx 10^8$ terms. In a realistic calculation, for $d = 3$ and SU(3), these would be $L^{4d}3^4 \approx 10^{14}$ for a meson-meson system, and even larger if baryons are involved. It is clear that an efficient numerical technique is called for to deal with computing quark propagators and correlator matrices.

A practical way is the use of random estimators [56, 57, 58, 59]. Let $R(x)$ be a vector of independent complex random deviates with

$$\langle R(x) R^*(y) \rangle_R = \delta_{x,y} \,, \tag{4.61}$$

where $\langle \cdots \rangle_R$ denotes the random R-average. Gaussian deviates come to mind, but Z^2 noise also works and may sometimes be more desirable [60]. The idea is to solve

$$\sum_y G^{-1}(x,y)\, X^{(R)}(y) = R(x) \tag{4.62}$$

for $X^{(R)}$, recalling that the fermion matrix G^{-1} is known, then the statistical average

$$\langle X^{(R)}(y) R^*(z) \rangle_R = G(y,z) \tag{4.63}$$

is an estimator for the inverse G of G^{-1}.

Since random sources R with various characteristics may be chosen the technique is very versatile. Let us expand our notation, where

$$R^{(r;\alpha)}(x) \,, \quad \text{with} \quad r = 1 \cdots N_R \,, \tag{4.64}$$

is a finite sample of N_R vectors drawn from the distribution. The label α distinguishes between various types of sources, for example we take $\alpha = t_0$ to mean

$$R^{(r;t_0)}(\vec{x}, t) = R^{(r;t_0)}(\vec{x}) \delta_t^{t_0} \,. \tag{4.65}$$

Here, random sources are placed on one time slice only. A more complicated example is $\alpha = (t_0, \vec{p})$ with

$$R^{(r;t_0,\vec{p})}(\vec{x}, t) = L^{-2}\, e^{i\vec{p}\cdot\vec{x}} R^{(r;t_0,\vec{p})}(\vec{x}) \delta_t^{t_0} \,. \tag{4.66}$$

These are Fourier-modified (boosted) sources on time slice t_0.

Given a finite sample (4.64), an estimator for the statistical average of (4.61) is

$$\frac{1}{N_R}\sum_{r=1}^{N_R} R^{(r;\alpha)}(x)R^{(r;\beta)^*}(y) \simeq \sum_{\langle r\rangle} R^{(r;\alpha)}(x)R^{(r;\beta)^*}(y) = \delta_{x,y}\delta^{\alpha,\beta}\,. \quad (4.67)$$

The above, intuitive, notation $\sum_{\langle r\rangle}$ for the statistical R-average $\langle\cdots\rangle_R$ is convenient in expressions of correlation functions. The second δ on the right hand side of (4.67) comes from drawing stochastically independent samples for each type of source.

For definiteness we continue with sources of type (4.65) set at a fixed time slice t_0. Then the solutions X of the linear equations (4.62)

$$\sum_{\vec{y}\,y_4} G^{-1}(\vec{x}x_4,\vec{y}y_4)\,X^{(r;t_0)}(\vec{y}y_4) = R^{(r;t_0)}(\vec{x})\delta^{t_0}_{x_4} \quad (4.68)$$

can be used to estimate columns of the propagator G. Applying G to both sides of (4.68) gives

$$X^{(r;t_0)}(\vec{z}z_4) = \sum_{\vec{x}} G(\vec{z}z_4,\vec{x}t_0)R^{(r;t_0)}(\vec{x})\,. \quad (4.69)$$

Multiplication with $R^{(r;t_0)^*}(\vec{y})$, taking the R-average and using (4.67) yields

$$G(\vec{z}z_4,\vec{y}t_0) = \sum_{\langle r\rangle} X^{(r;t_0)}(\vec{z}z_4)R^{(r;t_0)^*}(\vec{y})\,. \quad (4.70)$$

In this form the random estimator can be used directly in the expression (4.38) and (4.42) for $C^{(2)}$ and $C^{(4)}$, respectively.

Another, potentially useful, relation follows from (4.69)

$$\sum_{\langle r\rangle} X^{(r;t_0)}(\vec{z}z_4)X^{(r;t_0)^*}(\vec{y}y_4) = \sum_{\vec{x}} G(\vec{z}z_4,\vec{x}t_0)\,G(\vec{y}y_4,\vec{x}t_0)^*\,. \quad (4.71)$$

A more general version of (4.71) can be derived when Fourier-modified sources (4.66) are used. We have

$$\sum_{\langle r\rangle} X^{(r;t_0,\vec{p})}(\vec{z}z_4)X^{(r;t_0,\vec{q})^*}(\vec{y}y_4) = L^{-4}\sum_{\vec{x}} G(\vec{z}z_4,\vec{x}t_0)\,e^{i(\vec{p}-\vec{q})\cdot\vec{x}}\,G(\vec{y}y_4,\vec{x}t_0)^*\,. \quad (4.72)$$

Direct application of (4.72) gives the 2-point correlator (4.38) in the form

$$C^{(2)}_{\vec{p}\vec{q}}(t,t_0) = \langle L^{-2} \sum_{\vec{x}} e^{-i\vec{p}\cdot\vec{x}} \sum_{\langle r \rangle} X^{(r;t_0,\vec{q})}(\vec{x}t)\, X^{(r;t_0,\vec{0})^*}(\vec{x}t)\rangle \,. \qquad (4.73)$$

This equation is particularly useful for testing diagonality, see (4.40).

In the case of the 4-point correlator two independent sets of random sources are necessary. The four contributions $C^{(4A)}$ through $C^{(4D)}$, see (4.43), may be computed separately, for example

$$\begin{aligned} C^{(4A)}_{\vec{p}\vec{q}}(t,t_0) \quad = \quad & \Big\langle \sum_{\langle r_1 \rangle} \sum_{\vec{x}_1} \mathrm{e}^{-i\vec{p}\cdot\vec{x}_1}\, X^{(r_1;t_0,\vec{0})}(\vec{x}_1 t)\, X^{(r_1;t_0,\vec{q})^*}(\vec{x}_1 t) \\ & \sum_{\langle r_2 \rangle} \sum_{\vec{x}_2} \mathrm{e}^{+i\vec{p}\cdot\vec{x}_2}\, X^{(r_2;t_0,\vec{q})}(\vec{x}_2 t)\, X^{(r_2;t_0,\vec{0})^*}(\vec{x}_2 t) \Big\rangle \end{aligned} \qquad (4.74)$$

There are similar expressions for $C^{(4B)}$ and $C^{(4C)}$, $C^{(4D)}$. The noninteracting correlator $\overline{C}^{(4A)}$ is computed via (4.73) and the use of (4.56).

Having selected an even number N_R of random sources, in the above case a subset of $N_R/2$ of those can be used in $\sum_{\langle r_1 \rangle}$ and the other half in $\sum_{\langle r_2 \rangle}$. There are $\binom{N_R}{\frac{1}{2}N_R}$ possibilities to choose such a subdivision (*e.g.* 70 for $N_R = 8$ and 12870 for $N_R = 16$).

4.3 Effective Residual Interaction

Potentials are not unique in the sense that a class of potentials can produce the same phase shifts (phase-equivalent). Therefore an effective residual interaction extracted from the lattice is subject to definition. In this section we study the perturbative expansion of the time correlation matrix for an interacting elementary Bose field on a euclidean lattice. Our intention is to find guidance towards a sensible definition for the lattice simulation.

4.3.1 *Perturbative definition*

Let $\hat{\mathcal{L}}_0 = \hat{\mathcal{L}}_0(\hat{\phi}, \partial\hat{\phi})$ be the free Lagrangian for an elementary Bose field $\hat{\phi}(x)$ defined on the sites $x = (\vec{x}, t)$ of the lattice. It is understood that $\hat{\phi}$ is subject to the usual canonical quantization through commutators. Let $\hat{\mathcal{L}} = \hat{\mathcal{L}}_0 + \hat{\mathcal{L}}_I$ such that $\hat{\mathcal{L}}_I = \hat{\mathcal{L}}_I(\hat{\phi})$ is a (small) interaction. In the usual way $\hat{\mathcal{L}}$ gives rise to a Hamiltonian

$$\hat{H} = \hat{H}_0 + \hat{H}_I \,, \tag{4.75}$$

where $\hat{H}_0$ is the free part and $\hat{H}_I$ a perturbative interaction. In view of (4.28) and (4.30) define

$$\hat{\phi}_{\vec{p}}(t) = L^{-2} \sum_{\vec{x}} e^{i\vec{p}\cdot\vec{x}} \hat{\phi}(\vec{x}, t) \tag{4.76}$$

and

$$\hat{\Phi}_{\vec{p}}(t) = \hat{\phi}_{-\vec{p}}(t)\, \hat{\phi}_{+\vec{p}}(t) \,. \tag{4.77}$$

In the correlation matrix

$$\hat{C}^{(4)}_{\vec{p}\,\vec{q}}(t, t_c) = \langle 0 | \hat{\Phi}^{\dagger}_{\vec{p}}(t)\, \hat{\Phi}_{\vec{q}}(t_c) | 0 \rangle \tag{4.78}$$

the separable term of (4.41) has been dropped since it is zero for the flavored quark fields, like in Sec. 4.2.3. The (nondegenerate) vacuum state $|0\rangle$ satisfies $\hat{H}|0\rangle = W_0|0\rangle$. We will assume that its energy is zero, $W_0 = 0$. Thus the time dependence of the correlator (4.78) may be made explicit

$$\hat{C}^{(4)}_{\vec{p}\,\vec{q}}(t, t_c) = \langle 0 | \hat{\Phi}^{\dagger}_{\vec{p}}(t_c)\, e^{-\hat{H}(t-t_c)}\, \hat{\Phi}_{\vec{q}}(t_c) | 0 \rangle \,. \tag{4.79}$$

Switching to the interaction picture, we define

$$\hat{H}_I(t) = e^{\hat{H}_0(t-t_c)}\, \hat{H}_I\, e^{-\hat{H}_0(t-t_c)} \tag{4.80}$$

and the (euclidean) time evolution operator[‡]

$$\hat{U}(t, t_c) = e^{\hat{H}_0(t-t_c)}\, e^{-\hat{H}(t-t_c)} \,. \tag{4.81}$$

The perturbative expansion of the latter

$$\hat{U}(t, t_c) = \sum_{N=0}^{\infty} \frac{(-1)^N}{N!} \int_{t_c}^{t} dt_1 \cdots \int_{t_c}^{t} dt_N \, \mathrm{T}[\hat{H}_I(t_1) \cdots \hat{H}_I(t_N)] \tag{4.82}$$

then induces a perturbative expansion of the correlator

$$\begin{aligned} \hat{C}^{(4)}_{\vec{p}\,\vec{q}}(t, t_c) &= \langle 0 | \hat{\Phi}^{\dagger}_{\vec{p}}(t_c)\, e^{-\hat{H}_0(t-t_c)}\, \hat{U}(t, t_c)\, \hat{\Phi}_{\vec{q}}(t_c) | 0 \rangle & (4.83) \\ &= \sum_{N=0}^{\infty} \hat{C}^{(4;N)}_{\vec{p}\,\vec{q}}(t, t_c) \,. & (4.84) \end{aligned}$$

[‡]This is in analogy to the transfer matrix formalism.

The zero-order term $\hat{C}^{(4;N=0)}$ evidently describes noninteracting mesons.

Order $N = 0$

Let $|n\nu\rangle$ be a complete orthogonal set of eigenstates of $\hat{H}_0$

$$\hat{H}_0|n\nu\rangle = W_n^{(0)}|n\nu\rangle\,, \tag{4.85}$$

where $W_n^{(0)}$ are free two-meson energies on the lattice, and ν is a degeneracy index. Lattice effects set aside, those energies should be close to $2\sqrt{m^2+p^2}$ with m being the rest mass of one meson and $p = |\vec{p}|$ its momentum. Also, define the relative meson-meson momentum-space wave functions $\psi_{n\nu}^{(0)}(\vec{p})$ through

$$c_{n\nu}^{(0)}\psi_{n\nu}^{(0)}(\vec{p}) = \langle n\nu|\hat{\Phi}_{\vec{p}}(t_c)|0\rangle^*\,, \tag{4.86}$$

where $c_{n\nu}^{(0)}$ are normalization factors. The order $N = 0$ correlator then is

$$\hat{C}_{\vec{p}\vec{q}}^{(4;N=0)}(t,t_c) = \sum_{n\nu}|c_{n\nu}^{(0)}|^2 e^{-W_n^{(0)}(t-t_c)}\psi_{n\nu}^{(0)}(\vec{p})\psi_{n\nu}^{(0)*}(\vec{q})\,. \tag{4.87}$$

With properly chosen normalization factors $c_{n\nu}^{(0)}$ we expect orthonormality and completeness

$$\sum_{\vec{p}}\psi_{n\nu}^{(0)*}(\vec{p})\psi_{m\mu}^{(0)}(\vec{p}) = \delta_{nm}\delta_{\nu\mu} \tag{4.88}$$

$$\sum_{n\nu}\psi_{n\nu}^{(0)}(\vec{p})\psi_{n\nu}^{(0)*}(\vec{q}) = \delta_{\vec{p}\vec{q}}\,. \tag{4.89}$$

For a free elementary Bose field this is almost a trivial point since the $\psi_{n\nu}^{(0)}$ are merely plane (lattice) waves. A glance at (4.87) shows that, technically, those could be obtained as (normalized) eigenvectors from diagonalizing the correlation matrix $\hat{C}^{(4;N=0)}$, where $|c_{n\nu}^{(0)}|^2$ are the eigenvalues, at $t = t_c$. For the case considered here all eigenvalues of $\hat{C}^{(4;N=0)}$ will be nonzero.

In a lattice model, however, where the role of $\hat{\Phi}_{\vec{p}}(t_c)$ is assumed to be a composite operator made from fermion fields, see (4.28) and (4.30), it can not be *a priori* excluded that an operator matrix element, of the type as it occurs in (4.86), is identically zero for all $\vec{p}$. In this case the free correlator $\overline{C}^{(4)}$ would have an eigenvalue zero (for all t). Likewise, if, for some reason, the set of hadron operators used to construct the correlation matrices on the lattice is linearly dependent, one should expect $\overline{C}^{(4)}$ to have an eigenvalue zero.

<u>Order $N = 1$</u>

From (4.82)–(4.84) we obtain

$$\hat{C}^{(4;N=1)}_{\vec{p}\vec{q}}(t,t_c) = -\langle 0|\hat{\Phi}^{\dagger}_{\vec{p}}(t_c)\, e^{-\hat{H}_0(t-t_c)} \int_{t_c}^{t} dt_1 \hat{H}_I(t_1)\, \hat{\Phi}_{\vec{q}}(t_c)|0\rangle\,. \tag{4.90}$$

Upon inserting the complete set $|n\nu\rangle$ on both sides of $\hat{H}_I(t_1)$ and using (4.80) the t_1 integral over exponentials can be carried out explicitly. The result is

$$\begin{aligned}\hat{C}^{(4;N=1)}_{\vec{p}\vec{q}}(t,t_c) \;&=\; -\sum_{n\nu}\sum_{m\mu} \psi^{(0)}_{n\nu}(\vec{p})\psi^{(0)*}_{m\mu}(\vec{q}) \\ &\quad \langle n\nu|\hat{H}_I|m\mu\rangle c^{(0)*}_{n\nu} c^{(0)}_{m\mu} \exp\left[-\frac{W^{(0)}_n + W^{(0)}_m}{2}(t-t_c)\right] \\ &\quad \left\{(t-t_c)\delta_{nm} + \frac{\sinh\left[\frac{W^{(0)}_n - W^{(0)}_m}{2}(t-t_c)\right]}{\frac{W^{(0)}_n - W^{(0)}_m}{2}}(1-\delta_{nm})\right\}.\end{aligned} \tag{4.91}$$

Without loss of generality the normalization constants $c^{(0)}_{n\nu}$ may be chosen real and positive, with the phase factors being absorbed into $\psi^{(0)}_{n\nu}(\vec{p})$, as is evident from (4.86). A glance at (4.87) shows that the two normalization factors and the exponential in (4.91) may be removed by multiplying the correlation matrix $\hat{C}^{(4;N=1)}$ from both sides with the inverse square root of $\hat{C}^{(4;N=0)}$. Hence the matrix elements of

$$\hat{\mathfrak{C}}^{(4;N=1)}(t,t_c) = \hat{C}^{(4;N=0)}(t,t_c)^{-1/2}\; \hat{C}^{(4;N=1)}(t,t_c)\, \hat{C}^{(4;N=0)}(t,t_c)^{-1/2} \tag{4.92}$$

in the basis $\psi^{(0)}_{n\nu}(\vec{p})$ are products of $\langle n\nu|\hat{H}_I|m\mu\rangle$ and the expression inside $\{\cdots\}$ of (4.91). The t derivative of the latter is equal to one at $t = t_c$. Thus we have

$$\left[\frac{\partial\hat{\mathfrak{C}}^{(4;N=1)}_{\vec{p}\vec{q}}(t,t_c)}{\partial t}\right]_{t=t_c} = -\sum_{n\nu}\sum_{m\mu} \psi^{(0)}_{n\nu}(\vec{p})\langle n\nu|\hat{H}_I|m\mu\rangle\psi^{(0)*}_{m\mu}(\vec{q})\,. \tag{4.93}$$

Using (4.88) and (4.89), this translates into an explicit equation for $\hat{H}_I$

independent of the basis

$$\hat{H}_I = -\left[\frac{\partial \hat{\mathfrak{C}}^{(4;N=1)}_{\vec{p}\vec{q}}(t,t_c)}{\partial t}\right]_{t=t_c}, \tag{4.94}$$

which is valid to order $N = 1$. Finally, we may replace $\hat{C}^{(4;N=1)}$ in (4.92) with the full correlation matrix $\hat{C}^{(4)}$. The corresponding expression (4.94) for $\hat{H}_I$ will still be valid up to order $N = 1$ in perturbation theory. Thus, summarizing, define the effective correlator

$$\hat{\mathfrak{C}}^{(4)}(t,t_c) = \hat{C}^{(4;N=0)}(t,t_c)^{-1/2}\,\hat{C}^{(4)}(t,t_c)\,\hat{C}^{(4;N=0)}(t,t_c)^{-1/2} \tag{4.95}$$

understood as a matrix product, then the meson-meson interaction $\hat{H}_I$ satisfies

$$\hat{\mathfrak{C}}^{(4)}(t,t_c) = e^{-\hat{H}_I(t-t_c)} \tag{4.96}$$

up to (at least) order $N = 1$ in perturbation theory.

The utility of these results in the framework of a lattice simulation lies in the analogy which can be drawn between $\hat{C}^{(4;N=0)}$ and the free correlator $\overline{C}^{(4)}$, and between $\hat{C}^{(4)}$ and the full correlator $C^{(4)}$. The analogue of (4.96) may then be considered as the definition of an effective interaction.

Order $N = 2$

A calculation similar to the above reveals that (4.96) is valid to order $N = 2$. However, for $N > 2$ the situation is complicated by the presence of disconnected diagrams, see Fig. 4.8. The diagramatic series of connected insertions of $\hat{H}_I$ still gives rise to the exponential form (4.96). We conclude that, within the region of validity of (4.96), the operator

$$\hat{H}_I = -\frac{\partial \ln \hat{\mathfrak{C}}^{(4)}(t,t_c)}{\partial t} \tag{4.97}$$

is independent of t and t_c.

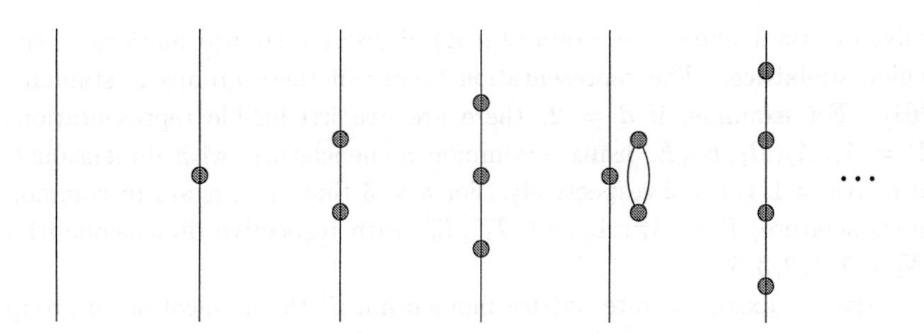

Fig. 4.8 Low-order diagrams for the elementary meson field correlator. The circles represent $\hat{H}_I$.

4.3.2 *Effective interaction for composite operators*

The above result is now easily transcribed to a system of composite hadrons. In this case the effective correlator (4.95) becomes

$$\begin{aligned}
\mathfrak{C}^{(4)}(t,t_0) &= \overline{C}^{(4)}(t,t_0)^{-1/2}\, C^{(4)}(t,t_0)\, \overline{C}^{(4)}(t,t_0)^{-1/2} \\
&= \overline{C}^{(4)}(t,t_0)^{-1/2}\, C_I^{(4)}(t,t_0)\, \overline{C}^{(4)}(t,t_0)^{-1/2} + \mathbf{1} \qquad (4.98)\\
&\stackrel{\text{def}}{=} \mathfrak{C}_I^{(4)}(t,t_0) + \mathbf{1}\,,
\end{aligned}$$

where (4.54) has been used. This then defines the effective residual interaction

$$\mathcal{H}_I \stackrel{\text{def}}{=} -\lim_{t\to\infty} \frac{\partial \ln \mathfrak{C}^{(4)}(t,t_0)}{\partial t}\,. \qquad (4.99)$$

In the lattice simulation the eigenvalues of $\mathfrak{C}_I^{(4)}(t,t_0)$ at asymptotic times (within lattice limits) select the low-energy part of the excitation spectrum of the two-body system. This will help to suppress intrinsic excitations of the composite particles and thus yields an effective residual interaction between the mesons (or hadrons) in their respective ground states. How well the excitation of intrinsic degrees of freedom is suppressed depends on the lattice size and on the design of the meson operators.

4.3.3 *Lattice symmetries*

Lattice symmetries should be utilized to reduce the size of the correlation matrices $C^{(2)}$ and $C^{(4)}$. For an $L^d \times T$ lattice the action is (usually) invariant under the group $O(d,\mathbb{Z})$ of discrete transformations of the cubic sublattice. The representation theory of these groups is standard [61]. For example, if $d = 2$, there are five irreducible representations $\Gamma = A_1, A_2, B_1, B_2, E$, using a common nomenclature, with dimensionalities $N_\Gamma = 1,1,1,1,2$ respectively. For $d = 3$ there are, again in common nomenclature, $\Gamma = A_1^\pm, A_2^\pm, E^+, T_1^\pm, T_2^\pm$ with respective dimensionalities $N_\Gamma = 1,1,2,3,3$.

Given a fixed, discrete, lattice momentum $\vec{p}\,'$ the application of group transformations $g \in O(d,\mathbb{Z})$ generates a set of momenta $\vec{p} = \mathcal{O}_g\,\vec{p}\,'$ that all have the same length $p = |\vec{p}|$. These transformations define a representation of $O(d,\mathbb{Z})$ with basis vectors $|\vec{p}>$ which is in general reducible. Let

$$|\vec{p}\rangle = \sum_\Gamma \sum_\epsilon |(\Gamma,p)\epsilon\rangle\langle(\Gamma,p)\epsilon|\vec{p}\rangle\,, \tag{4.100}$$

where $|(\Gamma,p)\epsilon\rangle$ denote a set of basis vectors, $\epsilon = 1\cdots$, of the subspace that belongs to Γ. Those can be constructed with group theoretical techniques [61]. A simple example, which will suffice for our present purposes, is given by choosing $\vec{p} = (p,0)$, assuming $d = 2$ and $pL/2\pi \in \mathbb{N}$, and the trivial one-dimensional representation $\Gamma = A_1$, thus

$$|(A_1,p)1\rangle = \frac{1}{2}\left(|(+p,0)\rangle + |(0,+p)\rangle + |(-p,0)\rangle + |(0,-p)\rangle\right)\,. \tag{4.101}$$

Since both the full and the free correlators $C^{(4)}$ and $\overline{C}^{(4)}$, respectively, commute with all group operations $g \in O(d,\mathbb{Z})$ there exist reduced matrices, say $C^{(4;\Gamma)}$ and $\overline{C}^{(4;\Gamma)}$, within each irreducible representation Γ such that (Schur's lemma)

$$\langle(\Gamma,p)\epsilon|C^{(4)}(t,t_0)|(\Gamma',q)\epsilon'\rangle = \delta_{\Gamma\Gamma'}\delta_{\epsilon\epsilon'}C^{(4;\Gamma)}_{p\,q}(t,t_0) \tag{4.102}$$

and similarly for $\overline{C}^{(4;\Gamma)}$. Above we identify $C_{\vec{p}\,\vec{q}}$ with $\langle\vec{p}|C|\vec{q}\rangle$.

Similar group theoretical considerations are of course also applicable on a coordinate space $|\vec{r}\rangle$ lattice.

An example from the QED_{2+1} model is shown in Fig. 4.9. The set $\{\overline{W}\}$ of energy levels, with labels $n = 1\cdots 6$, belongs to the free system. Those were extracted from the asymptotic time behavior of $\overline{C}^{(4)}_{\vec{p}\,\vec{q}}(t,t_0)$, see (4.59), and are degenerate (in general) for momenta of the same magnitude but different directions. The degeneracy is lifted by the residual interaction, the set $\{W\}$ being extracted from $C_{pq}^{(4;\Gamma)}(t, t_0)$. Since the angular momentum content of the various representations $\Gamma = A_1, A_2, B_1, B_2$ is different (A_1 mostly contains $\ell = 0$, B_1 and B_2 contain $\ell = 2$, *etc.*) the hyperfine splitting can be attributed to a spin-orbit like force.

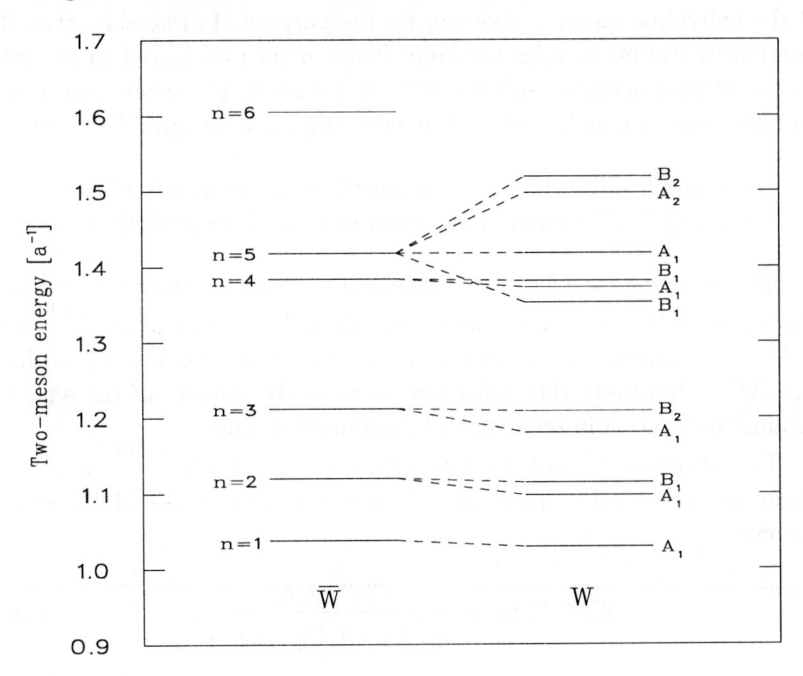

Fig. 4.9 Hyperfine splitting due to the residual interaction for the QED_{2+1} model [62]. The labels for $\{W\}$ refer to irreducible representations of the lattice symmetry group $O(2, \mathbb{Z})$.

4.3.4 *Truncated momentum basis*

It is obvious from (4.59) and (4.60) that the reduced matrix elements of the free correlator have the form

$$\overline{C}_{pq}^{(4;\Gamma)}(t, t_0) = \delta_{pq} \left| \bar{c}_p^{(\Gamma)}(t, t_0) \right|^2 . \tag{4.103}$$

The functions $\bar{c}_p^{(\Gamma)}(t, t_0)$ are linear combinations of 2-point correlator elements $C_{\vec{p}\vec{p}'}^{(2)}$ with momenta $\vec{p}$ and $\vec{p}'$ that differ only by their directions.

The eigenvalues of $\overline{C}^{(4;\Gamma)}(t, t_0)$ describe the time evolution of the eigenstates of a free meson-meson Hamiltonian, say $\mathcal{H}_0$. Since $\mathcal{H}_0$ relates to a composite system, it is useful to think of its degrees of freedom in terms of the relative motion of the two mesons (as clusters) and the intrinsic motion of the individual mesons. Assume, for the purpose of discussion, that the internal excitation energies are large (hard modes) compared to the relative excitation energies (soft modes). In principle, the separation of soft and hard modes may be enforced by choosing a big enough $L^d \times T$ lattice:

- Large L will yield a dense momentum spectrum $p \propto L^{-1}$.
- Large T will permit suppression of intrinsic excitations, at large t.

In practice there are, of course, numerical limitations. However, one may check how well those conditions are satisfied. We assume that $\overline{C}^{(4;\Gamma)}$ and $C^{(4;\Gamma)}$ are available on a truncated set of lattice momenta and both have size $N^{(\Gamma)}$. Naturally this will allow us to study features of the effective residual interaction in the long and intermediate range.

The advantage of working in momentum space is that $\overline{C}^{(4;\Gamma)}$ is already diagonal, see (4.103). Thus the effective correlator (4.95),(4.98) simply becomes

$$\mathfrak{C}^{(4;\Gamma)}_{pq}(t, t_0) = \frac{C^{(4;\Gamma)}_{pq}(t, t_0)}{\left|\bar{c}^{(\Gamma)}_p(t, t_0)\right| \left|\bar{c}^{(\Gamma)}_q(t, t_0)\right|} . \tag{4.104}$$

We envision numerical diagonalization, say

$$\mathfrak{C}^{(4;\Gamma)}_{pq}(t, t_0) = \sum_{n=1}^{N^{(\Gamma)}} v^{(\Gamma)}_n(p) \lambda^{(\Gamma)}_n(t, t_0)\, v^{(\Gamma)}_n(q) . \tag{4.105}$$

The eigenvectors $v^{(\Gamma)}_n$ are time independent [27], whereas the eigenvalues behave exponentially for asymptotic ($\simeq$) times,

$$\lambda^{(\Gamma)}_n(t, t_0) \simeq a^{(\Gamma)}_n \exp[-w^{(\Gamma)}_n (t - t_0)] , \tag{4.106}$$

or, if periodic boundary conditions across the time extent of the lattice are imposed,

$$\lambda^{(\Gamma)}_n(t, t_0) \simeq c^{(\Gamma)}_n \cosh[w^{(\Gamma)}_n (t - t_c)] . \tag{4.107}$$

Now, extraction of the effective residual interaction as defined in (4.99) becomes a trivial matter. Provided that the $v_n^{(\Gamma)}$ are orthonormal we have

$$\langle p|\mathcal{H}_I^{(\Gamma)}|q\rangle = \sum_{n=1}^{N^{(\Gamma)}} v_n^{(\Gamma)}(p)\, w_n^{(\Gamma)}\, v_n^{(\Gamma)}(q)\,. \tag{4.108}$$

These are the desired matrix elements of the effective residual interaction in the truncated momentum basis. They may be used in a variety of ways. For example, in the basis $|\vec{p}\rangle$ of lattice momenta we have

$$\langle\vec{p}|\mathcal{H}_I|\vec{q}\rangle = \sum_{\Gamma}\sum_{\epsilon}\langle\vec{p}|(\Gamma,p)\epsilon\rangle\langle p|\mathcal{H}_I^{(\Gamma)}|q\rangle\langle(\Gamma,q)\epsilon|\vec{q}\rangle\,. \tag{4.109}$$

This would require computation of the reduced matrix elements for all irreducible representations Γ of the cubic lattice symmetry group. However, depending on the system studied, it may well be that A_1 dominates the series.

The coordinate space matrix elements of the effective residual interaction are obtained from (4.109) by (discrete lattice) Fourier transformation

$$\langle\vec{r}|\mathcal{H}_I|\vec{s}\rangle = L^{-d}\sum_{\vec{p}}\sum_{\vec{q}} e^{i\vec{p}\cdot(\vec{r}-\vec{s})}e^{-i\vec{q}\cdot(\vec{r}+\vec{s})}\langle\vec{p}-\vec{q}|\mathcal{H}_I|\vec{p}+\vec{q}\rangle\,. \tag{4.110}$$

It is useful to write the matrix element of $\mathcal{H}_I$ as a sum of $\langle-\vec{q}|\mathcal{H}_I|+\vec{q}\rangle$ and a remainder. A reference system where the relative momenta before and after a scattering event are $\pm\vec{q}$, respectively, is known as the Breit frame [63]. Then, the sum over $\vec{p}$ in (4.110) gives rise to $\delta_{\vec{r}\,\vec{s}}$ and thus to a local potential $\mathcal{V}(\vec{r})$ for the Breit-frame contribution, and a genuinely nonlocal potential $\mathcal{W}(\vec{r},\vec{s})$ for the remainder

$$\langle\vec{r}|\mathcal{H}_I|\vec{s}\rangle = \delta_{\vec{r}\vec{s}}\,\mathcal{V}(\vec{r}) + \mathcal{W}(\vec{r},\vec{s})\,. \tag{4.111}$$

The potentials are

$$\mathcal{V}(\vec{r}) = \sum_{\vec{q}} e^{-i2\vec{q}\cdot\vec{r}}\langle-\vec{q}|\mathcal{H}_I|+\vec{q}\rangle \tag{4.112}$$

$$\begin{aligned}\mathcal{W}(\vec{r},\vec{s}) &= L^{-d}\sum_{\vec{p}}\sum_{\vec{q}} e^{i\vec{p}\cdot(\vec{r}-\vec{s})}e^{-i\vec{q}\cdot(\vec{r}+\vec{s})} \\ &\qquad (\langle\vec{p}-\vec{q}|\mathcal{H}_I|\vec{p}+\vec{q}\rangle - \langle-\vec{q}|\mathcal{H}_I|+\vec{q}\rangle)\,.\end{aligned} \tag{4.113}$$

Results from an actual simulation are displayed in Fig. 4.10. The (lattice) Fourier transform, see (4.110), of (4.109) is shown. An angular momentum projection was employed. Most of the $\ell = 0$ partial wave is contained in the A_1 sector. The corresponding scattering phase shifts, obtained from a Schrödinger equation, are shown in Fig. 4.11.

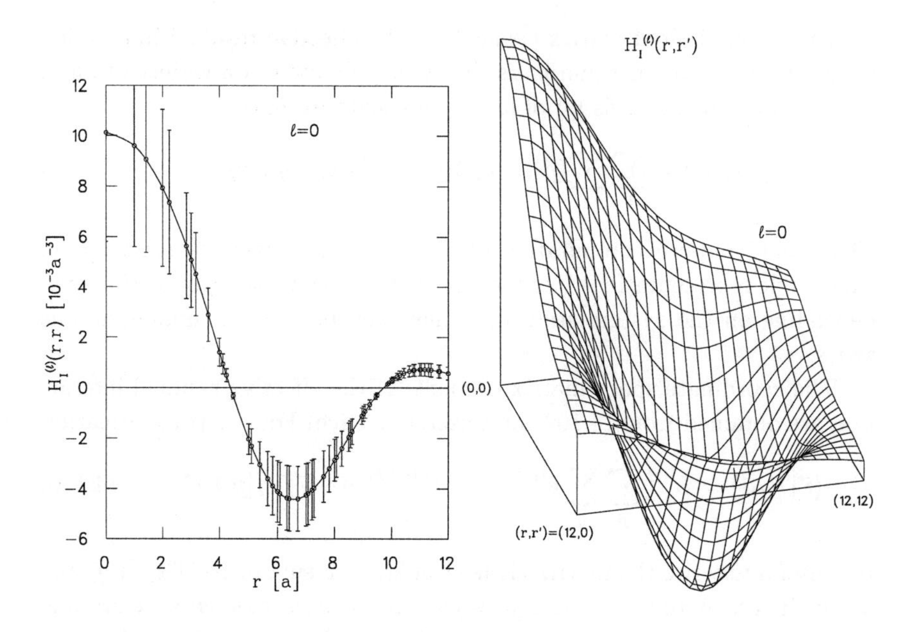

Fig. 4.10 Example of a meson-meson potential derived within the momentum-space approach. Local and nonlocal potentials are shown. The model is QED_{2+1} in 2+1 dimensions [62].

4.3.5 *Adiabatic approximation*

Probing the lattice with a set of coordinate space operators of the type

$$\phi_{\vec{x}}(t) = \sum_{\vec{p}} e^{-i\vec{p}\cdot\vec{x}} \phi_{\vec{p}}(t) = \bar{\chi}_d(\vec{x}, t)\, \chi_u(\vec{x}, t)\,, \tag{4.114}$$

where we have used (4.28), gives us an alternative window on the effective residual interaction. The Fourier transform of the two-meson operator

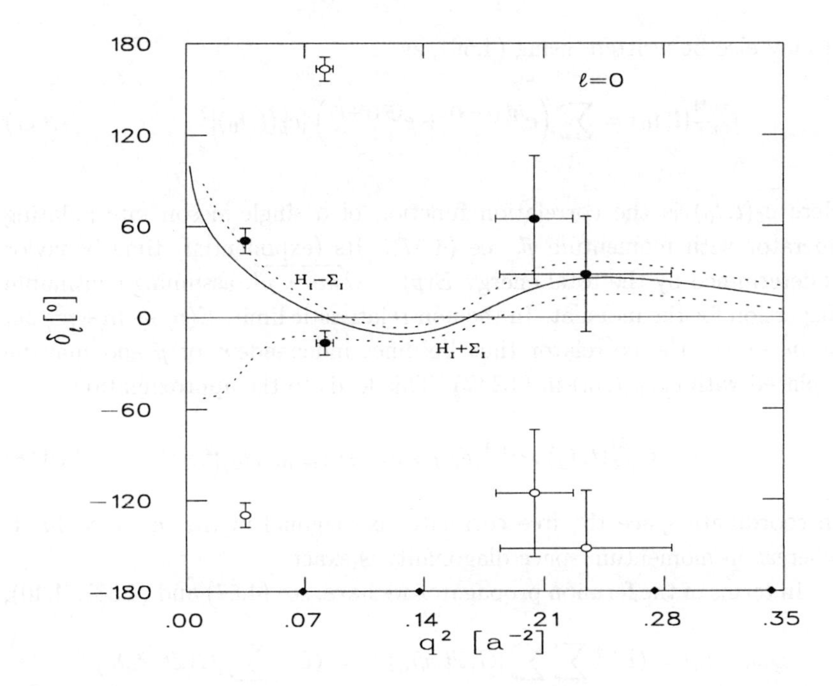

Fig. 4.11 Scattering phase shifts, for the s-wave, obtained from the effective interaction depicted in Fig. 4.10. The four data points (filled circles) were computed in Ref. [64], using Lüscher's method. The same set of phase shifts are also shown modulo 180° (open circles). The diamonds on the abscissa correspond to the values of $p^2 = (2\pi/L)^2k^2$ for an $L = 24$ lattice.

(4.30) becomes

$$\Phi_{\vec{r}}(t) = \sum_{\vec{p}} e^{-i\vec{p}\cdot\vec{r}}\Phi_{\vec{p}}(t) = \sum_{\vec{x}}\sum_{\vec{y}} \delta_{\vec{r},\vec{x}-\vec{y}}\,\phi_{\vec{y}}(t)\,\phi_{\vec{x}}(t)\,. \tag{4.115}$$

It corresponds to two composite mesons with relative distance $\vec{r}$ and total momentum zero.

Following Secs. 4.2.3 and 4.2.4 we consider the free meson-meson correlator

$$\overline{C}^{(4)}_{\vec{r}\,\vec{s}}(t,t_0) = \sum_{\vec{p}}\sum_{\vec{q}} e^{i\vec{p}\cdot\vec{r}}e^{-i\vec{q}\cdot\vec{s}}\;\overline{C}^{(4)}_{\vec{p}\,\vec{q}}(t,t_0)\,. \tag{4.116}$$

It may also be written, using (4.59), as

$$\overline{C}^{(4)}_{\vec{r}\,\vec{s}}(t,t_0) = \sum_{\vec{p}} \left(e^{i\vec{p}\cdot(\vec{r}-\vec{s})} + e^{i\vec{p}\cdot(\vec{r}+\vec{s})} \right) |c_{\vec{p}}(t,t_0)|^2 \,. \tag{4.117}$$

Here $c_{\vec{p}}(t,t_0)$ is the correlation function of a single-meson interpolating operator with momentum $\vec{p}$, see (4.57). Its (exponential) time behavior is determined by the total energy $E(p) = \sqrt{m^2+p^2}$, assuming continuum dispersion for the moment. In the non-relativistic limit, $E(p) \to m + p^2/2m$ as $m \to \infty$, the correlator thus becomes independent of $\vec{p}$ and may be replaced with $c_{\vec{p}=0}(t,t_0)$ in (4.117). This leads to the approximation

$$\overline{C}^{(4)}_{\vec{r}\,\vec{s}}(t,t_0) \approx L^2(\delta_{\vec{r},\vec{s}} + \delta_{\vec{r},-\vec{s}})|c_{\vec{p}=0}(t,t_0)|^2 \,. \tag{4.118}$$

In coordinate space the free correlator is diagonal in the $m \to \infty$ limit, whereas in momentum space diagonality is exact.

In terms of the fermion propagator we have, see (4.57) and (4.38),(4.40),

$$c_{\vec{p}=0}(t,t_0) = \langle L^{-4} \sum_{\vec{x}} \sum_{\vec{y}} |G(\vec{x}t,\vec{y}t_0)|^2 \rangle = \langle L^{-2} \sum_{\vec{x}} |G(\vec{x}t,\vec{x}_0 t_0)|^2 \rangle \,. \tag{4.119}$$

This is just the time correlation function for a single meson at rest.

Using lattice symmetry, see Secs. 4.3.3 and 4.3.4, the reduced correlator matrix has the form

$$\overline{C}^{(4;\Gamma)}_{r\,s}(t,t_0) \approx \delta_{r,s} |\overline{c}^{(\Gamma)}(t,t_0)|^2 \,, \tag{4.120}$$

where the asymptotic time behavior is given by

$$\overline{c}^{(\Gamma)}(t,t_0) \simeq \overline{a}^{(\Gamma)}\, e^{-m(t-t_0)} \,, \tag{4.121}$$

at least for $\Gamma = A_1$. The full, interacting, correlator matrix is built from (4.115)

$$C^{(4)}_{\vec{r}\,\vec{s}}(t,t_0) = \langle \Phi^{\dagger}_{\vec{r}}(t)\, \Phi_{\vec{s}}(t_0) \rangle - \langle \Phi^{\dagger}_{\vec{r}}(t) \rangle \langle \Phi_{\vec{s}}(t_0) \rangle \,. \tag{4.122}$$

Again, the separable term vanishes in our example (flavor assignment). Expressing $C^{(4)}$, via Wick's theorem, in terms of the fermion propagator and the diagrammatic classification proceeds just like in Secs. 4.2.3 and

4.2.4. Alternatively, we may simply Fourier transform (4.42),

$$C^{(4)}_{\vec{r}\,\vec{s}}(t,t_0) = L^{-4}\sum_{\vec{x}_1}\sum_{\vec{x}_2}\sum_{\vec{y}_1}\sum_{\vec{y}_2}\delta_{\vec{r},\vec{x}_1-\vec{x}_2}\delta_{\vec{s},\vec{y}_2-\vec{y}_1}$$
$$\langle\big(\,G^*(\vec{x}_2t,\vec{y}_2t_0)\,G(\vec{x}_2t,\vec{y}_2t_0)\,G^*(\vec{x}_1t,\vec{y}_1t_0)\,G(\vec{x}_1t,\vec{y}_1t_0)$$
$$+\,G^*(\vec{x}_1t,\vec{y}_2t_0)\,G(\vec{x}_1t,\vec{y}_2t_0)\,G^*(\vec{x}_2t,\vec{y}_1t_0)\,G(\vec{x}_2t,\vec{y}_1t_0)$$
$$-\,G^*(\vec{x}_2t,\vec{y}_1t_0)\,G(\vec{x}_2t,\vec{y}_2t_0)\,G^*(\vec{x}_1t,\vec{y}_2t_0)\,G(\vec{x}_1t,\vec{y}_1t_0)$$
$$-\,G(\vec{x}_2t,\vec{y}_1t_0)\,G^*(\vec{x}_2t,\vec{y}_2t_0)\,G(\vec{x}_1t,\vec{y}_2t_0)\,G^*(\vec{x}_1t,\vec{y}_1t_0)\big)\rangle\,. \tag{4.123}$$

In keeping with the $m\to\infty$ limit it is reasonable to assume that the relative distance between the (heavy) mesons will not change much during the propagation from t_0 to t. After all, the meson propagation is described by (4.119). This leads us to neglect the off-diagonal elements of $C^{(4)}$. More precisely, since it costs no energy to change the system according to a transformation that belongs to the lattice symmetry group $O(d,\mathbb{Z})$, the correct approximation is

$$C^{(4;\Gamma)}_{r\,s}(t,t_0)\approx\delta_{r\,s}\,C^{(4;\Gamma)}_{r\,r}(t,t_0)\,. \tag{4.124}$$

Now, the eigenvalues of $C^{(4;\Gamma)}(t,t_0)$ are just the diagonal elements. Those behave exponentially for asymptotic times.

$$C^{(4;\Gamma)}_{r\,r}(t,t_0)\simeq a^{(\Gamma)}_r\,e^{-W^{(\Gamma)}_r(t-t_0)}\,. \tag{4.125}$$

The effective correlator (4.104) is also diagonal in this approximation

$$\mathfrak{C}^{(4;\Gamma)}_{r\,s}(t,t_0)\approx\delta_{r\,s}\,\mathfrak{C}^{(4;\Gamma)}_{r\,r}(t,t_0)\,, \tag{4.126}$$

with

$$\mathfrak{C}^{(4;\Gamma)}_{r\,r}(t,t_0)\simeq\frac{a^{(\Gamma)}_r}{|\bar{a}^{(\Gamma)}|^2}e^{-[W^{(\Gamma)}_r-2m](t-t_0)}\,. \tag{4.127}$$

According to our definition (4.99) the effective residual interaction is

$$\langle r|\mathcal{H}^{(\Gamma)}_I|s\rangle\approx\delta_{r\,s}\mathcal{V}^{(\Gamma)}(r)\,,\quad\text{with}\quad\mathcal{V}^{(\Gamma)}(r)=W^{(\Gamma)}_r-2m\,. \tag{4.128}$$

The above line of arguments is somewhat similar in spirit to the Born-Oppenheimer approximation often used with systems of two atoms or molecules [65]. The time scale for the dynamics of the interaction is much faster than the time scale for the dynamics of the motion for the atoms. Borrowing a thermodynamics term, we speak of an adiabatic approximation. In our case we simply compute the total energy of the two-meson system at fixed (static) relative distance r and interpret the difference to the ground state energy as a potential [66].

An example from numerical work on the QED_{2+1} model [67] is shown in Fig. 4.12. The left part displays the static potential $\mathcal{V}^{(\Gamma)}$ according to (4.128). The distance $\vec{r} = 0$ is forbidden by the Pauli exclusion principle. This is the reason for the increase (repulsive core) of the potential as $r \to 0$.

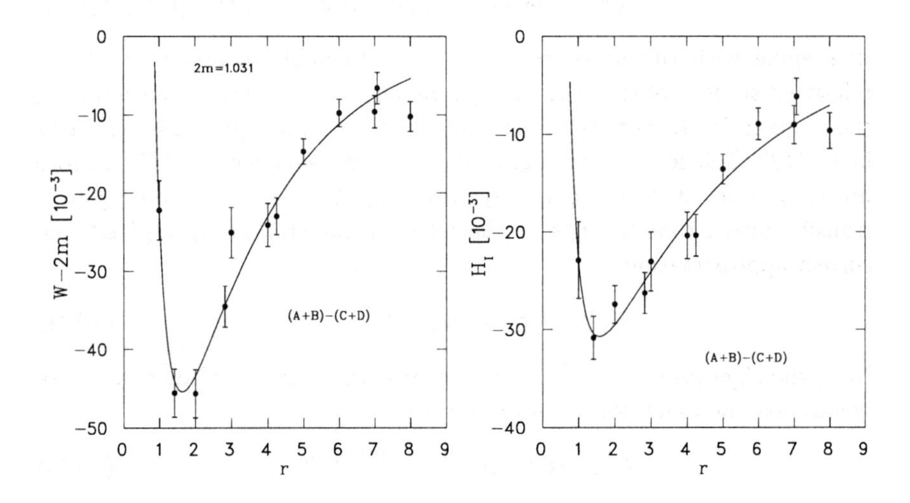

Fig. 4.12 Static meson-meson potential for the QED_{2+1} model according to (4.128), on the left. On the right a somewhat different approximation according to (4.129) serves to check the perturbative definition of $\mathcal{H}_I$.

The right part of Fig. 4.12 shows a potential computed slightly differently, for the sake of comparison. The perturbative version of (4.127) leads to

$$\mathcal{V}^{(\Gamma)}(r) \approx \frac{1}{t - t_0} \left[1 - \frac{\mathfrak{C}_{r\,r}^{(4;\Gamma)}(t, t_0)}{a_r^{(\Gamma)} |\bar{a}^{(\Gamma)}|^{-2}} \right] . \tag{4.129}$$

In this way the r-dependence of $a_r^{(\Gamma)}$ enters the analysis. The results in Fig. 4.12 are similar. They shed some light on the validity of the perturbative definition of $\mathcal{H}_I$ put forward in Sec. 4.3.1.

Of course, only the local part of the effective residual interaction is accessible in this way. Note, however, that a local potential

$$\mathcal{V}(\vec{r}) = \sum_{\Gamma} \sum_{\epsilon} \langle \vec{r} | (\Gamma, r)\epsilon \rangle \mathcal{V}^{(\Gamma)}(r) \langle (\Gamma, r)\epsilon | \vec{r} \rangle \,, \tag{4.130}$$

compared to (4.109), may not be the same as the approximate local potential of (4.112) even if the untruncated correlation matrices are used. The reason is that diagonal elements of the non-local part of the effective residual interaction may contribute to the adiabatic local potential.

The adiabatic approximation hinges on the assumption of heavy partners ($m \to \infty$). This should be reasonable if at least one of the quarks in each hadron is heavy, say an s-quark. However, in the chiral limit ($m_\pi \to 0$), using Wilson fermions, the adiabatic approximation is expected to fail if pseudoscalar mesons are involved. In this case momentum space methods like in Sec. 4.3.4 are preferable. Finally, the quality of the adiabatic approximation can, of course, be tested numerically by computing off-diagonal elements of the correlation matrices.

4.3.6 *Analysis on a periodic lattice*

Periodic boundary conditions across the space extent of the lattice are a common choice. The potentials $\mathcal{V}(\vec{r})$ and $\mathcal{W}(\vec{r}, \vec{s})$, see (4.112) and (4.113), extracted from the lattice simulation, in one or the other way, will reflect those conditions. In particular, the maximal usable relative distance r is $L/2$.

Suppose it is desired to make a fit to the lattice local potential with a class of functions $V^{(\alpha)}(\vec{r})$ depending on a set of parameters α. An example is

$$V^{(\alpha)}(\vec{r}) = \alpha_1 \frac{1 - \alpha_2 r^{\alpha_5}}{1 + \alpha_3 r^{\alpha_5 + 1} e^{\alpha_4 r}} + \alpha_0 \,. \tag{4.131}$$

This class (see Fig. 4.13) has enough flexibility to match heuristic features of the hadronic interaction, like short-range repulsion or attraction as the case may be, and a long range Yukawa form

$$V^{(\alpha)}(\vec{r}) \to -\frac{\alpha_1 \alpha_2}{\alpha_3} \frac{e^{-\alpha_4 r}}{r} + \alpha_0 \quad \text{as} \quad r \to \infty \,. \tag{4.132}$$

The periodic extension, defined as

$$V_L^{(\alpha)}(\vec{r}) = \sum_{\vec{n}} V^{(\alpha)}(\vec{r} - \vec{n}L)\,, \tag{4.133}$$

where $\vec{n}$ are vectors with d whole number components, can then be used to fit the lattice potentials with

$$\chi^2(\alpha) = \frac{1}{N}\sum_{\vec{r}} |\mathcal{V}(\vec{r}) - V_L^{(\alpha)}(\vec{r})|^2 \frac{1}{\sigma^2(\vec{r})}\,. \tag{4.134}$$

Here the sum runs over the set of N lattice distances $\vec{r}$ for which $\mathcal{V}(\vec{r})$ has been computed.

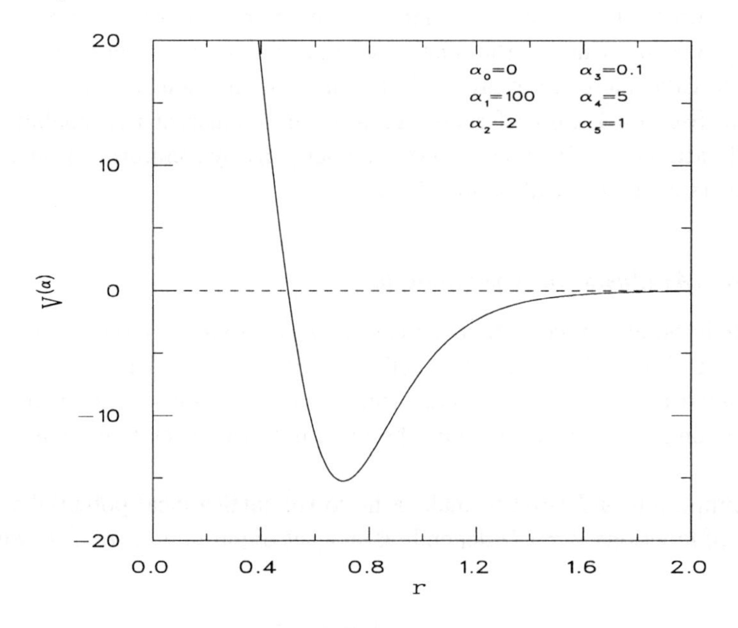

Fig. 4.13 Example of $V^{(\alpha)}$ as defined in (4.131). The parameters α are arbitrarily chosen, for the purpose of illustration.

The idea behind using a periodic extension, like (4.133), is to account for the fact that the lattice hadrons are interacting with their replicas in adjacent copies of the periodic lattice. With χ^2 minimized, we would take $V^{(\alpha)}(\vec{r})$ and $W^{(\beta)}(\vec{r}, \vec{s})$ obtained from a similar fit, as the results of the simulation.

4.4 Current State of QCD in 3+1 Dimensions

At the time of this writing, lattice work on the subject of hadronic interaction is in an exploratory phase. Nevertheless, it is useful to convey what has been done through some selected examples. We will not include in this section results other than QCD_{3+1}, for example $O(4)$ symmetric field theories and four-fermion models [68, 69, 70], models exploring static quark geometry[§] [71, 72], studies with an SU(2) gauge group [73], and work using the hamiltonian lattice formulation [74].

4.4.1 *Scattering lengths for π and N systems*

To date the most elaborate application within QCD_{3+1} has been via Lüscher's formula (4.9) by the CP-PACS collaboration aiming at scattering lengths for the π–π, π–N, K–N, $\bar{\text{K}}$–N, and N–N systems — see, for example, the Refs. [19, 18, 20, 21, 22, 23, 24, 25, 26]. The collaboration has applied considerable resources to this problem, recently using lattices as large as $L^3 \times T = 32^3 \times 60$ [24]. However, for most of their earlier explorations [21] lattices typically of size $12^3 \times 20$ with lattice spacings in the range of $a \approx 0.15 - 0.30$ fm were used.

Extracting the s-wave scattering length a_0 from Lüscher's formula

$$W_{h_1 h_2} - (m_{h_1} + m_{h_2}) = -\frac{2\pi(m_{h_1} + m_{h_2})a_0}{m_{h_1} m_{h_2} L^3}[1 + c_1\frac{a_0}{L} + c_2(\frac{a_0}{L})^2 + o(L^{-3})]\,, \tag{4.135}$$

where h_1 and h_2 denote hadrons 1 and 2, see (4.9), requires a standard mass calculation on the lattice. One therefore needs to compute correlation functions of suitable operators, say $\phi_{h_1}(t)$ and $\phi_{h_2}(t)$. For illustration, some examples using Wilson fermions are

$$\phi_{\pi^+}(t) = -\sum_{\vec{x}} \bar{\psi}_{dA}(\vec{x},t)\gamma_5\psi_{uA}(\vec{x},t) \tag{4.136}$$

$$\phi_{\pi^0}(t) = \frac{1}{\sqrt{2}}\sum_{\vec{x}} [\bar{\psi}_{uA}(\vec{x},t)\gamma_5\psi_{uA}(\vec{x},t) - \bar{\psi}_{dA}(\vec{x},t)\gamma_5\psi_{dA}(\vec{x},t)] \tag{4.137}$$

$$\phi_{K^+}(t) = \sum_{\vec{x}} \bar{\psi}_{sA}(\vec{x},t)\gamma_5\psi_{uA}(\vec{x},t)\,, \quad etc., \tag{4.138}$$

where $A = 1,2,3$ represents SU(3) color. Only zero-momentum inter-

[§]Chapter 5 discusses this subject in detail.

polating fields are needed here. Kogut-Susskind, or staggered, fermions [49, 50, 51] were also used by the CP-PACS collaboration. The construction of operators with specific hadron quantum numbers is technically more complicated, however see [75]. The authors of Ref. [21] chose to construct correlation functions where sources and sinks between h_1 and h_2 are one time slice apart. For the one-hadron systems this means

$$\begin{aligned} C_{h_1}(t,0) &= \langle \phi_{h_1}(t)\phi^\dagger_{h_1}(0)\rangle - \langle \phi_{h_1}(t)\rangle\langle \phi^\dagger_{h_1}(0)\rangle \quad (4.139)\\ &\simeq \bar{a}_{h_1} e^{-m_{h_1} t} \\ C_{h_2}(t+1,1) &= \langle \phi_{h_2}(t+1)\phi^\dagger_{h_2}(1)\rangle - \langle \phi_{h_2}(t+1)\rangle\langle \phi^\dagger_{h_2}(1)\rangle \quad (4.140)\\ &\simeq \bar{a}_{h_2} e^{-m_{h_2} t}\,, \end{aligned}$$

where $\simeq$ indicates the asymptotic behavior. The two-hadron interpolating fields are constructed from linear combinations of products of one-hadron operators, respecting the one-time-slice offset. Denoting those by $\Phi_{h_1 h_2}(t, t+1)$, illustrational examples in the π–π system with isospin I are

$$\begin{aligned} \Phi^{(I=0,I_3=0)}_{\pi\pi}(t,t+1) &= \frac{1}{\sqrt{3}}\left[\phi_{\pi^+}(t)\phi_{\pi^-}(t+1)\right. \quad (4.141)\\ &\quad \left.-\phi_{\pi^0}(t)\phi_{\pi^0}(t+1) + \phi_{\pi^-}(t)\phi_{\pi^+}(t+1)\right] \\ \Phi^{(I=2,I_3=2)}_{\pi\pi}(t,t+1) &= \phi_{\pi^+}(t)\phi_{\pi^+}(t+1)\,. \quad (4.142) \end{aligned}$$

In the s-wave $I = 1$ is not allowed (Bose symmetry). The corresponding two-hadron system correlator is

$$\begin{aligned} &C_{h_1 h_2}(t+1,t,1,0) = \quad (4.143)\\ &\quad \langle \Phi_{h_1 h_2}(t,t+1)\Phi^\dagger_{h_1 h_2}(0,1)\rangle - \langle \Phi_{h_1 h_2}(t,t+1)\rangle\langle \Phi^\dagger_{h_1 h_2}(0,1)\rangle\,. \end{aligned}$$

The measurement of the s-wave scattering length a_0 via Lüscher's formula (4.135) requires the energy difference $W_{h_1 h_2} - (m_{h_1} + m_{h_2})$, which can be extracted from the asymptotic time behavior of the ratio of correlation functions

$$R(t) \stackrel{\text{def}}{=} \frac{C_{h_1 h_2}(t+1,t,1,0)}{C_{h_1}(t,0)C_{h_2}(t+1,1)} \simeq \frac{a_{h_1 h_2}}{\bar{a}_{h_1}\bar{a}_{h_2}} e^{-[W_{h_1 h_2} - (m_{h_1}+m_{h_2})]t}\,. \quad (4.144)$$

Baryon interpolating field operators are more complicated. For the proton (N^+) and neutron (N^0) standard choices for zero-momentum operators are

$$\begin{aligned}\phi_{N^+}(t) &= \sum_{\vec{x}} \epsilon_{ABC} \Big[\left(\bar{\psi}^T_{uA}(\vec{x},t) \tilde{C} \psi_{dB}(\vec{x},t) \right) \psi_{uB}(\vec{x},t) \\ &\qquad - \left(\bar{\psi}^T_{dA}(\vec{x},t) \tilde{C} \psi_{uB}(\vec{x},t) \right) \psi_{uB}(\vec{x},t) \Big] \quad (4.145) \\ \phi_{N^0}(t) &= \sum_{\vec{x}} \epsilon_{ABC} \Big[\left(\bar{\psi}^T_{uA}(\vec{x},t) \tilde{C} \psi_{dB}(\vec{x},t) \right) \psi_{dB}(\vec{x},t) \\ &\qquad - \left(\bar{\psi}^T_{dA}(\vec{x},t) \tilde{C} \psi_{uB}(\vec{x},t) \right) \psi_{dB}(\vec{x},t) \Big] , \quad (4.146)\end{aligned}$$

where $\tilde{C}$ means charge conjugation and A, B, C are color indices. Operators for the $^3S_1$ and $^1S_0$ channels of the N–N system are constructed in Ref. [21]

$$\begin{aligned}\Phi_{^3S_1}(t, t+1) &= \frac{1}{\sqrt{2}} \left(\phi_{N^+}(t) \phi_{N^0}(t+1) - \phi_{N^0}(t) \phi_{N^+}(t+1) \right) \quad (4.147) \\ \Phi_{^1S_0}(t, t+1) &= \phi_{N^+}(t) \phi_{N^+}(t+1) \, . \quad (4.148)\end{aligned}$$

Again, ratios like (4.144) allow the extraction of the desired mass shifts.

Computation of the above time correlation functions requires matrix elements of the quark propagator $G = D^{-1}$, where $D(x, y)$ denotes the Dirac, or fermion, matrix and $x = (\vec{x}, t)$, are space-time lattice sites. Since a complete solution of

$$\sum_{x''} D(x', x'') \, G(x'', x) = \mathbf{1} \, \delta_{x', x} \, , \quad (4.149)$$

where $\mathbf{1}$ means the unit matrix in color-Dirac space, is not feasible in practice, source techniques are typically used to obtain at least some columns of G. The CP-PACS collaboration has employed so called wall sources. These are defined by placing a point source of value $\mathbf{1}$ on each spatial site of a certain time slice t, specifically

$$\sum_{\vec{x}'' t''} D(\vec{x}' t', \vec{x}'' t'') \, G_t(\vec{x}'' t'') = \mathbf{1} \sum_{\vec{x}} \delta_{(\vec{x}' t'), (\vec{x} t)} = \mathbf{1} \, \delta_{t' t} \, . \quad (4.150)$$

Note that the uniform nature of the wall source implies that the solution vector G_t must be independent of $\vec{x}'$. Indeed, employing (4.149) it is obvious that

$$G_t(\vec{x}'' t'') = \sum_{\vec{x}} G(\vec{x}'' t'', \vec{x} t) \, . \quad (4.151)$$

Loosely speaking, since the spatial source is structureless quark propagation to x'' proceeds in 'the same way' regardless of the spatial point of origin. However, this is not a gauge invariant mechanism. Under a gauge transformation $D(x,y) = U^\dagger(x)\tilde{D}(x,y)U(y)$, *etc.*, the wall source becomes $\mathbf{1}U(\vec{x}'t')\,\delta_{t',t}$. Thus, in contrast to $G(\vec{x}''t'',\vec{x}t)$, the (numerical) solution $G_t(\vec{x}''t'')$ of (4.150) does not transform covariantly. Gauge dependent noise in hadronic correlation functions caused by the wall sources is specifically created at time slices where two wall sources or one wall source and one sink are placed. This can be seen from Fig. 4.14 which shows a diagramatic analysis of the correlators for the π–π, π–N, and N–N systems in terms of quark propagators and the corresponding wall sources and sinks. Wall

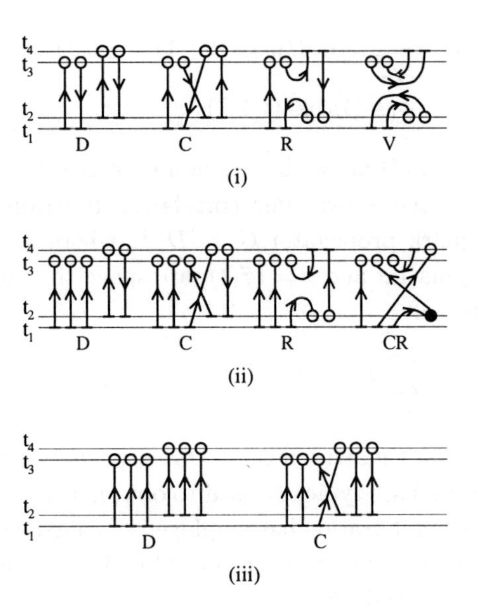

Fig. 4.14 Diagrams contributing to hadron-hadron 4-point functions, from Ref. [21]. Arrowed lines represent quark or antiquark propagators. Wall sources are depicted by short bars, and circles represent sinks for hadron operators. The diagrams in (i), (ii), and (iii) are for the π–π, π–N, and N–N systems, respectively. The decompositions (4.153) and (4.154), for example, refer to the labels D,C,R,V of diagrams (i).

sources are depicted as short lines and sinks as circles. The time slices refer to $t_1 = 0, t_2 = 1, t_3 = t, t_4 = t + 1$. For example the π–π rectangular diagram in Fig. 4.14(i) corresponds to

$$C^R(t_4, t_3, t_2, t_1) = \sum_{\vec{x}_2, \vec{x}_3} \langle \mathrm{ReTr}[G^\dagger_{t_1}(\vec{x}_2 t_2) G_{t_4}(\vec{x}_2 t_2) G^\dagger_{t_4}(\vec{x}_3 t_3) G_{t_1}(\vec{x}_3 t_3)] \rangle \,. \tag{4.152}$$

The paths involved in the computation of C^R are not all closed, they are interupted at time slices where a (nonlocal) wall source meets a (local) sink, or two wall sources meet. This happens, for example, at t_1 and t_4, see (4.152) and Fig. 4.14.

In order to deal with the gauge noise problem two strategies may be employed: The first one is gauge fixing, the other one is to rely on gauge fluctuations to cancel out in the final correlation function. For the latter the gauge noise is expected to decrease as $\sim L^{3/2}$ for sufficiently large L [21].

Correlator ratios (4.144) for various channels can be expressed in terms of linear combinations of certain diagrams which are referred to by labels D, C, R, V, *etc.*, in Fig. 4.14. For example, in the Wilson fermion scheme

$$R^{\pi\pi}_{I=0}(t) = R^D(t) + \frac{1}{2}R^C(t) - 3R^R(t) + \frac{3}{2}R^V(t) \tag{4.153}$$

$$R^{\pi\pi}_{I=2}(t) = R^D(t) - R^C(t) \,. \tag{4.154}$$

For staggered fermions the corresponding decomposition is slightly different.

Typical results for a ratio function (4.144) are shown in Fig. 4.15 for the $I = 0$ and $I = 2$ channels of the π–π system. A selection of scattering lengths, again for the π–π system, is displayed in Fig. 4.16. In Fig. 4.17 a comparison of scattering lengths for the systems π–π, π–N, and N–N is made. Table 4.1 contains the main results of the lattice simulation and current algebra predictions. A chiral extrapolation has not been made in [21]. Thus, perhaps not surprisingly, the lattice results agree best with the current algebra predictions that use the hadron masses supplied by the actual lattice simulation. The N–N system is special because of its large size. Realistic lattice results for the scattering length cannot be expected with contemporary resources.

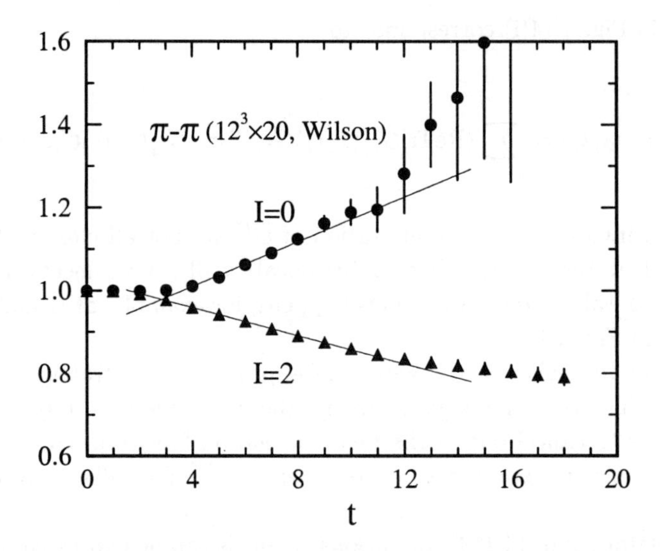

Fig. 4.15 Ratios (4.153) and (4.154) for the π–π four-point function, from Ref. [21]. This example is for Wilson fermions at $\beta = 5.7$, $\kappa = 0.164$, and wall sources without gauge fixing. The lattice size is $12^3 \times 20$.

4.4.2 *Static $N - N$ and $N - \bar{N}$ potentials*

Beyond the zero-momentum limit (scattering lengths) systems of hadrons with at least one heavy quark each pose the least technical difficulties. Going to the extreme, an infinitely heavy quark may be interpreted as a static color source. This idea goes back to Wilson's original work [37] and played a role in demonstrating confinement.

The time evolution operator of a static color source (a quark) located at fixed $\vec{x}$ is given by the product of link variables $U_4(\vec{x}t) \in SU(3)$ along a line $t_i \to t_i + 1 \to \cdots \to t_f$ in the time direction, $\mu = 4$. For a lattice $L^3 \times T$ with periodic boundary conditions in the time direction the line is closed, $t_i = t_f$. The trace of such an operator

$$Q(\vec{x}) = \frac{1}{3} \mathrm{Tr} \prod_{t=1}^{T} U_4(\vec{x}t) \tag{4.155}$$

is gauge invariant and is known as a Polyakov loop. It represents a static quark at $\vec{x}$, and $Q^\dagger(\vec{x})$ represents a static antiquark at $\vec{x}$. In the finite-

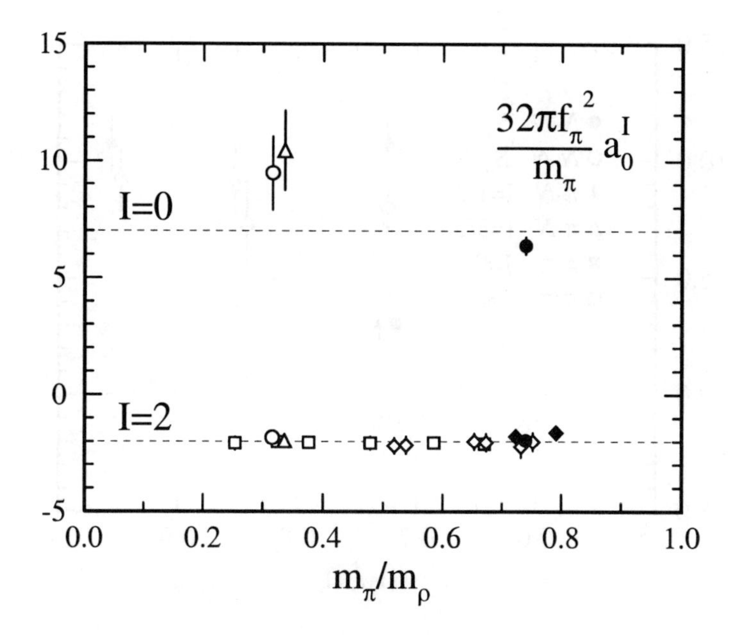

Fig. 4.16 Comparison of various scattering length results in the $I = 0$ and $I = 2$ channels for the π–π system, with units as given in the picture. Solid and open plot symbols refer to Wilson and Kogut-Susskind fermions, respectively. Results are from Ref. [21], except for the squares and diamonds, which are from Refs. [16, 17]. Dotted lines are current algebra predictions.

temperature formalism¶ the expectation value [66, 83, 84]

$$L^{-3} \sum_{\vec{x}} \langle Q(\vec{r} + \vec{x}) Q^\dagger(\vec{x}) \rangle = e^{-T F_{Q\overline{Q}}(\vec{r})} \tag{4.156}$$

is related to the free energy $F_{Q\overline{Q}}(\vec{r})$, loosely (but imprecisely) referred to as the heavy quark-antiquark potential. Adopting this interpretation it is interesting to compute the free energies of two clusters of two or three quarks [85, 86, 87]. Consider

$$O(\vec{x}) = \sum_{\vec{y}_1\vec{y}_2\vec{y}_3} \rho(\vec{y}_1\vec{y}_2\vec{y}_3) Q(\vec{x} + \vec{y}_1) Q(\vec{x} + \vec{y}_2) Q(\vec{x} + \vec{y}_3)\,, \tag{4.157}$$

where ρ is some function symmetric in $\vec{y}_1\vec{y}_2\vec{y}_3$, for example a Gaussian or

¶The temperature is the reciprocal of the time extent of the lattice, $1/Ta$. The path integral over fields becomes periodic in euclidean time [82].

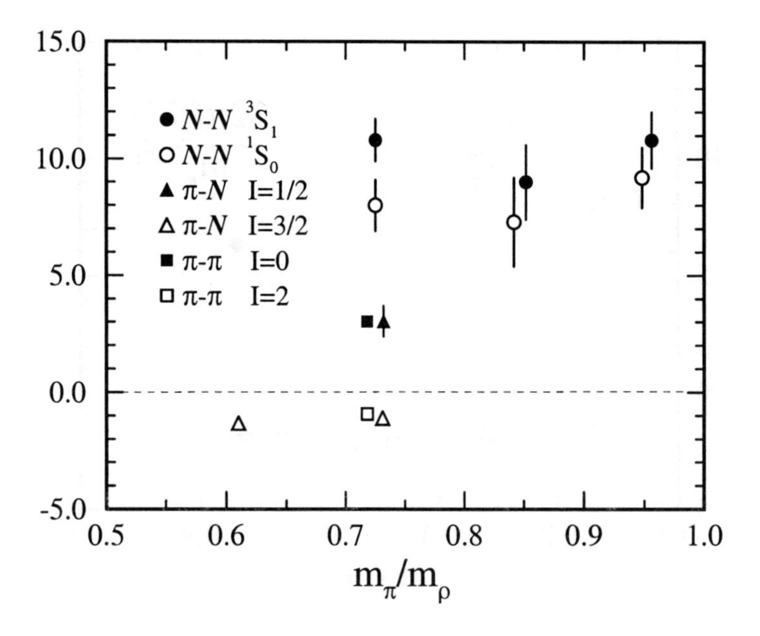

Fig. 4.17 Comparison of scattering lengths for the π–π, π–N, and N–N systems in physical units a_0 [fm] versus the π–ρ mass ratio. These are (quenched) results for Wilson fermions from Ref. [21] at a lattice spacing $a = 0.137(2)$ fm.

the characteristic function of a sphere with radius R, which serves as an *ad hoc* model for the three-quark cluster. The free energy of two three-quark clusters at separation $\vec{r}$ is then

$$F(\vec{r}) = -\frac{1}{T} \ln L^{-3} \sum_{\vec{x}} \langle O(\vec{r}+\vec{x}) O(\vec{x}) \rangle \,. \tag{4.158}$$

An example of $F(\vec{r})$ can be seen in Fig. 4.18. There is attraction in the overlap region of the clusters, somewhat reflecting the shape of ρ. The cluster interaction possesses no dynamics in this construction.

4.4.3 *Heavy-light meson-meson systems*

A step towards a more realistic calculation is to make some of the quarks dynamical. In a system of two mesons which consist of a heavy and a light quark each, we may still approximate the heavy quark by a Polyakov loop, as above. The role of the static quarks is then to localize the mesons. This

Table 4.1 Selected results from Ref. [21]. Scattering lengths a_0 for the s-wave in various channels are shown in physical units [fm], for the lattice (LAT), experiment (EXP), and current algebra (CUA). The lattice results are for Wilson fermions on a $12^3 \times 20$ lattice with $\beta = 5.7$. Light-quark masses are large as indicated by $m_\pi/m_\rho = 0.74$ and $m_N/m_\rho = 1.57$. Experimental and current algebra results are compiled from Refs. [76, 77] and [78], respectively. In the last column the masses from the lattice simulation enter the current algebra predictions.

		LAT	EXP	CUA	CUA(LAT)
π–π	$I = 0$	+3.02(17)	+0.37(7)	+0.222	+3.47(5)
	$I = 2$	−0.924(40)	−0.040(17)	−0.0635	−0.993(16)
π–N	$I = 1/2$	+3.04(66)	+0.245(4)	+0.221	+2.701(41)
	$I = 3/2$	−1.10(20)	−0.143(6)	−0.111	−1.350(20)
N–N	$^3S_1$	+10.8(9)	−5.432(5)		
	$^1S_0$	+8.0(1.1)	+20.1(4)		
K–N	$I = 0$	+0.55(47)	−0.0075	0	0
	$I = 1$	−1.56(13)	−0.225	−0.399	−2.701(41)
$\bar{\text{K}}$–N	$I = 0$	+4.64(37)	$-1.16 + i\,0.49$	+0.598	+4.051(61)
	$I = 1$	+2.63(64)	$+0.17 + i\,0.41$	+0.199	+1.350(20)

approach was pioneered by Richards *et al.* [30, 31]. The use of Wilson fermions simplifies the assignment of quantum numbers to the heavy-light meson operators. Writing $\bar{Q}\Gamma q$ for the latter and choosing $\Gamma = \gamma_5$ and $\Gamma = \gamma_{1,2,3}$ systems containing pseudoscalar (P) and vector (V) mesons, respectively, can be studied. According to Ref. [31] the PP–PP system allows to test ρ-exchange as an interaction mechanism. Likewise, the PV–VP operator probes for π meson exchange. Results to this effect are displayed in Fig. 4.19. Shown are effective masses of the meson-meson systems, extracted from large-time correlation functions, versus the relative distance R for π and ρ exchange mechanisms. At the quark masses used in Ref. [31] it is hard to discriminate between those two interaction mechanisms on numerical grounds. Aside from lattice artifacts (replicated lattices due to periodic boundary conditions) the results are consistent with an attractive force around $R \approx 2$. The significance of the work of Ref. [31] is that the strong interaction is probed from first principles, potentially aiming to distinguish between different physical mechanisms responsible for the forces

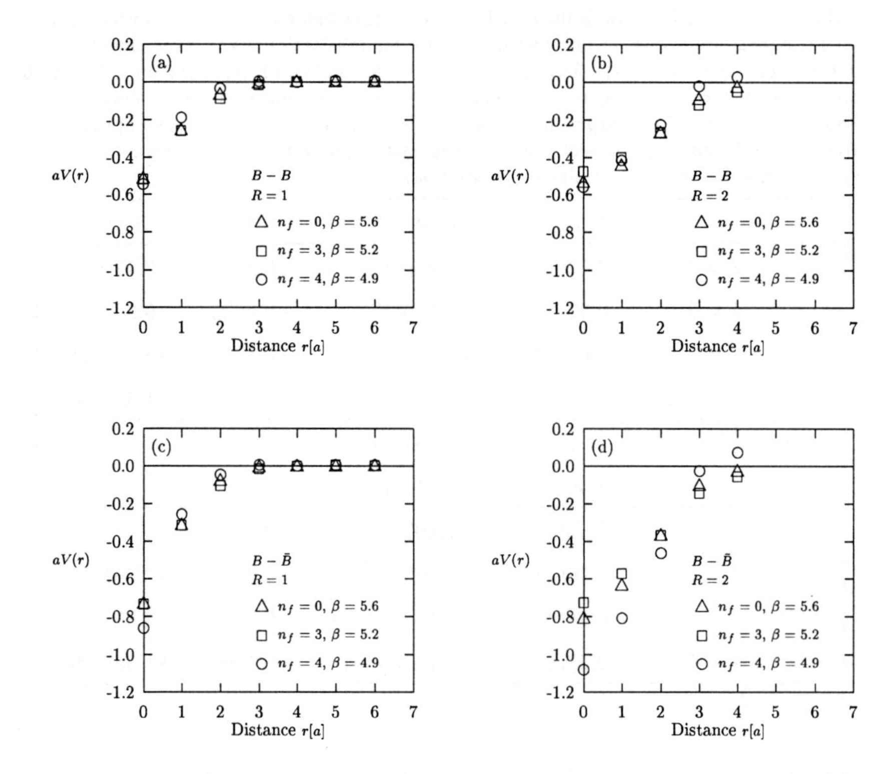

Fig. 4.18 Interaction energies of hadron-hadron systems versus the relative distance [85, 86] in the finite-temperature formalism. Hadronic structure is represented by a static cluster. The radius of the cluster is $R = 2$ in lattice units.

between hadrons.

We proceed to numerical investigations of heavy-light systems based on the staggered fermion scheme in a more detailed manner. Thus, in the coordinate-space approach of Sec. 4.3.5, we start with operators

$$\phi_{\vec{x}}(t) = \bar{\chi}_h(\vec{x}, t)\, \chi_u(\vec{x}, t)\,, \tag{4.159}$$

where h is the heavy flavor ($h = s, c, \cdots$). The two-meson interpolating field is defined just like (4.115). However, the heavy-flavor propagator, $G^{(h)}$, is taken in the limit $m_h \to \infty$. This is achieved by using the hopping parameter expansion [88, 82] and keeping the leading term only. We thus

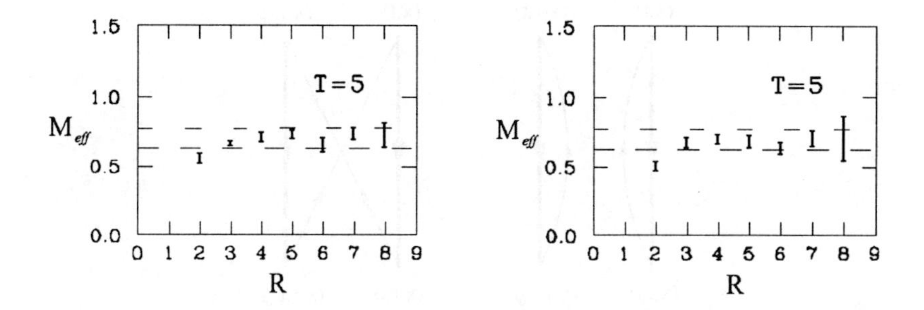

Fig. 4.19 Effective mass functions of a heavy-light meson meson system in the PP–PP channel (left) and the PV–VP channel (right) extracted at large time ($T = 5$), from [31]. The error bands (dashed lines) indicate systematic errors due to replicated lattices.

use

$$G^{(h)}(\vec{x}t, \vec{y}t_0) = \delta_{\vec{x},\vec{y}}\,[\frac{1}{2m_h}(-1)^{x_1+x_2+x_3}]^{t-t_0} \prod_{\tau=1}^{t-t_0} U_4(\vec{x}\tau)\,. \tag{4.160}$$

The product of link variables becomes proportional to the Polyakov loop (4.155) for $t - t_0 = T$. With $G^{(h)}$ diagonal in the space sites the 2-point and 4-point correlators become

$$C^{(2)}(t, t_0) = \langle L^{-6} \sum_{\vec{x}} G^{(h)*}_{BA}(\vec{x}t, \vec{x}t_0)\, G_{BA}(\vec{x}t, \vec{x}t_0)\rangle \tag{4.161}$$

$$\begin{aligned} C^{(4)}_{\vec{r}}(t, t_0) = \langle L^{-12} \sum_{\vec{x}} [&G^{(h)*}_{BA}(\vec{x}t, \vec{x}t_0)\, G_{BA}(\vec{x}t, \vec{x}t_0) \\ &\times G^{(h)*}_{DC}(\vec{x} - \vec{r}\,t, \vec{x} - \vec{r}\,t_0)\, G_{DC}(\vec{x} - \vec{r}\,t, \vec{x} - \vec{r}\,t_0) \\ &- G^{(h)*}_{BA}(\vec{x}t, \vec{x}t_0)\, G_{BC}(\vec{x} - \vec{r}\,t, \vec{x}t_0) \\ &\times G^{(h)*}_{DC}(\vec{x} - \vec{r}\,t, \vec{x} - \vec{r}\,t_0)\, G_{DA}(\vec{x}t, \vec{x} - \vec{r}\,t_0)]\rangle\,. \end{aligned} \tag{4.162}$$

Sums over color indices A, B, C, D are understood. The 4-point correlator (4.162) is a sum of two terms which are illustrated in Fig. 4.20. The light quarks can be exchanged between the mesons. Thus, besides gluon effects, also quark dynamics is included in the effective residual interaction.

For distances $\vec{r} \neq 0$ parallel to the coordinate axes, the A_1 sector pro-

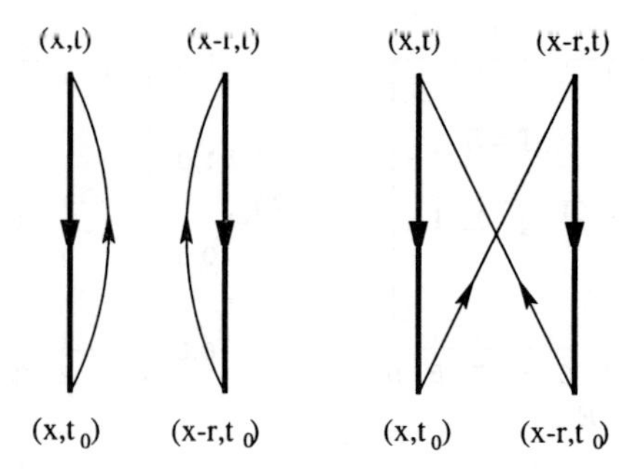

Fig. 4.20 Diagrams for the heavy-light meson-meson system. The thick and thin lines represent propagators for the heavy (static) and the light (dynamical) quark, respectively.

jection, see Sec. 4.3.3, is similar to (4.101), thus

$$C_r^{(4;A_1)}(t,t_0) = \frac{1}{4} \sum_{\vec{r},|\vec{r}|=r} C_{\vec{r}}^{(4)}(t,t_0)\,. \tag{4.163}$$

According to (4.127) the effective correlator

$$\mathfrak{C}_{r\,r}^{(4;A_1)}(t,t_0) \approx \frac{C_r^{(4;A_1)}(t,t_0)}{|C^{(2)}(t,t_0)|^2} \simeq c_r\, e^{-V^{(A_1)}(r)(t-t_0)} \tag{4.164}$$

serves to extract the (adiabatic) potential $V^{(A_1)}(r)$. Examples from an actual simulation [33] are shown in Fig. 4.21.

In the simple case of a one-hadron local operator it is possible to express the corresponding time correlation function in terms of propagator matrix elements from a source located at only one fixed lattice site, say $x = (\vec{x}_0 t_0)$. This is done by using translational invariance of the lattice action in the manner outlined in Sec. 4.2.2 leading to (4.40), for example. The cost of computing this is greatly reduced because only one column $G(\vec{x}t, \vec{x}_0 t_0)$ of the inverse fermion matrix (per color-Dirac index) is needed. In the case of a two-hadron system translational invariance might still be utilized to replace some propagator elements with $G(\vec{x}t, \vec{x}_0 t_0)$, for example via $\vec{x}_1 \to \vec{x}_1 + \vec{y}_1 - \vec{x}_0$ in (4.42). In the latter case elements $G(\vec{x}t, \vec{y}t_0)$ where $\vec{x}$ and $\vec{y}$ run separately over all spatial lattice sites still remain. Those are known as all-to-all propagators. Their computation poses an almost unavoidable problem in multi-hadron systems. The wall sources defined through (4.150) seemingly circumvent this difficulty, however, only at the

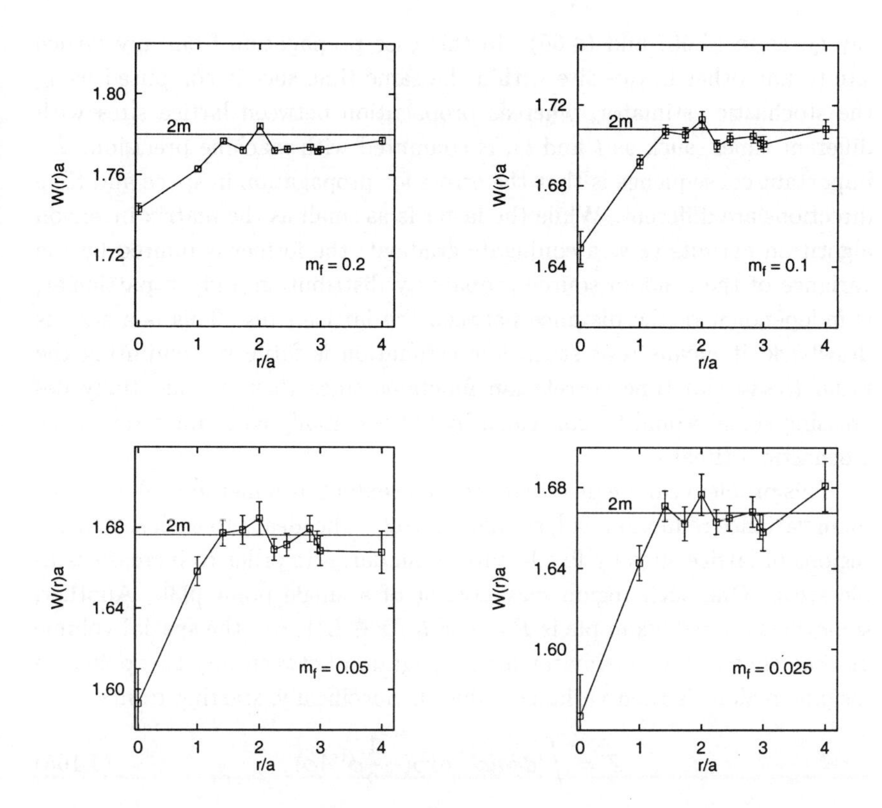

Fig. 4.21 Meson-meson energies $W(r) = V^{(A_1)}(r)$, see (4.164), for several light-quark masses of staggered fermions. Energies at distance $r/a = 3$ are degenerate, $|\vec{r}| = |(2,2,1)| = (3,0,0)|$. Those data points are slightly shifted [89].

cost of violating gauge invariance of the correlation functions, see Sec. 4.4.1. In a heavy-light hadron-hadron system, like (4.162) for example, the needed minimum number of sources is the number of relative spatial distances distances $\vec{r}$ that are probed (including those related by $O(3, \mathbb{Z})$ symmetry).

The use of stochastic methods for dealing with the all-to-all propagator problem was already discussed in some detail. Even in cases where translational invariance can be utilized it may be advantageous to employ random sources instead. The reason is that the additional (large) sums over lattice sites work to reduce statistical errors in the correlation functions. Besides, it may also be the only computationally viable strategy. In a typical application a random source is chosen to be nonzero on one time slice only, say t_0, as in (1.65) and (1.66). In this case propagation from any lattice site to any other lattice site within the same time slice is computed using the stochastic estimator, whereas propagation between lattice sites with different times, such as t and t_0, is computed with machine precision. An important consequence is that the errors for propagation in space and time directions are different. While the latter is as small as the matrix inversion algorithm permits (*e.g.* a conjugate gradient) the former is limited by the variance of the random source probability distribution and, in particular, is independent of the distance between the lattice sites. This is a serious drawback, it means that stochastic estimation is futile in computing the usual (hadronic) time correlation functions since their exponentially decreasing signal would be concealed by the essentially constant error of the stochastic estimator.

This problem can be dealt with to some extent by a method called maximum variance reduction [90], see also [91, 92]. The idea is to enclose disjoint regions of lattice sites by fixed-source boundaries in order to increase sample sizes. One such region may consist of a single point [93]. Another, somewhat related, example is $R = \{(\vec{x}, t_0) | \vec{x} \in L^3\}$, *i.e.* the spatial volume at time slice t_0. An estimator for propagation between any two points in disjoint regions is then variance reduced. Specifically, starting from

$$Z = \int [d\phi\, d\phi^*] \exp(-\frac{1}{2}\phi^\dagger A \phi)\,, \tag{4.165}$$

where ϕ is an auxiliary scalar complex lattice field and A is a hermitean positive definite matrix, one may generate stochastically independent ensembles of fields $\{\phi\}$ by way of common Monte Carlo methods with the gaussian probability distribution implied in (4.165). Writing $i = C,\ \mu,\ x$, *etc.*, for the color, Dirac, site,... indices it is then straightforward to show that

$$A^{-1}_{ij} = \frac{1}{Z} \int [d\phi\, d\phi^*] \exp(-\frac{1}{2}\phi^\dagger A \phi)\, \phi^*_j \phi_i \cong \langle \phi^*_j \phi_i \rangle\,, \tag{4.166}$$

where in this context $\langle \cdots \rangle$ means the average over the ensemble $\{\phi\}$. The above equations become useful for the choice

$$A = Q^\dagger Q\,, \tag{4.167}$$

with $Q = Q[U]$ being the fermion matrix for a given gauge field configuration U. In the Wilson scheme it has the form $Q[U] = \mathbf{1} - \kappa D[U]$. Quark propagator elements G_{ij} can then be estimated from the ensemble averages

$$G_{ij} \stackrel{\text{def}}{=} Q_{ij}^{-1} \cong \langle (Q\phi)_j^* \, \phi_i \rangle \,. \tag{4.168}$$

If the variance of the above estimator is σ_{ij} the statistical error is $\sigma_{ij}/\sqrt{N_\phi}$ for an ensemble of size N_ϕ.‖ However, the crucial point here is that the variance σ_{ij}, for $i = C$, μ, x and $j = B$, ν, y, is essentially independent of the space-time distance $|x - y|$, say $\sigma_{ij} \approx \sigma$ while σ is of the order of one inherited from its gaussian ancestry. This effect spoils using the stochastic estimator for the purpose of computing exponentially decreasing time correlation functions.

To alleviate the problem the authors of Ref. [90] proceed to consider a subset, or region, of contiguous lattice sites. Describe this region by sets of sites R and S, with $R \cap S = \emptyset$, where R contains all entirely interior points (having 8 neighbors $\in R \cup S$) and S is the boundary (having 1 through 7 neighbors $\in R \cup S$). In the process of generating ensembles $\{\phi\}$ from (4.165), consider a certain field ϕ. Then, the field components on the boundary S are kept fixed

$$s_l \stackrel{\text{def}}{=} \phi_l \quad \text{for} \quad l \in S \,, \tag{4.169}$$

and instead of (4.165) one uses

$$\mathcal{Z} = \int_R [d\phi \, d\phi^*] \exp\left(-\frac{1}{2} (\phi_i^\dagger \bar{A}_{ij} \phi_j + \phi_i^\dagger \tilde{A}_{il} s_l + s_l^\dagger \tilde{A}_{lj} \phi_j) \right) , \tag{4.170}$$

where $i, j \in R$ and $l \in S$ is understood for the index sum ranges and the notations $\bar{A}$ and $\tilde{A}$ are introduced for the corresponding blocks of A. The integral runs over the field components located on R. Now the R-averaged random source field

$$v_k = \frac{1}{\mathcal{Z}} \int_R [d\phi \, d\phi^*] \exp\left(-\frac{1}{2} (\phi_i^\dagger \bar{A}_{ij} \phi_j + \phi_i^\dagger \tilde{A}_{il} s_l + s_l^\dagger \tilde{A}_{lj} \phi_j) \right) \phi_k \,, \tag{4.171}$$

with $k \in R$, is called the variance reduced estimator for ϕ. Evaluation of the gaussian integral in (4.171) gives

$$v_k = -\bar{A}_{ki}^{-1} \tilde{A}_{il} s_l \,. \tag{4.172}$$

‖In fact, since a new estimator is computed for each of the gauge field configurations in the lattice simulation, the resulting error of a hadronic time correlation function is much smaller. For this reason a value of N_ϕ that is about $1/10 - 1/20$ of the number of gauge field configurations is usually sufficient.

Recall that in (4.168) single-site components of the random source field ϕ are employed to estimate the quark propagator. Replacing those with site-averaged variance reduced estimators v essentially amounts to using extended sources which of course reduce fluctuations. This is the main idea. To obtain a variance reduced estimator for quark propagator elements two disjoint regions R and R' are needed. Then, within the above framework, one may use (4.168) to show that

$$G_{k'k} \simeq \langle (Qv)^*_k \, v'_{k'} \rangle \quad \text{for} \quad k \in R, \; k' \in R' \,, \tag{4.173}$$

where v and v' are the variance reduced sources corresponding to the regions R and R' respectively [92]. Clearly, only propagator elements between any site in R and any site in R' can obtained in this way. Thus the all-to-all propagator problem is only partially solved. In order to maximize the variance reduction effect, R and R' should be equal in size and together with S and S' cover the entire lattice. This suggests the choices of all lattice sites with $0 < t < T/2$ and $T/2 < t < T$ for R and R', respectively. Further computational and technical details for the practitioner are given in Ref. [90].

Michael and Pennanen have utilized maximal variance reduced estimators in their work on heavy-light meson-meson systems [34] using improved Wilson fermions with a Sheikholeslami-Wohlert action [45]. Their main motivation was to explore the possibility of binding heavy-light systems. Since the kinetic energy of the quarks has a repulsive effect at shorter relative distances two hadrons containing a heavy quark each should have a better chance of forming a bound state. Concerning experiment, the simulation is expected to describe mesons with one (heavy) b-quark flavor. Using the static limit for the heavy flavor has the consequence that heavy-light pseudoscalar and vector mesons (B and B* respectively) are degenerate. However, experimentally their mass difference is less that 0.9% of the ground-state B-meson mass [94], so the static approximation is quite reasonable. The notation $\mathcal{B}$ is used in in Ref. [34] to refer to the degenerate heavy-light mesons. For the light quarks u and d flavors are employed, the corresponding Wilson hopping parameter κ puts their masses in the neighborhood of the strange quark mass.

Thus a $\mathcal{B}$–$\mathcal{B}$ system with relative distance r between the (static) quarks is classified in terms of the total spin S_q and isospin I_q of the light quarks, while imposing the usual particle exchange symmetries through appropriate heavy quark total spin S_b and spatial symmetries. The latter states are degenerate in the static quark limit. In the limit of an isotropic spatial wavefunction (L=0) there are only four degenerate ground state levels of the $\mathcal{B}$–$\mathcal{B}$ system characterized by $I_q = 0, 1$ and $S_q = 0, 1$. The situation is illustrated in Table 4.2.

Table 4.2 Allowed $\mathcal{B}$–$\mathcal{B}$ states with $L = 0$ according to [34]. Total isospin and spin of the light-quark subsystem are denoted by I_q and S_q, respectively. The total spin S_b of the heavy quarks only distinguishes degenerate states. Also shown are the physical heavy-light systems and their J^P coupling to these states.

I_q	S_q	S_b	J^P	BB	BB*	B*B*
1	1	1	0^+	✓		✓
1	1	1	1^+		✓	
1	1	1	2^+			✓
1	0	0	0^+	✓		✓
0	1	0	1^+		✓	✓
0	0	1	1^+		✓	✓

Selected results of the simulation [34] are shown in Fig. 4.22 exhibiting the energies of the $\mathcal{B}$–$\mathcal{B}$ system at fixed separation r of the static b-quarks. The authors suggest that there is evidence for deep binding at small r with a light di-quark configuration having $(I_q, S_q) = (0, 0)$ and $(1, 1)$ respectively. This binding energy is 400 – 200 MeV at $r = 0$ but is very short-ranged. It is essentially a gluonic effect and is rather insensitive to the light quark mass as shown by studies of static baryons with varying light quark masses [90]. Possibly, a configuration where the color of two heavy quarks combines to a color state symmetric under particle exchange (sextet) provides the interaction mechanism here. This situation is particular to relative distance $r = 0$ since two-quark color states other than local singlets are not suppressed by confinement dynamics.

At larger r, around 0.5 fm, evidence is seen for weak binding when the light quarks are in the $(I_q, S_q) = (0, 1)$ and $(1, 0)$ states. This can be related to meson exchange and one finds evidence of an interaction in the spin-dependent quark-exchange (cross) diagram which is compatible with the theoretical contribution from pion exchange in the study. Using lighter, more physical, light quark masses below the strange quark mass and, most

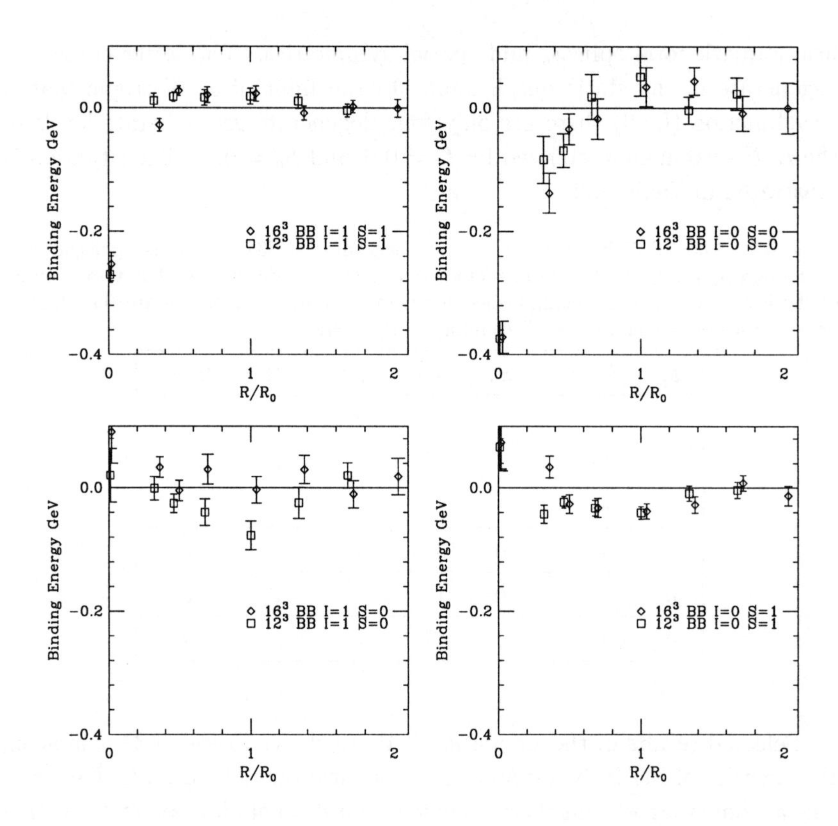

Fig. 4.22 Results for the binding energy $E(\mathcal{B}\text{–}\mathcal{B}) - 2E(\mathcal{B})$ between two $\mathcal{B}$ mesons with light quarks in $(I_q, S_q) = (1,1), (0,0), (1,0), (0,1)$ at separation R (called r throughout this chapter) in units of $R_0 \approx 0.5$ fm. The light quark mass used corresponds to strange quarks. Results at different spatial lattice sizes are displaced in R for legibility [34].

desirably a chiral extrapolation, will be necessary to confirm this interaction mechanism.

4.4.4 *Momentum-space work on the π–π system*

The calculation of euclidean correlation functions between different hadrons in the incoming and outgoing channel, like K$\rightarrow$ $\pi\pi$ decays, has to be treated with caution [95]. The program laid out in Secs. 4.2.2–4.2.5 and 4.3.4 has been put to the test in QCD_{3+1} for a π–π system of Wilson fermions [87]. The formalism is essentially unchanged, except for the appearance of color indices on the link variables $U_\mu(x) \in \mathrm{SU}(3)$ and, replacing staggered with Wilson fermions, the appearance of both a color $A, B \cdots = 1,2,3$ and a Dirac $\mu, \nu \cdots = 1,2,3,4$ index on the Grassmann quark fields $\psi_{fA\mu}(x)$. In addition, there is a rather technical, but important point:

A technique known as 'smearing' uses spatially extended field operators. For example, omitting flavor and Dirac indices for the moment, define the quark field recursively

$$\psi_A^{\{0\}}(\vec{x}t) = \psi_A(x), \qquad \psi_A^{\{k\}}(\vec{x}t) = \sum_{\vec{y}} \sum_{B} K_{AB}(\vec{x},\vec{y})\, \psi_B^{\{k-1\}}(\vec{y}t)\,, \quad (4.174)$$

with $k \in \mathbb{N}$, and the matrix

$$K_{AB}(\vec{x},\vec{y}) \;=\; \delta_{AB}\,\delta_{\vec{x},\vec{y}} + \alpha H_{AB}(\vec{x},\vec{y}) \quad (4.175)$$

$$H_{AB}(\vec{x},\vec{y}) \;=\; \sum_{m=1}^{3} \left[U_{m,AB}(\vec{x}\,t)\delta_{\vec{x},\vec{y}-\hat{m}} + U^{\dagger}_{m,AB}(\vec{y}\,t)\delta_{\vec{x},\vec{y}+\hat{m}} \right] \quad (4.176)$$

connecting (hopping) to next-neighbor sites. With this particular choice of the matrix K the procedure is known as Gaussian smearing [96]. The real number α and the maximal value for k are considered parameters. The advantage of smearing lies in the fact that interpolating operators constructed with smeared fields often have a larger overlap with the ground state of the hadron in question, see Fig. 4.23. The signal from the corresponding correlator is less contaminated by fluctuations from excited states and thus masses can be extracted from shorter euclidean time extents. This leads to better statistics and is often crucial for the numerical work. A similar technique, known as 'fuzzing' applies to the link variables [88] but will not be detailed here, since it can be found in Chapter 1 Sec. 1.2 and Chapter 5 Subsec. 5.4.1.3.

Thus, the equivalent of the one-meson field (4.28) is replaced by

$$\phi_{\vec{p}}(t) = L^{-3} \sum_{\vec{x}} e^{i\vec{p}\cdot\vec{x}}\, \bar{\psi}_{dA}^{\{k\}}(\vec{x},t)\gamma_5 \psi_{uA}^{\{k\}}(\vec{x},t) \quad (4.177)$$

for a π^+ pseudoscalar meson. As usual, the quark propagator is obtained from contractions between the (unsmeared) quark fields

$$\cdots \overset{n}{\psi}_{fA\mu}(x)\, \overset{n}{\bar{\psi}}_{gB\nu}(y) \cdots \;=\; \cdots G_{fA\mu,gB\nu}(x,y) \cdots \quad (4.178)$$

$$\cdots \overset{n}{\bar{\psi}}_{fA\mu}(x)\, \overset{n}{\psi}_{gB\nu}(y) \cdots \;=\; \cdots - \gamma_{5,\mu'\mu} G^{*}_{fA\mu',gB\nu'}(x,y)\gamma_{5,\nu\nu'} \cdots \quad (4.179)$$

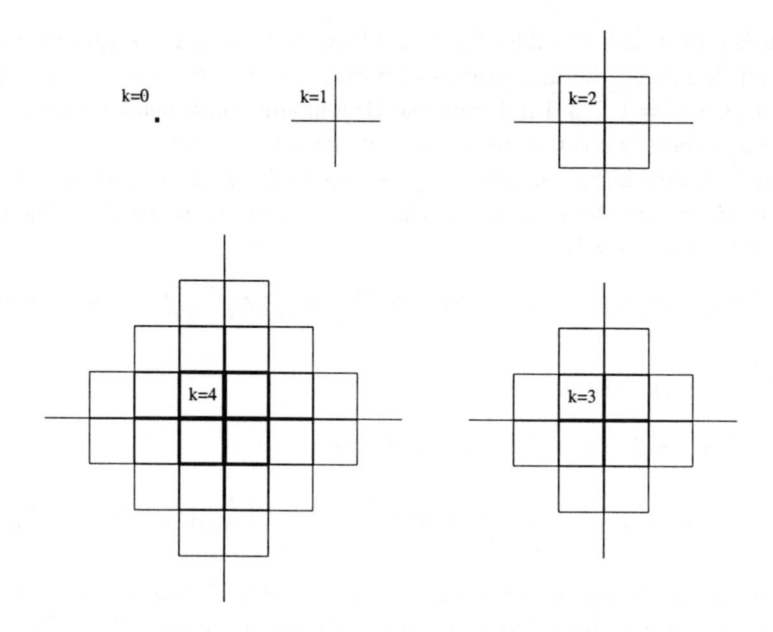

Fig. 4.23 Illustration of the 'smearing' technique for fermion operators. Various iterations ($k = 0$ local, *etc.*) are shown. The thickness of the lines indicates the 'multiplicity of use' of a particular link.

The latter equation is a consequence of the identity

$$G(y, x) = \gamma_5 G^\dagger(x, y)\gamma_5 \,. \tag{4.180}$$

The † refers to flavor-color-Dirac indices. The identity is valid if the lattice action is CPT invariant [97]. Further, if the mass matrix in the Wilson fermion action is flavor diagonal, then also

$$G_{fA\mu,gB\nu}(x, y) = \delta_{f\,g} G^{(f)}_{A\mu,B\nu}(x, y) \,. \tag{4.181}$$

Typically, one would choose equal Wilson hopping parameters κ (those determine the quark masses) for u and d quarks and different ones for the strange quark.

We now give a list of essential things to do, according to the plan of Secs. 4.3 and 4.2.

Consider the linear equation

$$\sum_{\vec{y}y_4}\sum_{B\nu} G^{-1(f)}_{A\mu,B\nu}(\vec{x}x_4,\vec{y}y_4)\, X^{(f;\,A'\mu' r\, x_4')}_{B\nu}(\vec{y}y_4) = \delta_{AA'}\,\delta_{\mu\mu'}\, R^{(A'\mu' r\, x_4')}(\vec{x})\,\delta_{x_4x_4'}\,. \tag{4.182}$$

The meaning of the indices are $A, B\cdots = 1,2,3$ color, $\mu,\nu\cdots = 1,2,3,4$ Dirac, $\vec{x},\vec{y}\cdots$ space ($d = 3$), x_4, y_4 time, and $r = 1,\ldots,N_R$ labels the random sources R for each source point. A prime $'$ denotes a source point. There is some freedom in choosing the latter. In (4.182) the sources are nonzero on one time slice only. The same set of sources is used for different flavors (Wilson hopping parameters κ), but a separate source is chosen for each color, Dirac and time index. The features of the random sources are somewhat linked to the choice of the solution algorithm [98]. Also, Gaussian, $\mathbb{Z}_2$, or other random sources may be employed. In any case, the random-source average, which we approximate numerically as

$$\sum_{\langle r\rangle}\cdots \approx \frac{1}{N_R}\sum_{r=1}^{N_R}\cdots\,, \tag{4.183}$$

must satisfy

$$\sum_{\langle r\rangle} R^{(A'\mu' r\, x_4')}(\vec{x})R^{(B'\nu' r\, y_4')*}(\vec{y}\,) = \delta_{A'B'}\,\delta_{\mu'\nu'}\,\delta_{\vec{x}\vec{y}}\,\delta_{x_4'y_4'}\,. \tag{4.184}$$

A conjugate-gradient algorithm [99], or a variant, is suitable for solving the linear Eq. (4.182). An estimator for the propagator matrix elements is then

$$G^{(f)}_{B\nu,A\mu}(\vec{y}y_4,\vec{x}x_4) = \sum_{\langle r\rangle} X^{(f;\,A\mu\, r\, x_4)}_{B\nu}(\vec{y}y_4)R^{(A\mu\, r\, x_4)*}(\vec{x})\,. \tag{4.185}$$

For simple hadron-hadron systems it may be sufficient to place sources on one time slice x_4 only. This is the situation for the π^+–π^+ system ($I = 2$ channel). Matching the notation in the examples discussed in previous sections we set $x_4 = t_0$. Putting things together, the correlation matrices are then computed through the following sequence of steps:

$$R_C^{\{0\}(B\nu\, r\, t')}(\vec{x}) = \delta_{CB} R^{(B\nu\, r\, t')}(\vec{x}) \quad \text{(no sum over } B) \tag{4.186}$$

$$R_C^{\{k\}(B\nu\, r\, t')}(\vec{x}) = \sum_{\vec{y}} \sum_A K_{CA}(\vec{x}, \vec{y}) R_A^{\{k-1\}(B\nu\, r\, t')}(\vec{y}) \tag{4.187}$$

$$X_{C\mu}^{\{0\}(f;\, B\nu\, r\, t')}(\vec{x}t) = X_{C\mu}^{(f;\, B\nu\, r\, t')}(\vec{x}t) \tag{4.188}$$

$$X_{C\mu}^{\{k\}(f;\, B\nu\, r\, t')}(\vec{x}t) = \sum_{\vec{y}} \sum_A K_{CA}(\vec{x}, \vec{y}) X_{A\mu}^{\{k-1\}(f;\, B\nu\, r\, t')}(\vec{y}t) \tag{4.189}$$

$$\mathcal{R}^{\{k\}(AB\, \mu\nu\, r' r\, t')}(\vec{p}) = \sum_{\vec{x}} L^{-3} e^{-i\vec{p}\cdot\vec{x}} \sum_C \times R_C^{\{k\}(A\mu\, r'\, t')*}(\vec{x}) R_C^{\{k\}(B\nu\, r\, t')}(\vec{x}) \tag{4.190}$$

$$\mathcal{X}_{\nu\mu}^{\{k\}(gf;\, AB\, \mu'\nu'\, r' r\, t')}(\vec{p}t) = \sum_{\vec{x}} L^{-3} e^{-i\vec{p}\cdot\vec{x}} \sum_C \times X_{C\nu}^{\{k\}(g;\, A\mu'\, r'\, t')*}(\vec{x}t) X_{C\mu}^{\{k\}(f;\, B\nu'\, r\, t')}(\vec{x}t) \tag{4.191}$$

$$C_{\vec{p}\vec{p}}^{(2)}(t, t_0) = -\gamma^*_{5,\nu\mu} \mathcal{X}_{\nu\mu}^{\{k\}(ud;\, AB\, \mu'\nu'\, r' r\, t_0)*}(\vec{p}t) \times \gamma_{5,\mu'\nu'} \mathcal{R}^{\{k\}(ud;\, AB\, \mu'\nu'\, r' r\, t_0)} . \tag{4.192}$$

Summation over doubly-occuring indices is understood, except where indicated otherwise. The expressions for the 4-point correlators $\overline{C}^{(4)}$ and $C^{(4)}$ are more complicated, but similar.

Figure 4.24 shows energy-level diagrams of the non-interacting (from $\overline{C}^{(4)}$), and the interacting (from $C^{(4)}$), π–π systems for two values of the hopping parameter, $\kappa^{-1} = 5.888$ and $\kappa^{-1} = 5.972$. These are from a set of six values κ^{-1} which have been used to perform the chiral extrapolation $m_\pi^2 \to 0$ (equivalently $\kappa \to \kappa_c$). The other lattice parameters [100] were $L = 9$, $T = 13$, and $\beta = 6.2$ with the next-nearest-neighbor $o(a^2)$ tadpole-improved action [11, 46]. The lattice spacing, matched to the string tension, is $a = 0.4$ fm or $a^{-1} = 500$ MeV. An interesting feature of the spectra is that high-momentum levels are lowered by the interaction indicating attraction at short distances, and some of the lower levels move up indicating some repulsion in the system.

The corresponding approximate local potentials, according to (4.112) of

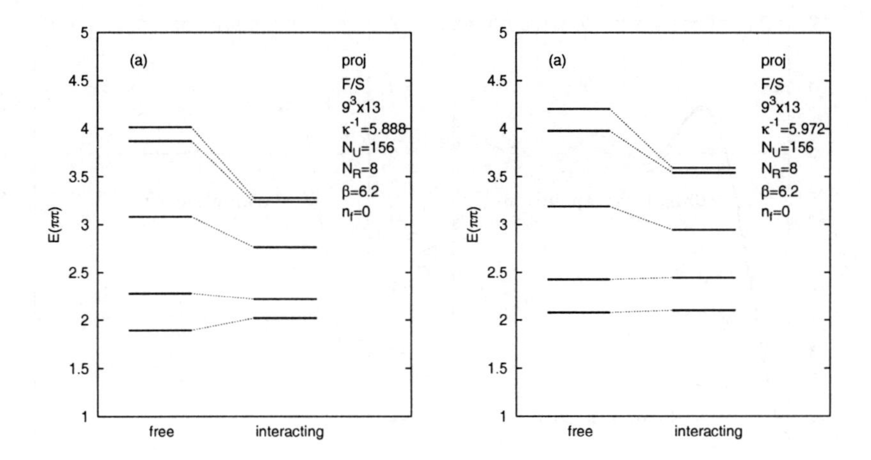

Fig. 4.24 Meson-meson energy level splitting due to the residual interaction. Two different values of the hopping parameter (quark masses) are shown [87].

Sec. 4.3.4, were projected onto the partial wave $\ell = 0$

$$\mathcal{V}_0(r) = \frac{1}{4\pi} \int d\Omega_{\vec{r}}\, \mathcal{V}(\vec{r}) = \sum_q j_0(2qr) \langle q|\mathcal{H}_I^{(A_1)}|q\rangle , \tag{4.193}$$

with $q = \frac{2\pi}{L}k, k = 0 \cdots k_{\rm max}$. Results of the chiral extrapolation plot [35] are shown in Fig. 4.25. The Fourier nature of the resultant potentials is apparent. A parametric fit, using (4.131), (4.133) and (4.134), is also shown. Of course, short distance r's are beyond the scope of the truncated momentum basis and should be interpreted with caution. Thus only small relative momenta make sense when calculating scattering phase shifts with this potential. Results using standard (non-relativistic) scattering theory [36] with parameterized potentials (4.131) are displayed in Fig. 4.26.

Recently, scattering phase shifts for the π–π system in the $I = 2$ channel have been obtained in full QCD [102].

4.4.5 *Coordinate-space work on the π–π system*

We turn to another exploration in coordinate space with staggered fermions [87]. The final interaction potential is shown in Figs. 4.27(a) and (c) for $m_f = 100$ MeV and $m_f = 500$ MeV, respectively. The fits of the periodic potential were stable for the data with $m_f = 100$ MeV. The data

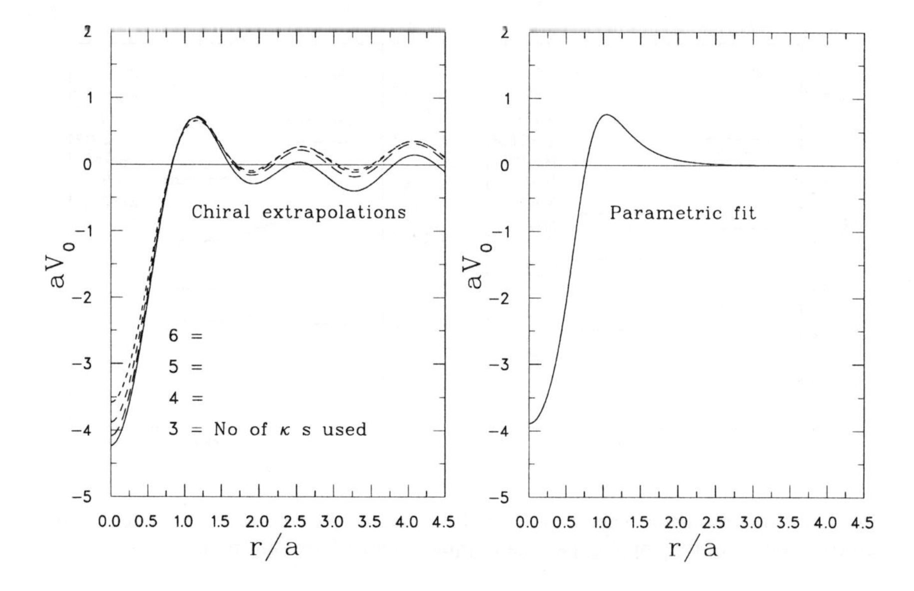

Fig. 4.25 Approximate s-wave projected local meson-meson potentials according to (4.112) and (4.193). Shown is the result of a chiral extrapolation ($m_\pi^2 \to 0$) of the potential based on various hopping parameter values; curves resulting from the use of 3 to 6 κ-values are compared. A parametric fit with $V^{(\alpha)}(r)$ based on (4.131), using 3 values of κ, is presented [35].

for $m_f = 500$ MeV suffer from very small energy differences which makes a fit unreliable. The potentials exhibit attraction at intermediate range and may have a repulsive behavior at short distances. The potential at $m_f = 500$ MeV is compatible with zero interaction. The analytic form (4.131) of the interaction potential was used as an input to the Schrödinger equation for a phase shift calculation according to scattering theory [36]. The resulting phase shifts are displayed in Figs. 4.27(b) and (d) for different values of the meson mass parameter used in the Schrödinger equation.

One aim of this study was the comparison of results obtained from lattice QCD with results calculated from inverse scattering theory. There is a wealth of inversion results from hadron-hadron scattering [110]. We here restrict ourselves to cases with partial wave $L = 0$ meson-meson scattering. The basis for any inversion is the existence of experimental data which, after a suitable interpolation and extrapolation, leads to a solution of the underlying ill-posed problem. For the π–π system the experimental situation is acceptable. The main source of experimental information for the inversion

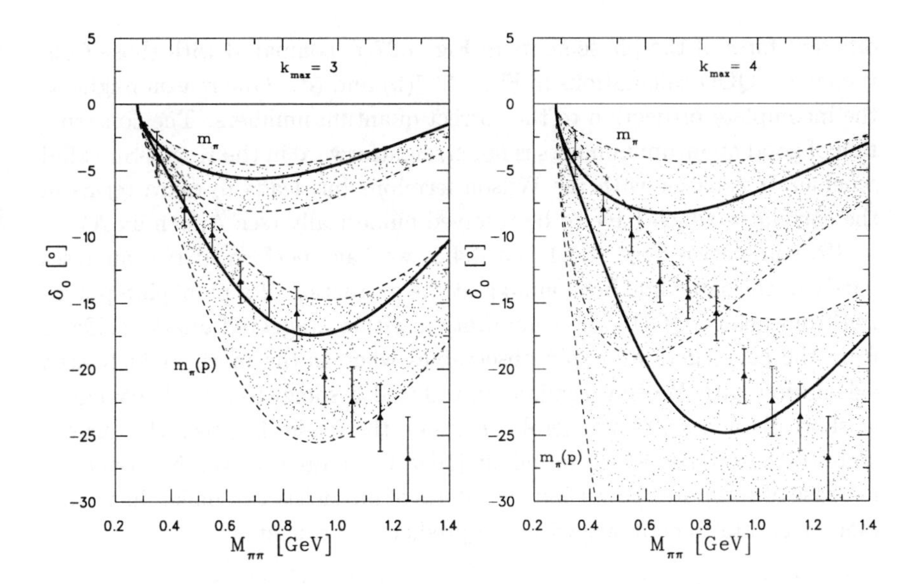

Fig. 4.26 Scattering phase shifts $\delta_{\ell=0}^{I=2}$ calculated from the π–π potential of the lattice QCD simulation (thick lines) [35]. Influence of momentum cutoff $k_{\max} = 3$ (left) and $k_{\max} = 4$ (right) is shown. Results using the classical and relativistic dispersion relation are distinguished by m_π and $m_\pi(p)$, respectively. Errors are represented by the dotted regions. Their boundaries (dashed lines) correspond to the phase shifts calculated with extremal (bootstrap) potentials. The experimental data are from Ref. [101].

procedure is the analysis of the CERN–Munich experiment $\pi^- p \to \pi^- \pi^+ n$ [103, 107, 111]. Even though there are ambiguities in the analysis of the $\pi\pi$ final state, all phase shift analyses reach comparable results, which can be described by both meson exchange [8] and chiral perturbation theory (χPT) [104, 105]. This experimental situation is depicted in Fig. 4.27(f). The inversion results for several sources of phase shifts are shown in Fig. 4.27(e). While the isospin $I = 2$ channel is purely repulsive, the isospin $I = 0$ channel shows an attractive component. A resonance is supported in this channel [110].

The lattice mesons are coupled to isospin $I = 2$ but constructed without a specific angular momentum projection. Thus there is no one-to-one correspondence between the quantum numbers of the lattice mesons and the experimental data. The lattice potentials in Figs. 4.27(a) and (c) are attractive with repulsion at short distances, whereas the potentials obtained from inverse scattering in Fig. 4.27(e) are only repulsive. This is expressed in the different form of the phase shifts in Fig. 4.27(f) compared with those from the lattice QCD calculations in Figs. 4.27(b) and (d). One reason might be the incomplete projection to the correct quantum numbers. The construction of good quantum numbers is not straightforward in the Kogut-Susskind scheme. It is easier with the Wilson fermions but leads to extra terms in the correlator which have to be resolved numerically (see Appendix A).

To summarize Secs. 4.4.4 and 4.4.5, we have performed two analyses partly in the same line both aiming at the extraction of a pion–pion potential from lattice QCD. The computation within staggered quarks suffered from the construction of spin observables whereas the more sophisticated calculation with Wilson fermions introduced a somewhat arbitrary cutoff at short distances. As a general conclusion the reader has seen the current state of these trials which he might judge to be qualitative. Not only the increase of computer power but the efforts of physicists can make the results more accurate to compare with and predict experiments.

4.5 Conclusion and Outlook

The field of *hadronic physics* has experienced a change of paradigm in recent years. From a modern perspective its primary goal now is the understanding of hadronic phenomena in terms of the fundamental degrees of freedom of the underlying theory, quantum chromodynamics [112]. Historically, phenomenological descriptions of hadron-hadron interactions have been with us for more than six decades while new questions related to the quark-gluon structure of hadrons have emerged as our insight has deepened over time. Thus hadronic physics now is closely intertwined with QCD. Dramatic advances in computational technology, lattice field theory, and algorithms of computational physics, have moved unprecedented opportunities for fundamental cognition within our reach. New accelerators and advanced detector design should of course also be mentioned in this context. Among the many topics tackled by lattice QCD – the hadron mass spectrum, hadronic structure, heavy flavor and exotic hadrons, to name just a few – the physics of hadron-hadron interactions is one of the more challenging subjects.

It is thus not surprising that the topic of this chapter is only in a nascent state. Concluding words with a quality of closure would be truly misleading. The current stage of hadron-hadron interactions on the lattice is driven by the tools of lattice technology with mass calculations being directly accessible from the euclidean formulation. All methods discussed in this chapter essentially use the energy spectra of certain systems as a starting point. Three ways to utilize those have been visited: finite volume methods and direct calculation of scattering phase shifts (Lüscher's method), construction of an effective interaction from two-hadron states (perturbative potentials), and direct computation of the interaction in heavy-light systems (adiabatic potential).

Of those methods the first one addresses measurable data directly, so far its most prominent application being aimed at scattering lengths. The other variants leave the realm of field theory in the hope of unveiling physical mechanisms of the interaction. Perhaps, in the latter case the usual lattice technology constraints (related to chiral symmetry, quark masses, finite volume, continuum limit, *etc.*) are a lesser impediment to progress in the sense that qualitative insights into the physics of the interaction can still be gained with 'less than perfect' lattice parameters.

Precision simulations reproducing experimental scattering data should probably not be the first priority at this time. To achieve this the field may not be advanced enough, particularly in the nucleon-nucleon case. Also, fermion methods dealing properly with chiral symmetry [113] should certainly be incorporated.

Understanding hadronic interaction mechanisms, for example the importance of gluon or quark degrees of freedom at certain relative distances, is probably more desirable at this stage. Emphasizing this aspect in future studies, for example heavy-light meson-baryon and baryon-baryon systems, seems a promising line of work that should shed some light on the QCD aspects of the hadron-hadron interaction. A related aspect is the understanding of the origin of effective field theories [114].

From the viewpoint of lattice technology the computation of multi-quark correlation functions, which requires all-to-all fermion propagators, is certainly the biggest problem. Considerable space was devoted to random source methods. Those are likely to remain pillars of hadron-hadron systems simulations. In another development, not mentioned before, Baysian analysis methods for lattice generated time correlation functions promise access to excited states [115, 116, 117], an aspect crucial for the topic of this chapter.

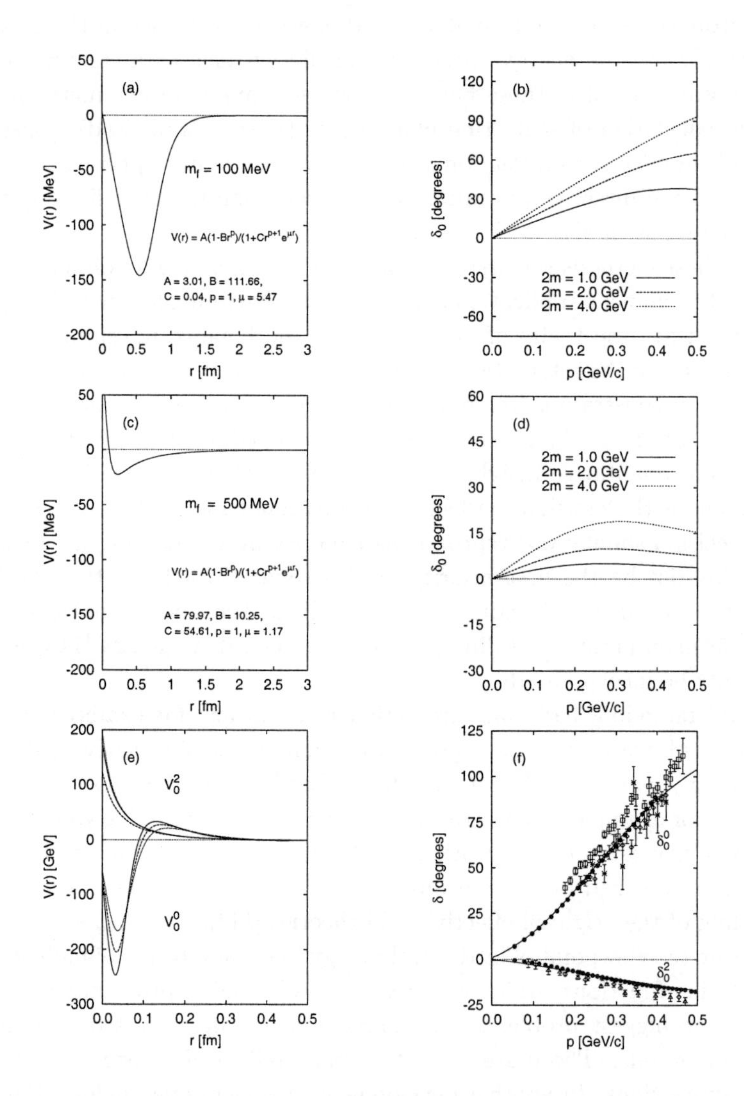

Fig. 4.27 (a)–(d): Lattice results for meson-meson interaction potentials $V(r)$ after the subtraction of the effects of the mirror particles, and phase shifts δ, for quark masses $m_f = 100$ MeV and $m_f = 500$ MeV. (e): $\pi\pi$ $L = 0$, $I = 0$ and 2 inversion potentials. Froggatt [103] (solid line), χPT [104, 105] (dashed line) and meson exchange [8] (dotted). (f): from $\pi\pi$ $L = 0$, $I = 0$ and 2 phase shifts of Froggatt data [103] and interpolation (dots and solid line), Estabrooks *et al.* [106] (boxes), Grayer *et al.* [107] (diamonds), Männer [108] (triangles) and Baillon *et al.* [109] (asterixes).

Finally, although it is true that the euclidean formulation of lattice QCD in imaginary time currently dominates the field, one should realize that it may not be the framework best suited to the task. It is not clear how to overcome the difficulty of describing asymptotic states, in the sense of the LSZ scattering formalism, other than trying to solve the problem via spectrum computations at multiple lattice volumes [118]. In the context of scattering, the alternative hamiltonian formulation in real time is largely unexplored [74].

Acknowledgements

It is a great pleasure to acknowledge the many significant contributions of (in alphabetical order) J. Canosa, A. Dominguez, O. Linsuain, A. Mihály, K. Rabitsch, R.M. Woloshyn, who have at one time or another advanced the subject matter of this review in various ways. The support and hospitality of the Thomas Jefferson National Laboratory JLab during three summers and by the European Center for Theoretical Studies ECT* are gratefully acknowledged. This material is based upon work supported by the U.S. National Science Foundation under Grant No. 0073362. Resources made available through the Lattice Hadron Physics Collaboration LHPC were used. This work was further supported by the Austrian Science Foundation FWF under Project P10468-PHY.

Appendix A: Remarks on Staggered Fields

The two most widely used schemes for discretizing fermion fields are commonly known as 'Wilson fermions' [119] and 'Kogut-Susskind' or 'staggered fermions' [49, 50, 120]. Wilson fermions are computationally more demanding than staggered fermions. While Wilson fermions explicitly break chiral symmetry, though they do have a chiral limit which in practice can be recovered via a $m_\pi^2 \to 0$ extrapolation of observables, staggered fermions retain a discrete remnant of chiral symmetry [121]. In the Wilson scheme, the construction of interpolating fields for hadrons with definite quantum numbers is transparent, and works more or less as in the continuum theory. This is opposite for staggered fermions, which entail complicated mathematics when constructing interpolating fields [51, 75, 121].

A case in point is an ordinary π^+ interpolating field. First, the quark fields, q_f, are nonlocal operators which live on elementary hypercubes of the lattice [120, 121]. For a U(1) gauge model (no colors) we have

$$q_f(x) = C^{-3/2} \sum_{\eta} \Gamma_\eta U(\eta, x) \chi_f(2x + \eta) \tag{A.1}$$

built from the staggered fields χ_f. The sum is over the set of the d-dimensional vectors η that point to the corners of the hypercube, $\eta_\mu = 0, 1$ for $\mu = 1, \cdots, d$, with $d = 4$, see Ref. [120] for an odd number of dimensions. Further

$$\Gamma_\eta = \gamma_1^{\eta_\mu} \cdots \gamma_d^{\eta_d} \quad \text{and} \quad U(\eta, x) = [U_1(2x)]^{\eta_1} \cdots [U_d(2x + \eta_1 \cdots \eta_{d-1})]^{\eta_d}\,, \tag{A.2}$$

and C, in (A.1), is the dimension of the Clifford algebra. A π^+ interpolating field is then

$$\phi(x) = \bar{q}_d(x)(\gamma_5 \otimes \gamma_5^*) q_u(x)\,. \tag{A.3}$$

The direct product $\otimes$ refers to row and column indices, respectively, of Clifford algebra matrices. Now, turning to the corresponding correlation function it can be shown that the latter reduces to a sum of local terms

$$\begin{aligned} \langle \sum_x \phi^\dagger(\vec{x}, t)\phi(\vec{x}_0, t_0)\rangle \;&=\; \langle 2\sum_{\vec{x}} [2G(\vec{x}, 2t; \vec{x}_0, t_0)G^*(\vec{x}, 2t; \vec{x}_0, t_0) \\ &\quad + G(\vec{x}, 2t+1; \vec{x}_0, t_0)G^*(\vec{x}, 2t+1; \vec{x}_0, t_0) \\ &\quad + G(\vec{x}, 2t-1; \vec{x}_0, t_0)G^*(\vec{x}, 2t-1; \vec{x}_0, t_0)]\rangle\,. \end{aligned} \tag{A.4}$$

For a concise exposition see Ref. [122]. This shows that the following equivalent correlator may be used

$$C^{(2)}(t, t_0) = \langle \sum_{\vec{x}} G(\vec{x}, t; \vec{x}_0, t_0)G^*(\vec{x}, t; \vec{x}_0, t_0)\rangle\,. \tag{A.5}$$

For our purposes a useful observation is that the local field

$$(-1)^{x_1 + \cdots + x_{d-1} + t} \bar{\chi}_d(\vec{x}, t)\chi_u(\vec{x}, t) \tag{A.6}$$

supplemented by a phase factor as shown gives rise to the correlator (A.5). Hence, rather than using (A.1) and going through an involved derivation, we may as well work with the interpolating field

$$\phi_{\vec{p}}(t) = L^{-d/2} \sum_{\vec{x}} e^{i\vec{p}\cdot\vec{x}} (-1)^{x_1 + \cdots + x_{d-1} + t}\, \bar{\chi}_d(\vec{x}, t)\, \chi_u(\vec{x}, t) \tag{A.7}$$

from the outset. The construction of correlators requires contractions between the staggered Grassmann fields

$$\cdots \overset{n}{\chi}_f(x)\, \overset{n}{\bar{\chi}}_{f'}(x') \cdots = \cdots \delta_{ff'} G(x,x') \cdots , \tag{A.8}$$

where n indicates the partners of the contraction, and G is the inverse fermion matrix, see (4.26). We also need

$$\cdots \overset{n}{\bar{\chi}}_f(x)\, \overset{n}{\chi}_{f'}(x') \cdots = \cdots - \overset{n}{\chi}_{f'}(x')\, \overset{n}{\bar{\chi}}_f(x) \cdots = \cdots - \delta_{f'f} G(x',x) \cdots . \tag{A.9}$$

In our application x' lives on a fixed time slice t_0 whereas x runs over all lattice sites. This would require computing the entire inverse fermion matrix, as opposed to only one column. Here a symmetry can be employed. In the staggered fermion scheme the propagator satisfies [75]

$$G(x',x) = (-1)^{x'} G^\dagger(x,x')\,(-1)^{-x} \quad \text{with} \quad (-1)^x = (-1)^{x_1+\cdots+x_d} , \tag{A.10}$$

which is actually obvious from (4.26). (The Wilson fermion version of this is (4.180).) Thus

$$\cdots \overset{n}{\bar{\chi}}_f(x)\, \overset{n}{\chi}_{f'}(x') \cdots = \cdots - \delta_{ff'} (-1)^{x'} G^\dagger(x,x') (-1)^{-x} \cdots . \tag{A.11}$$

It is straightforward to check that the correlation matrices $C^{(2)}$ and $C^{(4)}$ of Sec. 4.2.2 constructed with the fields (A.7) give exactly the same results as stated in (4.38) and (4.42) respectively. The reason is the cancellation of the phase factors appearing in (A.7) and (A.11).

For the purpose of concise presentation we omitted the phase factors in Sec. 4.2.2.

Appendix B: Improved Lattice Actions

The history of improved lattice actions originates with work by Symanzik [39, 40, 123]. The idea is to cancel discretization errors through adding additional, carefully tuned, terms to the lattice action. Quantum effects (radiative corrections) play an important role in their construction. Symanzik's improvement program has received considerable attention — see, for example, [124, 125, 126, 127, 128] since its inception, but has only recently evolved into a form which is finding widespread application through the work of Lepage and Mackenzie [38]. Through a combination of beyond-elementary geometric structures in the lattice action and accounting for part of the quantum fluctuations through numerically self-consistent renormalization, lattice simulations now operate considerably closer to the continuum limit than ever before. With improved lattice actions results, which can be compared to experiment, are now within reach in many cases. With regard to the current topic (hadronic interactions) improved actions are important because they allow us to use large lattices, say $L \approx 4$ fm or so, with relatively moderate computing facilities.

Improvement of the Lepage-Mackenzie flavor has been well publicized [42]. Educational articles (for beginners) detailing the basics, ramifications, and physical significance of this development have been written by its originators [129, 130]. There is no point in duplicating those writings. However, given the importance of large lattices for hadronic interactions it is appropriate to devote at least an appendix to the basics of this development. We will do so by way of selected examples.

B.1: A scalar example

In the lattice discretization of the space-time continuum derivatives are approximated by finite differences. For example, considering a scalar field theory,

$$\partial^2 \phi(x) \approx \sum_\mu \frac{\phi(x + a\hat{\mu}) + \phi(x - a\hat{\mu}) - 2\phi(x)}{a^2} . \tag{B.1}$$

In the classical limit, this approximation is 'good' if $\phi(x)$ is smooth on a scale larger than the lattice spacing a. We may give a more precise meaning to this statement by saying that the Fourier transform

$$\tilde{\phi}(p) = (2\pi)^{-2} \int d^4x \, e^{ip \cdot x} \phi(x) \tag{B.2}$$

is zero (within some bound ϵ) for momenta $|p| > \pi/a$. The error of the finite-difference approximation (B.1) is $o(a^2)$, as a Taylor expansion shows. In fact we may systematically improve upon the classical discretization by including next-nearest-neighbor finite differences. Writing

$$\phi(x + a\hat{\mu}) = \sum_{n=0}^{\infty} \frac{a^n}{n!} \partial_\mu^n \phi(x) = e^{a\partial_\mu} \phi(x) \tag{B.3}$$

we may solve the four equations

$$e^{\pm a\partial_\mu} = 1 \pm a\partial_\mu + \frac{1}{2}a^2\partial_\mu^2 \pm \frac{1}{6}a^3\partial_\mu^3 + \frac{1}{24}a^4\partial_\mu^4 \pm \cdots \tag{B.4}$$

$$e^{\pm 2a\partial_\mu} = 1 \pm 2a\partial_\mu + 2a^2\partial_\mu^2 \pm \frac{4}{3}a^3\partial_\mu^3 + \frac{2}{3}a^4\partial_\mu^4 \pm \cdots \tag{B.5}$$

for ∂_μ^n, $n = 1 \cdots 4$, in terms of powers of single-step shift operators

$$\Delta_\mu^{(\pm)} = e^{\pm a\partial_\mu}\,. \tag{B.6}$$

Specifically we find

$$a\partial_\mu = \frac{2}{3}(\Delta_\mu^{(+)} - \Delta_\mu^{(-)}) - \frac{1}{12}(\Delta_\mu^{(+)2} - \Delta_\mu^{(-)2}) + o(a^5) \tag{B.7}$$

$$a^2\partial_\mu^2 = \frac{4}{3}(\Delta_\mu^{(+)} + \Delta_\mu^{(-)} - 2) - \frac{1}{12}(\Delta_\mu^{(+)2} + \Delta_\mu^{(-)2} - 2) + o(a^6). \tag{B.8}$$

Introducing the finite-difference derivatives

$$\Delta_\mu^{(1)} = \frac{\Delta_\mu^{(+)} - \Delta_\mu^{(-)}}{2a} \quad \text{and} \quad \Delta_\mu^{(2)} = \frac{\Delta_\mu^{(+)} + \Delta_\mu^{(-)} - 2}{a^2} \tag{B.9}$$

we may also write

$$\partial_\mu = \Delta_\mu^{(1)} - \frac{1}{12}a^2(\Delta_\mu^{(1)}\Delta_\mu^{(2)} + \Delta_\mu^{(1)}\Delta_\mu^{(2)}) + o(a^4) \tag{B.10}$$

$$\partial_\mu^2 = \frac{4}{3}\Delta_\mu^{(2)} - \frac{1}{3}\Delta_\mu^{(1)2} + o(a^4)\,. \tag{B.11}$$

Note that $\Delta_\mu^{(1)}$ and $\Delta_\mu^{(2)}$ commute. Also, it is straightforward to generalize the expansions to higher orders in a in a systematic way. Thus the continuum action, say for a scalar Klein-Gordon field with quartic self interaction

$$S[\phi] = \int d^4x[-\frac{1}{2}(\partial_\mu\phi)^2 + \frac{1}{2}m^2\phi^2 + \frac{g}{4!}\phi^4] \tag{B.12}$$

may then be discretized, alternatively, as

$$S[\phi] \approx \beta a^4 \sum_x \left[-\frac{1}{2}\sum_\mu \left(\Delta_\mu^{(1)}\phi(x) - \frac{1}{6}a^2\Delta_\mu^{(1)}\Delta_\mu^{(2)}\phi(x) \right)^2 + \frac{1}{2}m^2\phi^2(x) + \frac{g}{4!}\phi^4(x) \right], \tag{B.13}$$

or

$$S[\phi] \approx \quad \beta a^4 \sum_x \left[-\frac{1}{2} \sum_\mu \phi(x) \left(\frac{4}{3} \Delta_\mu^{(2)} - \frac{1}{3} \Delta_\mu^{(1)2} \right) \phi(x) \right.$$
$$\left. + \frac{1}{2} m^2 \phi^2(x) + \frac{g}{4!} \phi^4(x) \right] . \tag{B.14}$$

The above lattice actions are equivalent up to $o(a^4)$. Evidently the advantage of $o(a^4)$ improvement is that one works close to the continuum limit ($a \to 0$), or from a numerical point of view, at the same 'closeness' to the continuum theory on can afford a coarser (larger a) lattice which translates into less computational cost.

Since a classical field $\phi(x)$ is smooth on some scale, say a, improvement should be expected to work at some level of accuracy.

However, upon quantizing the theory, smoothness of the field is lost. In the generating functional

$$Z[J] = \int [d\phi] \, e^{-S[\phi] + J \cdot \phi} \tag{B.15}$$

$\phi(x)$ and $\phi(y)$ are independent integration variables for $x \neq y$ even if x and y are infinitesimally close. At a given scale, say a, those fluctuations $|\phi(x) - \phi(y)|$ can in principle be large for $|x - y| \ll a$. Their importance is determined by the Boltzmann weight in the euclidean path integral. In a manner of speaking we say that the quantum fields are rough on any scale, see Fig. 4.28. This means that the Fourier transform (B.2) is unbounded and, generally speaking, momentum-space integrals diverge. The well-known solution is renormalization. In a nutshell, introducing a momentum cutoff Λ, which makes the momentum-space integrals finite, then renormalizing the fields

$$\phi(x) = Z_\Lambda^{-1} \phi_\Lambda(x) , \tag{B.16}$$

couplings *etc.*, and, at the level of physical quantities (observables) removing the cutoff Λ, becomes an integral part of the quantum field theory considered.

Clearly, lattice discretization introduces a natural cutoff momentum

$$\Lambda = \frac{\pi}{a} . \tag{B.17}$$

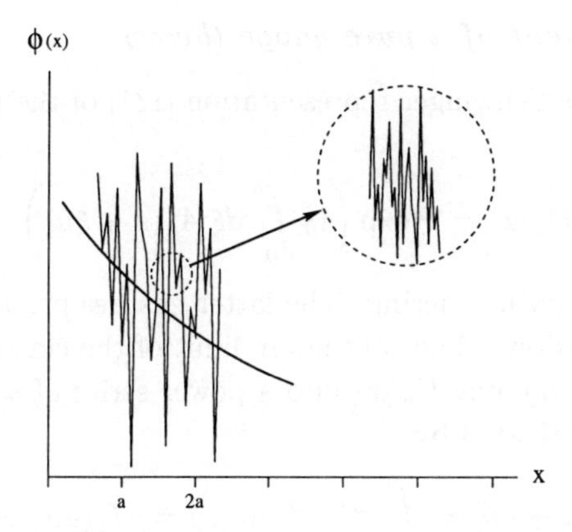

Fig. 4.28 Roughness of the quantum field on any scale. The smooth curve illustrates a Fourier average with a finite momentum cutoff Λ.

In a way, the roughness of the fields at scales less than the lattice spacing a is rendered harmless at the expense of introducing renormalization. (One may think of Z_Λ, for example, as playing a role similar to a dielectric constant κ_e for a medium of electric dipoles. The small-scale fluctuations of the electric field on the scale of the dipoles are lumped into κ_e.)

The field renormalization constant is without consequence in the present example because it can be made to disappear by changing integration variables $\phi(x) \to \phi_\Lambda(x)$ in the generating functional (B.15). Thus

$$Z[J_\Lambda] = \text{const} \,.\int [d\phi_\Lambda]\, e^{-S_\Lambda[\phi_\Lambda]+J_\Lambda\cdot\phi_\Lambda}, \tag{B.18}$$

where, S_Λ is as in (B.13, B.14) except that $\beta_\Lambda = \beta Z_\Lambda^{-2}$ and $g_\Lambda = g Z_\Lambda^{-4}$. Improvement does not change the physics of the quantum theory in this particular example of the scalar interacting Klein-Gordon field.

This culture of thinking has, perhaps, impeded a much earlier discovery of the type of improvement now routinely used in lattice gauge simulations. The above line of thought does lead to a genuinely different situation in the case of gauge theories.

B.2: Improvement of a pure gauge theory

We start from the Schwinger representation [131] of the (lattice) link variables

$$U_\mu(x) = \mathcal{P}\exp\left(ig\int_0^a d\xi\, A_\mu(x+\xi\hat{\mu})\right), \tag{B.19}$$

where $\mathcal{P}$ denotes path ordering. The latter ensures proper behavior under gauge transformations. The continuum limit of the classical, smooth, field is obtained by expanding $U_\mu(x)$ into a power series of a. Some formalism proves useful. Motivated by

$$\int_0^a d\xi\, A_\nu(x+\xi\hat{\mu}) = \int_0^a d\xi\, e^{\xi\partial_\mu} A_\nu(x) = aT_1(a\partial_\mu)A_\nu(x) \tag{B.20}$$

we define the set of (integrated) translators

$$T_n(X) = \sum_{k=0}^{\infty} \frac{n!}{(k+n)!} X^k\,. \tag{B.21}$$

Those satisfy

$$T_0(X) = e^X \quad , \quad T_n(X) = \mathbf{1} + \frac{1}{n+1} T_{n+1}(X)X\,. \tag{B.22}$$

A straight path along m links in direction $\hat{\mu}$ involves

$$\int_0^{ma} d\xi\, A_\nu(x+\xi\hat{\mu}) = maT_1(ma\partial_\mu)A_\nu(x). \tag{B.23}$$

These can be used as building blocks for various loop operators, for example, the planar $m\times n$ plaquettes, see Fig. 4.29,

$$\begin{aligned} P_{\mu\nu}^{(m\times n)}(x) = \quad & \frac{1}{3}\mathrm{ReTr}\mathcal{P}\exp\left[+ig\int_0^{ma} d\xi\, A_\mu(x+\xi\hat{\mu})\right. \\ & +ig\int_0^{na} d\xi\, A_\nu(x+ma\hat{\mu}+\xi\hat{\nu}) \\ & -ig\int_0^{ma} d\xi\, A_\mu(x+na\hat{\nu}+\xi\hat{\mu}) \\ & \left. -ig\int_0^{na} d\xi\, A_\nu(x+\xi\hat{\nu})\right] \qquad \text{(B.24)} \\ = \quad & \frac{1}{3}\mathrm{ReTr}\mathcal{P}\exp\left[igmna^2 T_1(na\partial_\nu)T_1(ma\partial_\mu)\right. \\ & \left.(\partial_\mu A_\nu(x) - \partial_\nu A_\mu(x))\right]\,. \qquad \text{(B.25)} \end{aligned}$$

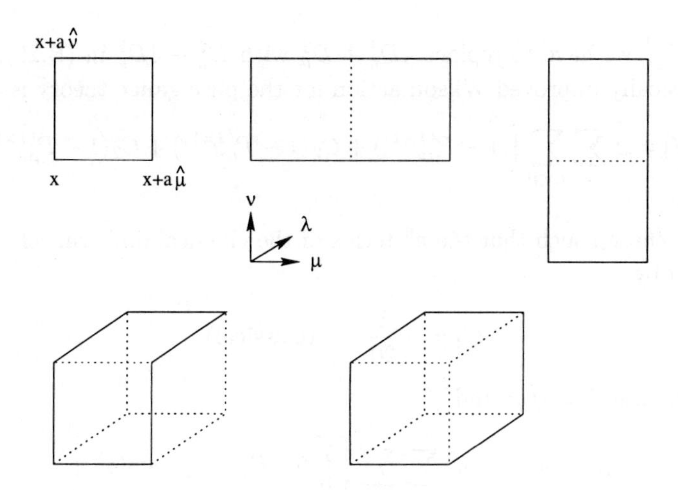

Fig. 4.29 Geometry of some loop operators, elementary 4-link plaquette, 6-link planar rectangles, bent rectangle, and parallelogram, considered for improved gauge field actions.

The combination of path ordering $\mathcal{P}$ and taking the color trace Tr ensures that $P_{\mu\nu}^{(m\times n)}$ is gauge invariant. Manifest gauge invariance may be recovered by replacing ∂_μ with the covariant derivative

$$D_\mu = \partial_\mu - igA_\mu(x)\,. \tag{B.26}$$

As a consequence, the expansion of (B.25) into a power series of a only contains gauge-invariant terms. Specifically, for the 4-link elementary plaquette we obtain

$$\begin{aligned} P_{\mu\nu}^{(1\times1)}(x) &= 1 - \frac{1}{6}a^4\mathrm{Tr}(gF_{\mu\nu}(x))^2 \\ &\quad -\frac{1}{72}a^6\mathrm{Tr}\left(gF_{\mu\nu}(x)(D_\mu^2 + D_\nu^2)gF_{\mu\nu}(x)\right) + o(a^8) \end{aligned} \tag{B.27}$$

and for the 6-link planar rectangle

$$\begin{aligned} P_{\mu\nu}^{(2\times1)}(x) &= 1 - \frac{2}{3}a^4\mathrm{Tr}(gF_{\mu\nu}(x))^2 \\ &\quad -\frac{1}{18}a^6\mathrm{Tr}\left(gF_{\mu\nu}(x)(4D_\mu^2 + D_\nu^2)gF_{\mu\nu}(x)\right) + o(a^8). \end{aligned} \tag{B.28}$$

For $P_{\mu\nu}^{(1\times2)}$ we have to replace $4D_\mu^2 + D_\nu^2$ with $D_\mu^2 + 4D_\nu^2$ in (B.29). Thus the classically improved Wilson action for the pure gauge theory is

$$S_G[U] = \beta \sum_x \sum_{\mu<\nu} \left[(1 - P_{\mu\nu}^{(1\times1)}) + C_2(1 - P_{\mu\nu}^{(2\times1)}) + C_2(1 - P_{\mu\nu}^{(1\times2)})\right] \tag{B.29}$$

with C_2 chosen such that the a^6 terms of the classical limit cancel. This is achieved by

$$C_2 = -\frac{1}{20} \qquad \text{(classical)} \tag{B.30}$$

in which case $\beta = 5/g^2$ and

$$S_G[U] \to \qquad \beta a^4 \sum_x \sum_{\mu<\nu} \left[\frac{g^2}{10}\mathrm{Tr}(F_{\mu\nu}(x))^2 + o(a^3)\right] \tag{B.31}$$

$$= a^4 \sum_x \left[\frac{1}{4}\sum_{\mu,\nu} \mathrm{Tr}(F_{\mu\nu}(x))^2 + o(a^3)\right] . \tag{B.32}$$

In the quantized theory, expansion of the gauge fields into a power series may no longer be justified. Nevertheless, we expect the expansion to be 'good' if the expectation values of the plaquette operators (B.28) and (B.29) are 'close' to one,

$$\langle \frac{1}{6L^3T} \sum_x \sum_{\mu<\nu} P_{\mu\nu}^{(m\times n)} \rangle \approx 1 \,. \tag{B.33}$$

This condition can be easily tested in practical situations. In Fig. 4.30 for example, are shown the lattice averages of the 4-link plaquette and the planar 6-link rectangle. The deviations from one are sizeable and increase if the gauge field coupling ($g^2 \propto \beta^{-1}$) is made larger. This effect is, of course, due to quantum fluctuations. Its significance is that:

- *Quantum fluctuations spoil the classical continuum limit.*

Again, the standard procedure to deal with short-ranged (ultraviolet) fluctuations is renormalization. An ultraviolet momentum cutoff is in place through lattice discretization. The gauge fields are represented by the link variables (B.19), thus replace

$$U_\mu(x) \to u^{-1} U_\mu(x) \tag{B.34}$$

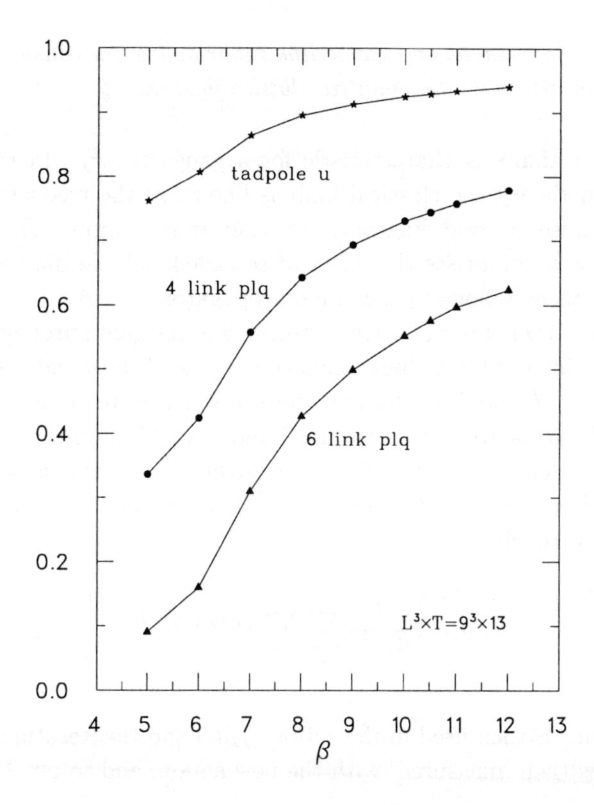

Fig. 4.30 Expectation values of the 6-link plaquette (planar rectangle), the 4-link (elementary) plaquette, and the corresponding tadpole renormalization factor u for versus β.

where u is a renormalization constant. Note that without the 6-link terms in the action $S_G[U]$, see (B.29), the factor u would just renormalize the gauge field coupling, and be harmless (and useless) very much like in the example of Sec. 4.5. Link variable renormalization does not do anything if only the standard Wilson action is used. The crucial point here is that the 6-link improvement terms in the action carry two more factors of link variables. Their fluctuations (in the mean-field sense) can be divided out by two renormalization factors. Hence, using

$$C_2 = -\frac{1}{20u^2} \tag{B.35}$$

in the action results in the cancellation of the $o(a^6)$ terms, as above, between the 4-link and 6-link terms in the action *i.e.*

• *Classical improvement and gauge link renormalization conspire to cancel discretization errors in the quantized lattice field theory.*

This state of affairs is characteristic for a gauge theory. In the lingo of perturbation theory the classical limit is known as the tree-level approximation, because no loop diagrams are taken into account. The renormalization factor u comprises the effect of so-called tadpole diagrams, which are singly-attached loops to the gluon propagator.

To be sure, there are radiative corrections to the gluon propagator which are not contained in the tadpole diagrams. Some of those induce $o(a^2)$ corrections $\propto \alpha_s(\pi/a)$ and, containing loops, are harder to come by [41, 128].

The tadpole factor, on the other hand, can be monitored during an actual simulation. Starting with an arbitrary value, say $u = 1$, after a number of iterations we simple 'measure' the elementary plaquette (for example) and redefine

$$u = \left\langle \frac{1}{6L^3T} \sum_x \sum_{\mu<\nu} P_{\mu\nu}^{(1\times 1)} \right\rangle^{1/4} . \tag{B.36}$$

The new value is then used in the action $S_G[U]$ and the iterations continue. Another u is then 'measured' with the new action, and so on. In this way:

• *The tadpole renormalization factor is determined self consistently during the simulation.*

Numerical work has shown that tree-level and tadpole improvement (even without loop corrections) already account for much of the discretization errors [47, 48]. The effect of improvement is quite dramatic. A well-known example [130] is the restoration of rotational invariance of the static quark-antiquark potential, often parametrized as $V(r) = Kr - \pi/12r + C$. On-axis distances $r/a = 1, 2, 3, 4 \cdots$ and off-axis distances $r/a = \sqrt{2}, \sqrt{3}, ..,$ which are supposed to fall on a smooth curve $V(r)$, are off by as much as 38%. With a tree-level and tadpole improved action errors are reduced to the one-percent level! Improvement schemes are not unique. Various loop geometries (bent, twisted rectangles, *etc.*) may be used to cancel discretization errors. A corresponding statement, of course, holds for improving the fermionic part of the action.

B.3: Improvement of a fermionic action

A simple fermion action, which is improved up to the same order as the gauge field action, can be constructed by employing next-nearest-neighbor terms [46, 132, 133]. Consider the following

$$S_F^{(0)} = a^3 \sum_x \bar{\psi}(x)\psi(x) \tag{B.37}$$

$$\begin{aligned} S_F^{(1)} &= a^3 \sum_x \sum_\mu [\bar{\psi}(x)(r - \gamma_\mu)U_\mu(x)\psi(x + a\hat{\mu}) \\ &\quad + \bar{\psi}(x + a\hat{\mu})(r + \gamma_\mu)U_\mu^\dagger(x)\psi(x)] \end{aligned} \tag{B.38}$$

$$\begin{aligned} S_F^{(2)} &= a^3 \sum_x \sum_\mu [\bar{\psi}(x)(r' - \gamma_\mu)U_\mu(x)U_\mu(x + a\hat{\mu})\psi(x + 2a\hat{\mu}) \\ &\quad + \bar{\psi}(x + 2a\hat{\mu})(r' + \gamma_\mu)U_\mu^\dagger(x + a\hat{\mu})U_\mu^\dagger(x)\psi(x)] \end{aligned} \tag{B.39}$$

where r, r' are parameters. Knowing about the essential ingredients for tree-level and tadpole improvement we construct a fermionic action as a linear combination

$$S_F = c_0 S_F^{(0)} + \frac{c_1}{u} S_F^{(1)} + \frac{c_2}{u^2} S_F^{(2)} . \tag{B.40}$$

The factor u is the tadpole renormalization (B.36) which reflects the relative number of gauge links in $S_F^{(0)}$ and $S_F^{(1)}$, $S_F^{(2)}$. We need the following classical continuum limits

$$\begin{aligned} S_F^{(1)} &= 8r\, a^3 \sum_x \bar{\psi}\psi - 2a^4 \sum_x \bar{\psi}\gamma_\mu D_\mu \psi \\ &\quad + r\, a^5 \sum_x \bar{\psi} D_\mu D_\mu \psi - \frac{1}{3} a^6 \sum_x \bar{\psi}\gamma_\mu D_\mu D_\nu D_\nu \psi + o(a^7) \end{aligned} \tag{B.41}$$

$$\begin{aligned} S_F^{(2)} &= 8r' a^3 \sum_x \bar{\psi}\psi - 4a^4 \sum_x \bar{\psi}\gamma_\mu D_\mu \psi \\ &\quad + 4r' a^5 \sum_x \bar{\psi} D_\mu D_\mu \psi - \frac{8}{3} a^6 \sum_x \bar{\psi}\gamma_\mu D_\mu D_\nu D_\nu \psi + o(a^7). \end{aligned} \tag{B.42}$$

Remarkably it is possible to simultaneously cancel the $o(a^5)$ and $o(a^6)$ terms. Cancellations must be realized on the tree level, $u = 1$. The requirement is

$$\begin{pmatrix} r & 4r' \\ -\frac{1}{3} & -\frac{8}{3} \end{pmatrix} \begin{pmatrix} c_1 \\ c_2 \end{pmatrix} = \begin{pmatrix} 0 \\ 0 \end{pmatrix} . \tag{B.43}$$

A nontrivial solution exists if

$$r' = 2r \qquad \text{with} \qquad c_1 + 8c_2 = 0 . \tag{B.44}$$

We solve the last equation writing

$$c_1 = -\frac{4}{3}c' \qquad c_2 = \frac{1}{6}c', \tag{B.45}$$

where c' is a normalization. Still on the tree level, we then have

$$\begin{aligned} S_F \quad \to \quad & (c_0 - 8c'r)a^3 \sum_x \bar{\psi}\psi + 2c'a^4 \sum_x \bar{\psi}\gamma_\mu D_\mu \psi + o(a^7) \\ & = 2c'a^4 \sum_x \left[\frac{c_0 - 8c'r}{2c'a} \bar{\psi}\psi \; + \bar{\psi}\gamma_\mu D_\mu \psi + o(a^3) \right] \end{aligned} \tag{B.46}$$

for $a \to 0$. Thus, in the continuum limit, the bare fermion mass is $m_f = (c_0 - 8c'r)/2c'a$. The choice

$$r = 1 \tag{B.47}$$

for the Wilson parameter takes care of the doubling problem (violating chiral symmetry). In order to meet with the usual conventions we introduce the 'hopping' parameter κ through

$$c' = \kappa u c_0 \tag{B.48}$$

and choose the normalization

$$c_0 = \frac{1}{a^3} . \tag{B.49}$$

Thus, finally, the improved lattice action (B.40) becomes

$$
\begin{aligned}
S_F = \quad & \sum_x \bar{\psi}(x)\psi(x) - \frac{4\kappa}{3}\sum_x\sum_\mu \left[\bar{\psi}(x)(1-\gamma_\mu)U_\mu(x)\psi(x+a\hat{\mu})\right. \\
& \left. +\bar{\psi}(x+a\hat{\mu})(1+\gamma_\mu)U_\mu^\dagger(x)\psi(x)\right] \\
& +\frac{\kappa}{6u}\sum_x\sum_\mu \left[\bar{\psi}(x)(2-\gamma_\mu)U_\mu(x)U_\mu(x+a\hat{\mu})\psi(x+2a\hat{\mu})\right. \\
& \left. +\bar{\psi}(x+2a\hat{\mu})(2+\gamma_\mu)U_\mu^\dagger(x+a\hat{\mu})U_\mu^\dagger(x)\psi(x)\right] \ . \qquad \text{(B.50)}
\end{aligned}
$$

With this improved action we are hopefully 'close' to the continuum limit, so the fermion mass is 'close' to $m_f = (1-8\kappa u)/2\kappa u a$ with a critical value of $\kappa_c = 1/8u$ for $m_f = 0$.

B.4: More on highly improved actions

An example of a dispersion function computed with the $o(a^2)$ improved action (B.29, B.35) and (B.50) is shown in Fig. 4.31. The filled circles are the energies of a pseudoscalar meson with momentum p from a $5^3 \times 7$ lattice at $\beta = 5.0$. The data are close to a smooth (fitted) curve, as opposed to the open circles, which are from the standard lattice dispersion relation for bosons [88].

Another example of 'closeness to the continuum' achievable with improved lattice actions is shown in Fig. 4.32. The nucleon-rho mass ratio is plotted versus the rho mass. Open plot symbols correspond to the standard, unimproved, Wilson action (gauge and fermion fields). The filled symbols refer to various 'flavors' of improved actions. The experimental value of m_N/m_ρ is 1.22. The two curves in Fig. 4.32, meant to 'guide the eye', illustrate dramatically the 'quantum leap' provided to us by improved actions: Physical results, close to the continuum limit, can be obtained from simulations on relatively coarse lattices with lattice spacings a about a factor of ≈ 4 larger compared to the standard Wilson action. Systematic errors still are large though, at this time.

The lattice actions discussed in Secs. 4.5 and 4.5 are improved up to two leading orders in the lattice constant. Those actions are commonly referred to as $o(a^2)$ improved, or highly improved actions. It should be mentioned that highly improved actions are not without problems. The dispersion function $E(p)$ for free massless quarks, as calculated from the zeros of the inverse lattice propagator, exhibit unphysical branches that describe ad-

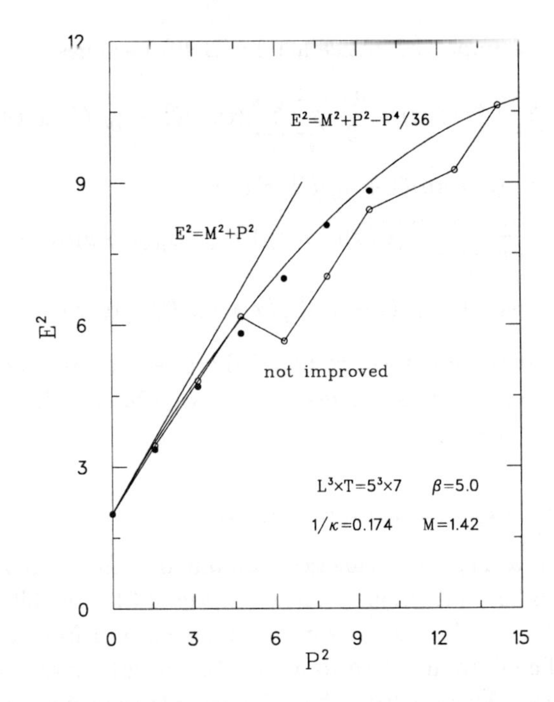

Fig. 4.31 Energy-momentum relation for a pseudoscalar meson obtained from a lattice simulation with the improved action (filled circles), and a fitted curve. Also shown are the unimproved lattice dispersion relation for a free Bose field [88], and the continuum relation.

ditional particles, or ghosts. Hopefully their masses are sufficiently large, so as not to interfere with the propagation of the physical quarks on the lattice. The effect of those particles reveals itself in the early-time-slice behaviour of hadronic correlation functions. Thus, in practice, their effect can be isolated and hardly poses a problem.

A class of highly improved actions that has received much attention recently is known as D234 actions, the nomenclature taken from the way derivative operators are used [137]. Here, in addition to the techniques discussed above, one utilizes a construct known as the Sheikholeslami-Wohlert (SW), or clover, action [45] as well as anisotropic lattices. The latter have different lattice spacings a_s and a_t for space and time respectively, typically $a_s > a_t$. The purpose of combining those features is to push doublers, or ghosts, into the high-mass region where they can do little harm to hadronic physics. Actions of this class are well documented, technical issues are in

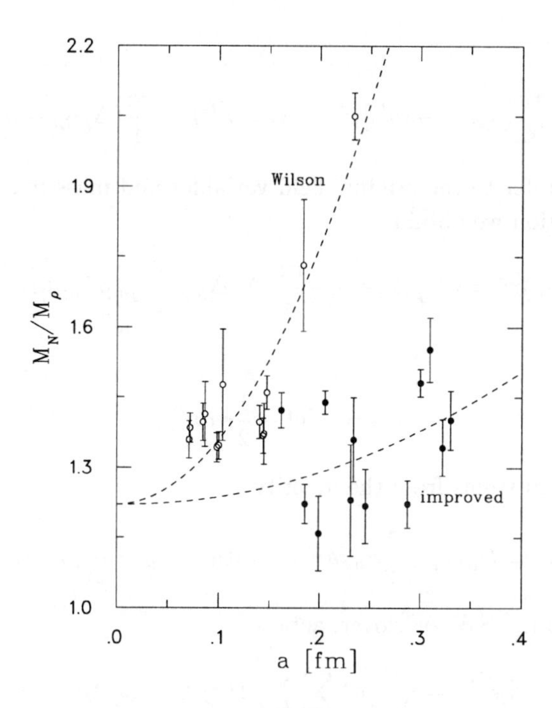

Fig. 4.32 Nucleon ρ-meson mass ratio versus the lattice constant a. The scale for a was obtained from $M_\rho = 770$ MeV. See Ref. [134, 135, 136] to trace the references for the data points.

the foreground at this time [137, 138, 139, 140]. We refrain from discussing these further.

B.5: Clover leaf fermion action

The SW action deserves being mentioned here because of its widespread use and relatively simple implementation.

The idea is to consider (integration) variable transformations of the fermion fields in the path integral that defines the lattice QFT. This will, of course, not alter the physics of the theory but, in general, change the appearance of the lattice action written in terms of the transformed fields. The additional terms are called redundant. Sheikholeslami and Wohlert [45] considered (among an extensive list of other possibilities) the transformation

$$\psi = [\mathbf{1} - \frac{ra}{4}(\gamma_\mu \Delta_\mu - m')]\psi' \qquad \bar{\psi} = \bar{\psi}'[\mathbf{1} - \frac{ra}{4}(\Delta_\mu \gamma_\mu - m')] \tag{B.51}$$

where ψ',$\bar{\psi}'$ refer to the original field variables and mass m'. Working out the Dirac action we obtain

$$\bar{\psi}'(\gamma_\mu \Delta_\mu + m')\psi' = \bar{\psi}[\gamma_\mu \Delta_\mu + m - \frac{ra}{2}(\Delta_\mu \Delta_\mu + \frac{1}{2}\sigma_{\mu\nu} g F_{\mu\nu})]\psi + o(a^2) \tag{B.52}$$

with

$$m = m'(1 + \frac{1}{2} ram'). \tag{B.53}$$

The $\sigma \cdot F$ term stems from the identity

$$(D_\mu \gamma_\mu)^2 = D_\mu D_\mu + \frac{1}{2}\sigma_{\mu\nu} g F_{\mu\nu} \quad \text{with} \quad \sigma_{\mu\nu} = \frac{1}{2i}[\gamma_\mu, \gamma_\nu]. \tag{B.54}$$

This leads to the SW, or clover, action

$$S_F^{(SW)} = \frac{\kappa}{2iu^3} a^4 \sum_x \sum_{\mu,\nu} \bar{\psi}(x) \sigma_{\mu\nu} P_{\mu\nu}(x) \psi(x), \tag{B.55}$$

which is an additive part to the standard Wilson fermion action. The name clover action is apparent from Fig. 4.33 which shows the geometry of link variables involved in

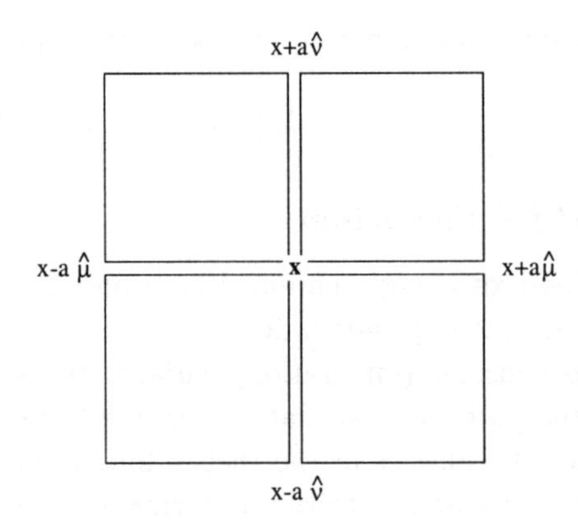

Fig. 4.33 Geometry of the paths used in $P_{\mu\nu}(x)$ for the clover action.

$$
\begin{aligned}
P_{\mu\nu}(x) &= \frac{1}{4}\Big(U_\mu(x)U_\nu(x+a\hat{\mu})U_\mu^\dagger(x+a\hat{\nu})U_\nu^\dagger(x) \\
&\quad - U_\nu^\dagger(x-a\hat{\nu})U_\mu^\dagger(x-a\hat{\mu}-a\hat{\nu})U_\nu(x-a\hat{\mu}-a\hat{\nu})U_\mu(x-a\hat{\mu}) \\
&\quad + U_\nu(x)U_\mu^\dagger(x-a\hat{\mu}+a\hat{\nu})U_\nu^\dagger(x-a\hat{\mu})U_\mu(x-a\hat{\mu}) \\
&\quad - U_\mu(x)U_\nu^\dagger(x+a\hat{\mu}-a\hat{\nu})U_\mu^\dagger(x-a\hat{\nu})U_\nu(x-a\hat{\nu})\Big) . \qquad \text{(B.56)}
\end{aligned}
$$

The action is $o(a)$ improved on the tree level, and local, meaning that no 'hopping' to next neighbors or beyond occurs. This feature is computationally desirable since it leads to simpler and faster codes.

The clover action and the action discussed in Sec. 4.5 are related [141].

Bibliography

[1] H. Yukawa. *Proc. Phys. Math. Soc. Japan*, 17:48, 1935.
[2] H. Yukawa. *Proc. Phys. Math. Soc. Japan*, 19:712, 1937.
[3] M. Taketani, S. Nakamura, and M. Sasaki. *Proc. Theor. Phys.*, 6:581, 1951.
[4] K. Holinde. *Phys. Rept.*, 68:121, 1981.
[5] M. Lacombe *et al. Phys. Rev.*, C23:2405, 1981.
[6] R. Machleidt, K. Holinde, and C. Elster. *Phys. Rept.*, 149:1, 1987.
[7] R. Machleidt. *Adv. Nucl. Phys.*, 19:189, 1989.
[8] D. Lohse, J. W. Durso, K. Holinde, and J. Speth. *Nucl. Phys.*, A516:513, 1990.
[9] G. Janssen, B. C. Pearce, K. Holinde, and J. Speth. *Phys. Rev.*, C52:2690, 1995.
[10] Virginia Tech Partial-Wave Analysis Facility (SAID), CNS DAC Services, http://gwdac.phys.gwu.edu/, 1997-2000.
[11] H. W. Hamber, E. Marinari, G. Parisi, and C. Rebbi. *Nucl. Phys.*, B225:475, 1983.
[12] M. Lüscher. *Commun. Math. Phys.*, 104:177, 1986.
[13] M. Lüscher. *Commun. Math. Phys.*, 105:153, 1986.
[14] M. Guagnelli, E. Marinari, and G. Parisi. *Phys. Lett.*, B240:188, 1990.
[15] S. R. Sharpe. *Nucl. Phys. Proc. Suppl.*, 20:406, 1991.
[16] S. R. Sharpe, R. Gupta, and G. W. Kilcup. *Nucl. Phys.*, B383:309, 1992.
[17] R. Gupta, A. Patel, and S. R. Sharpe. *Phys. Rev.*, D48:388, 1993.
[18] Y. Kuramashi, M. Fukugita, H. Mino, M. Okawa, and A. Ukawa. *Phys. Rev. Lett.*, 71:2387, 1993.
[19] Y. Kuramashi, M. Fukugita, H. Mino, M. Okawa, and A. Ukawa. *Nucl. Phys. Proc. Suppl.*, 34:117, 1993.
[20] M. Fukugita, Y. Kuramashi, H. Mino, M. Okawa, and A. Ukawa. *Phys. Rev. Lett.*, 73:2176, 1994.

[21] M. Fukugita, Y. Kuramashi, M. Okawa, H. Mino, and A. Ukawa. *Phys. Rev.*, D52:3003, 1995.
[22] Y. Kuramashi. *Prog. Theor. Phys. Suppl.*, 122:153, 1996.
[23] S. Aoki *et al. Nucl. Phys. Proc. Suppl.*, 83:241, 2000.
[24] S. Aoki *et al. Nucl. Phys. Proc. Suppl.*, 106:230, 2002.
[25] S. Aoki *et al. hep-lat/0110151*, 2002.
[26] S. Aoki *et al. hep-lat/0206011*, 2002.
[27] M. Lüscher and U. Wolff. *Nucl. Phys.*, B339:222, 1990.
[28] M. Lüscher. *Nucl. Phys.*, B354:531, 1991.
[29] H. Markum, M. Meinhart, G. Eder, M. Faber, and H. Leeb. *Phys. Rev.*, D31:2029, 1985.
[30] D. G. Richards. *Nucl. Phys. Proc. Suppl.*, 9:181, 1989.
[31] D. G. Richards, D. K. Sinclair, and D. W. Sivers. *Phys. Rev.*, D42:3191, 1990.
[32] K. Rabitsch, H. Markum, and W. Sakuler. *Phys. Lett.*, B318:507, 1993.
[33] A. Mihály, H. R. Fiebig, H. Markum, and K. Rabitsch. *Phys. Rev.*, D55:3077, 1997.
[34] C. Michael and P. Pennanen. *Phys. Rev.*, D60:054012, 1999.
[35] H. R. Fiebig, K. Rabitsch, H. Markum, and A. Mihály. *Few Body Syst.*, 29:95, 2000.
[36] J. R. Taylor. *Scattering Theory.* Wiley, New York, 1972.
[37] K. G. Wilson. *Phys. Rev.*, D10:2445, 1974.
[38] G. P. Lepage and P. B. Mackenzie. *Phys. Rev.*, D48:2250, 1993.
[39] K. Symanzik. *Nucl. Phys.*, B226:187, 1983.
[40] K. Symanzik. *Nucl. Phys.*, B226:205, 1983.
[41] M. Lüscher and P. Weisz. *Phys. Lett.*, B158:250, 1985.
[42] G. P. Lepage. *Nucl. Phys. Proc. Suppl.*, 47:3, 1996.
[43] C. DeTar and S. Gottlieb, Physics Today, February 2004, 45.
[44] H. Neuberger, "Lattice Field Theory: past, present and future", hep-ph/0402148.
[45] B. Sheikholeslami and R. Wohlert. *Nucl. Phys.*, B259:572, 1985.
[46] T. Eguchi and N. Kawamoto. *Nucl. Phys.*, B237:609, 1984.
[47] M. G. Alford, W. Dimm, G. P. Lepage, G. Hockney, and P. B. Mackenzie. *Nucl. Phys. Proc. Suppl.*, 42:787, 1995.
[48] M. G. Alford, W. Dimm, G. P. Lepage, G. Hockney, and P. B. Mackenzie. *Phys. Lett.*, B361:87, 1995.
[49] J. B. Kogut and L. Susskind. *Phys. Rev.*, D11:395, 1975.
[50] L. Susskind. *Phys. Rev.*, D16:3031, 1977.
[51] N. Kawamoto and J. Smit. *Nucl. Phys.*, B192:100, 1981.
[52] P. D. Coddington, A. J. G. Hey, A. A. Middleton, and J. S. Townsend. *Phys. Lett.*, B175:64, 1986.
[53] E. Dagotto, J. B. Kogut, and A. Kocić. *Phys. Rev. Lett.*, 62:1083, 1989.
[54] H. R. Fiebig and R. M. Woloshyn. *Phys. Rev.*, D42:3520, 1990.
[55] S. Ma. *Statistical Mechanics.* World Scientific, Philadelphia - Singapore,

1985.
[56] G. G. Batrouni, G. R. Katz, A. S. Kronfeld, G. P. Lepage, B. Svetitsky, and K. G. Wilson. *Phys. Rev.*, D32:2736, 1985.
[57] S. Duane and J. B. Kogut. *Nucl. Phys.*, B275:398, 1986.
[58] R. T. Scalettar, D. J. Scalapino, and R. L. Sugar. *Phys. Rev.*, B34:7911, 1986.
[59] S. Gottlieb, W. Liu, D. Toussaint, R. L. Renken, and R. L. Sugar. *Phys. Rev.*, D35:3972, 1987.
[60] S.-J. Dong and K.-F. Liu. *Phys. Lett.*, B328:130, 1994.
[61] M. Hamermesh. *Group Theory and its Application to Physical Problems.* Addison-Wesley, Reading, 1964.
[62] J. Canosa and H. R. Fiebig. *Phys. Rev.*, D55:1487, 1997.
[63] H. Feshbach. *Theoretical Nuclear Physics.* Wiley, New York, 1992.
[64] H. R. Fiebig, A. Dominguez, and R. M. Woloshyn. *Nucl. Phys.*, B418:649, 1994.
[65] A. Messiah. *Quantum Mechanics - Vol. II.* North-Holland, Amsterdam, 1966.
[66] H. R. Fiebig, H. Markum, A. Mihály, K. Rabitsch, W. Sakuler, and C. Starkjohann. *Nucl. Phys. Proc. Suppl.*, 47:394, 1996.
[67] H. R. Fiebig, O. Linsuain, H. Markum, and K. Rabitsch. *Phys. Lett.*, B386:285, 1996.
[68] F. Zimmermann, J. Westphalen, M. Göckeler, and H. A. Kastrup. *Nucl. Phys. Proc. Suppl.*, 30:879, 1993.
[69] M. Göckeler, H. A. Kastrup, J. Westphalen, and F. Zimmermann. *Nucl. Phys.*, B425:413, 1994.
[70] M. Göckeler, H. A. Kastrup, J. Viola, and J. Westphalen. *Nucl. Phys. Proc. Suppl.*, 47:831, 1996.
[71] A. M. Green, C. Michael, J. E. Paton, and M. E. Sainio. *Int. J. Mod. Phys.*, E2:479, 1993.
[72] A. M. Green, C. Michael, and J. E. Paton. *Nucl. Phys.*, A554:701, 1993.
[73] C. Stewart and R. Koniuk. *Phys. Rev.*, D57:5581, 1998.
[74] A. M. Chaara, H. Kröger, L. Marleau, K. J. M. Moriarty, and J. Potvin. *Phys. Lett.*, B336:567, 1994.
[75] G. W. Kilcup and S. R. Sharpe. *Nucl. Phys.*, B283:493, 1987.
[76] M. M. Nagels *et al.* *Nucl. Phys.*, B147:189, 1979.
[77] O. Dumbrajs *et al.* *Nucl. Phys.*, B216:277, 1983.
[78] S. Weinberg. *Phys. Rev. Lett.*, 17:616, 1966.
[79] C. Liu, J.-h. Zhang, Y. Chen, and J. P. Ma. *Nucl. Phys.*, B624:360–376, 2002.
[80] C.-a. Liu, J.-h. Zhang, Y. Chen, and J. P. Ma. hep-lat/0109010, 2001.
[81] G.-w. Meng, C. Miao, X.-n. Du, and C. Liu. hep-lat/0309048, 2003.
[82] I. Montvay and G. Münster. *Quantum Fields on the Lattice.* Cambridge University Press, Cambridge, UK, 1994.
[83] L. D. McLerran and B. Svetitsky. *Phys. Lett.*, B98:195, 1981.

[84] J. Kuti, J. Polonyi, and K. Szlachanyi. *Phys. Lett.*, B98:199, 1981.
[85] K. Rabitsch. *Zur Nukleon-Nukleon Wechselwirkung aus der Gitter–Quantenchromodynamik.* Diplomarbeit, Technische Universität Wien, Vienna, 1993.
[86] W. Bürger, M. Faber, H. Markum, K. Rabitsch, and W. Sakuler. *Int. J. Mod. Phys.*, C5:387, 1994.
[87] K. Rabitsch. *Hadron-Hadron Potentials from Lattice Quantum Chromodynamics.* PhD thesis, Technische Universität Wien, Vienna, 1997.
[88] H. J. Rothe. *Lattice Gauge Theories – An Introduction*, Volume 43 of *Lecture Notes in Physics.* World Scientific, Singapore, 1992.
[89] A. Mihály. *Studies of Meson-Meson Interactions within Lattice QCD.* PhD thesis, Lajos Kossuth University, Debrecen, 1998.
[90] C. Michael and J. Peisa. *Phys. Rev.*, D58:034506, 1998.
[91] C. Michael and J. Peisa. *Nucl. Phys. Proc. Suppl.*, 60A:55, 1998.
[92] J. Peisa and C. Michael. *Nucl. Phys. Proc. Suppl.*, 63:338, 1998.
[93] G. M. de Divitiis, R. Frezzotti, M. Masetti, and R. Petronzio. *Phys. Lett.*, B382:393, 1996.
[94] D. E. Groom *et al. Eur. Phys. J.*, C15:1, 2000.
[95] L. Maiani and M. Testa. *Phys. Lett.*, B245:585, 1990.
[96] C. Alexandrou, S. Güsken, F. Jegerlehner, K. Schilling, and R. Sommer. *Nucl. Phys.*, B414:815, 1994.
[97] D. Weingarten. *Phys. Lett.*, B109:57, 1982.
[98] U. Glässner, S. Güsken, T. Lippert, G. Ritzenhöfer, K. Schilling, and A. Frommer. *Int. J. Mod. Phys.*, C7:635, 1996.
[99] F. S. Beckman. In A. Ralston and H. S. Wilf, editors, *Mathematical Methods for Digital Computers.* Wiley, New York, 1960.
[100] H. R. Fiebig, H. Markum, A. Mihály, K. Rabitsch, and R. M. Woloshyn. *Nucl. Phys. Proc. Suppl.*, 63:188, 1998.
[101] W. Hoogland *et al. Nucl. Phys.*, B126:109, 1977.
[102] T. Yamazaki *et al.* hep-lat/0309155, 2003.
[103] C. D. Froggatt and J. L. Petersen. *Nucl. Phys.*, B129:89, 1977.
[104] J. Gasser and H. Leutwyler. *Phys. Lett.*, B125:325, 1983.
[105] J. Gasser and H. Leutwyler. *Ann. Phys.*, 158:142, 1984.
[106] P. Estabrooks and A. D. Martin. *Nucl. Phys.*, B79:301, 1974.
[107] G. Grayer *et al. Nucl. Phys.*, B75:189, 1974.
[108] W. Männer. In D. A. Garelick, editor, *Experimental Meson Spectroscopy - 1974*, page 22. AIP Conf. Proc. 21, Boston, Mass., 1974.
[109] P. Baillon *et al. Phys. Lett.*, B38:555, 1972.
[110] M. Sander. *Quantum Inversion and Hadron Hadron Interactions. (in German).* Shaker, Aachen, 1997. ISBN 3–8265–2129–3.
[111] W. Ochs. *PiN Newslett.*, 3:25, 1991.
[112] S. Capstick *et al.* Key issues in hadronic physics. hep-ph/0012238, 2000. APS Division of Nuclear Physics Town Meeting on Electromagnetic and Hadronic Physics, Newport News, Virginia, 1-4 Dec 2000.

[113] H. Neuberger. *Ann. Rev. Nucl. Part. Sci.*, 51:23, 2001.
[114] G. Colangelo, J. Gasser, and H. Leutwyler. *Nucl. Phys.*, B603:125, 2001.
[115] Y. Nakahara, M. Asakawa, and T. Hatsuda. *Phys. Rev.*, D60:091503, 1999.
[116] G. P. Lepage, B. Clark, C. T. H. Davies, K. Hornbostel, P. B. Mackenzie, C. Morningstar, and H. Trottier. *Nucl. Phys. Proc. Suppl.*, 106:12, 2002.
[117] H. R. Fiebig. *Phys. Rev.*, D65:094512, 2002.
[118] M. Lüscher. *Nucl. Phys.*, B364:237, 1991.
[119] K. G. Wilson. In A. Zichichi, editor, *New Phenomena in Subnuclear Physics*, page 69. Plenum Press, New York, 1975.
[120] C. Burden and A. N. Burkitt. *Europhys. Lett.*, 3:545, 1987.
[121] H. Kluberg-Stern, A. Morel, O. Napoly, and B. Peterson. *Nucl. Phys.*, B220:447, 1983.
[122] W. Wilcox and R. M. Woloshyn. *Phys. Rev.*, D32:3282, 1985.
[123] K. Symanzik. In R. Schrader, R. Seiler, and D. A. Uhlenbrock, editors, *Mathematical Problems in Theoretical Physics – Lecture Notes in Physics 153*. Springer, Berlin, 1982.
[124] G. Curci, P. Menotti, and G. Paffuti. *Phys. Lett.*, B130:205, 1983. Erratum-ibid.B135:516,1984.
[125] P. Weisz. *Nucl. Phys.*, B212:1, 1983.
[126] P. Weisz and R. Wohlert. *Nucl. Phys.*, B236:397, 1984.
[127] M. Lüscher. Improved lattice gauge theories. In K. Osterwalder and R. Stora, editors, *Critical Phenomena, Random Systems, Gauge Theories - Les Houches, Session XLIII, 1984*, pages 359–374. Elsevier, Amsterdam, 1986.
[128] M. Lüscher and P. Weisz. *Commun. Math. Phys.*, 97:59, 1985. Erratum-ibid.98:433,1985.
[129] G. P. Lepage. Lattice QCD for Small Computers. In S. Raby and T. Walker, editors, *The Building Blocks of Creation*. World Scientific, Singapore, 1994.
[130] G. P. Lepage. Redesigning Lattice QCD. In H. Latal and W. Schweiger, editors, *Perturbative and Nonperturbative Aspects of Quantum Field Theory - Springer Lecture Notes in Physics 479*, pages 1–48. Springer, Heidelberg, 1997.
[131] C. Itzykson and J.-B. Zuber. *Quantum Field Theory*. McGraw-Hill, New York, 1980.
[132] H. W. Hamber and C. M. Wu. *Phys. Lett.*, B133:351, 1983.
[133] W. Wetzel. *Phys. Lett.*, B136:407, 1984.
[134] H. R. Fiebig and R. M. Woloshyn. *Phys. Lett.*, B385:273, 1996.
[135] S. Collins, R. G. Edwards, U. M. Heller, and J. H. Sloan. *Nucl. Phys. Proc. Suppl.*, 53:877, 1997.
[136] F. X. Lee and D. B. Leinweber. *Phys. Rev.*, D59:074504, 1999.
[137] M. G. Alford, T. Klassen, and G. P. Lepage. *Nucl. Phys. Proc. Suppl.*, 47:370, 1996.
[138] M. G. Alford, T. R. Klassen, and G. P. Lepage. *Nucl. Phys. Proc. Suppl.*, 53:861, 1997.

[139] M. G. Alford, T. R. Klassen, and G. P. Lepage. *Nucl. Phys.*, B490:377, 1997.

[140] M. G. Alford, T. R. Klassen, and G. P. Lepage. *Nucl. Phys. Proc. Suppl.*, 63:862, 1998.

[141] G. Martinelli, C. T. Sachrajda, and A. Vladikas. *Nucl. Phys.*, B358:212, 1991.

Chapter 5

Bridges from Lattice QCD to Nuclear Physics

A.M. Green

Helsinki Institute of Physics, P.O. Box 64, FIN–00014, Finland
E-mail:anthony.green@helsinki.fi

A review is given of attempts to bridge the gap between everyday particle and nuclear physics — involving many quarks — and the basic underlying theory of QCD that can only be evaluated exactly for few quark systems. Even the latter requires the original theory of QCD to be discretised to give Lattice QCD — but this modification can still yield *exact* results for the original theory. These LQCD results can then be considered on a similar footing to experimental data — namely as cornerstones that must be fitted by phenomenological models. In this way, the hope is that "QCD inspired" models can become more and more "QCD based" models, by fixing — in the few-quark case where LQCD can be carried out — the form of these models in such a way that they can be extended to multi-quark systems.

5.1 Introduction

Even though for over 30 years QCD has been thought to be the theory of strong interactions, it has had a rather limited impact on most other branches of physics — except for few-quark hadron physics. Of course, the reason is well known — to write down the Lagrangian that describes *exactly* the quark and gluon interactions is easy, but actually performing calculations *directly* with this Lagrangian has turned out to be extremely difficult. The one exception to this last statement is at high energies, where — due to asymptotic freedom — the interactions become sufficiently weak for perturbation theory to be applicable and this has had much success [1]. However, most of "everyday" physics is far from this limit. Furthermore, some quantities such as masses depend on the interaction strength

g as $\exp(-1/g^2)$, which immediately rules out a perturbation expansion in powers of g.

This inability to treat the QCD Lagrangian directly has led to several different types of approximation being made. Essentially these fall into two broad categories: the numerical and the effective theory approaches. Unfortunately, the latter — especially the Effective Potential Theories (EPTs), which are the main subject of this chapter — often have little overlap with the numerical approaches. Those approaches that use the direct numerical way concentrate mainly on the description of single mesons or baryons *i.e.* $q\bar{q}$ or qqq states, whereas most of the EPTs tend to concentrate more on multiquark systems. However, even though these EPTs are often advertized as being "QCD motivated", in most cases they are simply based on many-body ideas and techniques that are well founded in nuclear physics — but are not necessarily justified for the description of multiquark states. The main purpose of this chapter is to see to what extent these "nuclear physics motivated" methods are justified. But first a few general words should be said about these two approaches.

5.1.1 *Numerical treatment of QCD*

The numerical treatment of the QCD Lagrangian — Lattice QCD — has been the main subject of this volume and so, to avoid too much repetition, only points relevant to this chapter will be mentioned. In Lattice QCD the original exact Lagrangian is replaced by an approximate form that is discretized on a 4-dimensional lattice with links of length a. This discretization can be done in many ways, but in all cases the original QCD Lagrangian must be recovered as $a \to 0$. The discretization also reduces the subject to being mainly numerical. However, as emphasized by Lüscher [2]: "In general, numerical simulations have the reputation of being an approximate method that mainly serves to obtain qualitative information on the behaviour of complex systems. This is, however, not so in lattice QCD, where the simulations produce results that are exact (on a given lattice) up to statistical errors. The systematic uncertainties related to the non-zero lattice spacing and the finite lattice volume then still need to be investigated, but these effects are theoretically well understood and can usually be brought under control." Therefore, in order to recover results that are appropriate to the original continuum Lagrangian, two main limits need to be studied:

- **Limit 1. Are the results stable as $\mathbf{a} \to \mathbf{0}$?**
Of course, this limit must be approached with consideration of the number of spatial sites in the lattice N^3, since the volume (V) of the system being studied, *e.g.* 2-, 3- or 4-quarks, should be much smaller than the physical lattice size $L^3 = (aN)^3$. Furthermore, the Euclidean time — needed to extract observables — should be much greater than L. In practice, usually $T \approx 2L$ suffices. Therefore, the two inequalities that need to be satisfied can be combined as $V^{1/3} \ll L \ll T$. This can instantly lead to problems for a meson–meson system ($q^2\bar{q}^2$) with mesons of size ≈ 1 fm, since this would need $V^{1/3} > 2$ fm, if the gradual separation of the mesons is of interest. In this case, a lattice spacing of ≈ 0.1 fm would require $N > 20$ and so $T > 40$. Such large lattices are used by some groups *e.g.* $32^3 \times 60$ in the study of scattering lengths in Ref. [3] — with even larger lattices $64^3 \times 128$ now becoming feasible [2]. In fact, the progress in computer technology allows the lattice extents to be doubled in all directions roughly every 8 years [2]. Unfortunately, at present many of us have to be satisfied with sizes more like $16^3 \times 24$, which rules out the study of completely separated mesons. One way of partially overcoming this problem is to use so called Improved Actions. These incorporate modifications to the standard lattice Lagrangian in order to remove the lowest order dependences on a. This enables coarser lattices to be used, so that, in some cases, $a \approx 0.5$ fm suffices compared with more usual values of $a \approx 0.1$ fm — see Refs. [4]. In this way, the inequality $V^{1/3} \ll L$ is satisfied by increasing a and not N. Of course, there is often a price to pay. Since improved actions are more complicated, they take more computer time to implement compared with the standard actions. Sometimes this is sufficient to remove the advantage of using a smaller lattice. Having said that, there are some improved actions that seem to be always advantageous. Probably the most common of these is the so-called clover action for improving the quark part of the QCD Lagrangian [5, 6] — see Subsec. 5.1.2.1 and also Appendix B.5 of Chapter 4.

Often the $a \to 0$ limit is checked in three steps:

  - i) First a **benchmark calculation** is performed with, say, a lattice $16^3 \times 24$ and $a = 0.12$ fm. This is the least time and storage consuming of these three steps.

- ii) Then the **finite size** effect $V^{1/3} \ll L$ is checked with a larger lattice, say, $24^3 \times 32$ but with the same $a = 0.12$ fm. This is much more time and storage consuming by a factor of about 4, but will show whether or not the process being treated "fits" into the lattice. It is concluded that the finite size effect is no problem, if these latter results are directly the same, within error bars, as the benchmark results.
- iii) Finally, the calculation is repeated with the above larger lattice but using, say, $a' = 0.08$ fm and the results compared with the above by now including appropriate factors of a'/a. These final **scaled** results should now agree, within error bars, with the results from the above two $a = 0.12$ fm lattices.

A specific example of this **scaling** procedure appears later in Subsec. 5.4.3.3. This raises the question: Given a fixed number of "computer units" for a calculation, then what is the most efficient way of spending these units? As pointed out by Kronfeld [7], it is much more efficient to run at several lattice spacings than to put all the resources onto the finest conceivable lattice. Kronfeld gives the example of a computer budget of 100 units and suggests using 65, 25 and 10 units on a series of coarser lattices with spacings a_0, $a_1 = 2^{1/4}a_0$ and $a_2 = 2^{1/2}a_0$. The time needed to create statistically independent lattice gauge fields grows as $\tau_g \propto a^{-(4+z)}$, where the 4 in the exponent arises because the number of variables to process grows as a^{-4} in a 3+1 dimensional world and $z \approx 1-2$ depending on the algorithm for updating the lattice. In this case the three sets of lattices would have comparable error bars that would only be slightly larger — by a factor of about 1.25 (*i.e.* $1/\sqrt{0.65}$) — than the case of using all 100 units on the finest lattice with a_0. However, using all three sets enables an estimate to be made of the discretization effect. As Kronfeld says "The slightly larger statistical error seems a small price to pay".

- **Limit 2. The mass of the light quarks should be realistic.** This seems to be a much more difficult limit to achieve. In practice, light quarks (u, d) with a mass ~ 100 MeV are often used instead of the true values of less than 10 MeV. This is reflected in the computed ratio of the $\pi-$ and $\rho-$masses $R_{\pi\rho} = m_\pi/m_\rho$. The experimental masses give $R_{\pi\rho} = 0.2$, whereas $R_{\pi\rho} \approx 0.7$, if $m_{u,d} \approx m_s$ — the strange quark mass (a value used in many works). Unfortunately, extrapolating from results using different bare quark masses to get the observed $R_{\pi\rho} = 0.2$ is not straightforward. At sufficiently low values of the bare quark masses the effective field theory of Chiral Perturbation Theory becomes applicable and shows that in these quark mass extrapolations logarithmic terms arise in addition to simple power-law behaviour — see Subsec. 5.1.2.1.

The above limits are discussed in more detail in Ref. [8]. There it is pointed out that QCD is a multiscale problem. Not only is there a characteristic scale of QCD ($\Lambda_{\rm QCD} \sim 200 - 250$ MeV [2]) but also a wide range of quark masses with light quark masses $m_q \approx 10$ MeV up to $m_Q \approx 5$ GeV for the b-quark, leading to the hierarchy

$$m_q \ll \Lambda_{\rm QCD} \ll m_Q. \tag{5.1}$$

But two more scales are needed before QCD can be put on a lattice. Firstly, for light quarks the lattice size (L) must be larger than the size of a light quark *i.e.* $m_q^{-1} \ll L$. Secondly, for heavy quarks the lattice spacing (a) must be finer than the size of such quarks *i.e.* $a \ll m_Q^{-1}$. The hierachy in Eq. 5.1 then becomes

$$L^{-1} \ll m_q \ll \Lambda_{QCD} \ll m_Q \ll a^{-1}. \tag{5.2}$$

However, for numerical reasons it is not possible to satisfy all these conditions. So that, in practice, finite computer resources force the hierachy

$$L^{-1} < m_q \ll \Lambda_{QCD} \ll m_Q \sim a^{-1} \tag{5.3}$$

instead of the idealized one.

In Ref. [9] the dependence on a and the appropriate value of m_π for the work required to obtain a "new" configuration are combined into the single approximate expression

$$\frac{\text{Gflops}}{\text{config}} \approx 0.157 \left(\frac{L}{a}\right)^{3.41} \left(\frac{T}{a}\right)^{1.14} \left(\frac{1}{am_\pi}\right)^{2.77}, \tag{5.4}$$

where 1 Gflop is 10^9 computer operations per second. This shows the $\sim a^{-8}$ dependence that is the main numerical problem for Lattice QCD to overcome and is the reason for being interested in using coarser lattices with improved actions to be discussed in the next Subsection. Estimates similar to Eq. 5.4 are also made in Refs. [2, 10].

For those readers who would like a detailed development of Lattice QCD the text books by Creutz [11], Montvay and Münster [12], and Rothe [13] are recommended. Also there are many review articles and summer school lecture series — see Ref. [7] for a partial listing of these. At the time of writing, some of the most recent reviews for a general audience are listed in Ref. [2].*

5.1.2 *Effective Field/Potential Theories*

Effective theory formulations describing quark–gluon systems fall into distinct categories. On the one extreme are the Effective Field Theories (EFTs) that have a rigorous basis, whereas at the other extreme we have the Effective Potential Theories (EPTs), which are essentially phenomenological being based on models with potentials in differential equations. It is important to discuss these two types of theory separately, since they play very different rôles in the present chapter and should not be confused with each other.

5.1.2.1 *Effective Field Theories (EFTs)*

Effective field theories play a crucial part in extracting continuum (physical) results from the purely numerical lattice techniques. A review of this topic has been given by Kronfeld [7].

In these theories an energy scale (Λ) is introduced. This essentially separates the *short* distance effects (*i.e.* less than $1/\Lambda$), which are lumped into the coupling constants of the theory, from *long* distance effects, which are described explicitly by the operators of the theory. Such theories are then only applicable to processes involving energies less than Λ. In QCD the energy scale characteristic of non-perturbative effects is ~ 1 GeV. There are several theories that fall into this category, examples of which are:

(1) **Symanzik effective field theory**
The most obvious difference between Lattice QCD and the real life

*A popular level review in the February 2004 edition of Physics Today[14] has not been well received by everyone [15].

situation of continuum QCD is the presence of the lattice with spacing a. Only the $a \to 0$ limit has a physical meaning. A systematic way of studying this lattice artifact was developed by Symanzik [16]. He showed how $O(a)$ effects could be removed in a systematic way from lattice results by assuming Λa is small and treating lattice artifacts as perturbations. He achieves this through creating an effective field theory by adding terms with increasing powers of a (and containing parameters c_i) to the basic lattice QCD Lagrangian of Wilson. The c_i parameters are then adjusted (tuned) to kill off the offending $O(a)$ effects. This has now been developed into an industry for generating improved actions that only contain $O(a^2)$ lattice spacing corrections — see Subsec. 4.1.7 and Appendix B of Chapter 4 for a more detailed description and for references.
A simple example of this is the quark–gluon coupling $\Gamma_\mu(p,p')$, where p, p' are the 4-momenta of the initial and final quark. The above strategy is to first replace the continuum coupling $\Gamma^C_\mu(p,p')$ by its lattice counterpart $\Gamma^L_\mu(p,p')$ and then to expand the latter in powers of a *i.e.*

$$\begin{aligned} \Gamma^C_\mu(p,p') &= c\gamma_\mu \longrightarrow \\ \Gamma^L_\mu(p,p') &= c\{\gamma_\mu \cos[\tfrac{1}{2}(p+p')_\mu a] - i\sin[\tfrac{1}{2}(p+p')_\mu a]\} \\ &\longrightarrow c\{\gamma_\mu - \tfrac{i}{2}a(p+p')_\mu + O(a^2)\}. \end{aligned}$$

To remove the $O(a)$ term, Sheikholeslami and Wohlert [5] suggested adding a lattice form of $\sigma_{\mu\nu}F^{\mu\nu}$ to the Wilson action so that $\Gamma^L_\mu(p,p')$ became

$$\begin{aligned} \Gamma^L_\mu(p,p') = c\{\gamma_\mu \cos[\frac{1}{2}(p+p')_\mu a] - i\sin[\frac{1}{2}(p+p')_\mu a] + \\ \frac{1}{2}c_{SW}\sigma_{\mu\nu}\cos[\frac{1}{2}k_\mu a]\sin[\frac{1}{2}k_\nu a]\}. \end{aligned} \quad (5.5)$$

Expanding $\Gamma^L_\mu(p,p') \longrightarrow c\{\gamma_\mu - \frac{i}{2}a[(p+p')_\mu + c_{SW}i\sigma_{\mu\nu}k^\nu] + O(a^2)\}$, where $k = p' - p$ and $\sigma_{\mu\nu} = i[\gamma_\mu, \gamma_\nu]/2$. On the mass shell, if c_{SW} is now "tuned" to unity, then the two terms of $O(a)$ cancel to leave corrections of only $O(a^2)$. When this procedure is applied to Wilson's fermion action the outcome is usually referred to as the **clover** action.

(2) **Chiral Perturbation theory of Gasser and Leutwyler** [17]
Light quarks have a mass $m_l \approx 10$ MeV *i.e.* $m_l \ll \Lambda$. This makes it

numerically impractical to perform lattice QCD calculations with such masses since the algorithms for computing the quark propagators become slower and slower — as seen from Eq. 5.4. Therefore, the procedure to reach this physical region is to first carry out the lattice calculations with a sequence of masses (m_q) in a range of, say, $0.2m_s < m_q < m_s$, where $m_s \sim 100$ MeV – appropriate for the strange quark. Given this sequence of results (masses or matrix elements), the task is then to use a reliable method for extrapolating these results to quark masses appropriate for the light quarks. By far the most successful method for this extrapolation is based on Chiral Perturbation Theory (χPT), which can be viewed as an expansion in m_q/Λ. The numerical data, with m_q in the range $0.2m_s < m_q < m_s$, can then be tested against the leading order (next-to-leading order or next-to-next-to-leading order) prediction of χPT. If this is successful, then it gives confidence in extrapolating m_q to the light quark masses [18].

(3) **Heavy-quark effective theory and Non-relativistic QCD**
For a brief review of Heavy-quark effective theory (HQET) and Non-relativistic QCD (NRQCD) see Ref. [19]. In situations involving heavy quarks — such as B-physics, where some of the quarks have a mass of $m_Q \approx 5$ GeV and are non-relativistic — it is appropriate to make expansions in terms of Λ/m_Q or ν, the relative velocity between the quark and the antiquark in the B-meson. These two expansions are usually referred to, respectively, as HQET — for systems containing a single heavy-quark — and NRQCD for a heavy-quark heavy-antiquark system. One way of deriving these effective theories is to write down the heavy–quark theory as an expansion in terms of Λ/m_Q or ν in the continuum and then replace the derivatives that arise by their lattice counterparts to give Lattice HQET and Lattice NRQCD. These ideas are still being developed. For example in Ref. [20] some of the irrelevant degrees of freedom in NRQCD are integrated out to yield a theory called potential NRQCD (or pNRQCD), which is much simpler to treat.

EFTs are not only used with few-quark systems but also have a long history in few- and many-nucleon systems — see the works of van Kolck *et al.* [21, 22]. This approach was first advocated by Weinberg [23], who illustrated how the nucleon–nucleon and many–nucleon potentials could be qualitatively understood. For example, these arguments show that, if the strength of the NN-potential is $\sim$ 10 MeV, then those of the NNN- and NNNN-potentials are $\sim$ 0.5 and $\sim$ 0.02 MeV respectively — numbers that are in accord with detailed few-nucleon phenomenology based on realistic potentials such as that of Argonne [24]. However, it is not clear that this approach — in spite of its impressive **qualitative** results — could ever compete **quantitatively** with standard meson exchange models for describing the NN-potential, where baryon resonances such as the $\Delta(1236)$ and $N^*(1535)$ are included explicitly. As pointed out by van Kolck himself, even the inclusion of the ρ- and ω-mesons give rise to interactions that are "at present an insurmountable obstacle for a systematic approach".

In multi-particle systems EFTs are often converted into Mean Field Theories (MFTs), in which the emphasis is on single particle properties with all other particles being treated "on the average". Unfortunately, in some applications — such as the equation of state of high density nuclear matter, as encountered in relativistic heavy ion collisions or in neutron stars — there are serious questions concerning their validity. Even advocates of the MFT approach (see, for example, Glendenning on pages 127 and 287 in Ref. [25]) express reservations by writing "In many ways it is not as good a theory as the Schrödinger-based theory of nuclear physics" and "The status of an exotic solution of an effective theory is more tenuous than from a fundamental theory". Others are even more critical — see for example Ref. [26]: "The Relativistic Mean Field approximation is very elegant and pedagogically useful, but is not valid in the context of what is known about nuclear forces $\cdots$. It requires $\mu \langle r \rangle \ll 1$, where $\langle r \rangle$ is the average interparticle distance and $1/\mu$ a meson range. However, for pions $\mu_\pi \langle r \rangle \sim 0.8 - 1.4$ but for vector mesons $\mu_v \langle r \rangle \sim 4.7 - 5.8$."

5.1.2.2 *Effective Potential Theories (EPTs)*

The reason for briefly describing the above Effective Field Theories (EFTs) is to emphasize their difference from Effective Potential Theories (EPTs), which are the main interest in most of this chapter. The above EFTs are an integral part of the development of Lattice QCD and play a crucial rôle in extracting precise continuum results for few (2 or 3) quark systems *i.e.*

Lattice QCD+**EFTs** $\longrightarrow$ Continuum results for few quark systems.

On the other hand, the EPTs attempt to understand (interpret) these continuum results for few quark systems in such a way that the theory can be extended to the multiquark systems of interest to nuclear physics *i.e.*

Continuum results for few quark systems + **EPTs** $\longrightarrow$
Descriptions of multiquark systems (Hadron Physics).

This step is here referred to as "Bridges from Lattice QCD to Nuclear Physics" — the title of this chapter. In most cases this step is mainly phenomenological.

The EFTs mainly concentrate on the properties of a *single* particle, so that, for example, the energy of a multiparticle system is expressed as the sum of the effective masses of the separate particles with the effect of all other particles being treated in an average manner. In this way symmetries of the fundamental Lagrangian can be preserved. However, as said above, the extension of these ideas to more complicated Lagrangians or many-body systems presents problems. On the other hand, in the less ambitious and more phenomenological approach of Effective Potential Theories (EPTs), the emphasis is first on the *two-body* system. In this case, two-body potentials are the main ingredient.

For multi-nucleon systems the NN-potential can be mainly phenomenological or based on EFTs as with the Argonne and Bonn potentials respectively [24, 27]. Those based on EFTs can be generated with varying degrees of ability to describe the two-body system. We have already mentioned the works of Weinberg [23] and van Kolck [21], which follow the procedure — referred to by van Kolck as Weinberg's "theorem":

(1) Identify the relevant degrees of freedom and symmetries involved.
(2) Construct the most general Lagrangian consistent with item (1).
(3) Do standard quantum field theory with this Lagrangian.

The outcome is qualitatively correct being within $\approx 10\%$ of the two-nucleon data — but this is an accuracy that is often insufficient for understanding nuclear phenomena. At the other extreme we have the one-boson-exchange (OBE) potentials in which various meson–baryon couplings are tuned to ensure a good fit to the NN experimental data. These OBE potentials, even though they are based on EFT-like Lagrangians, often incorporate couplings such as $N\Delta(1236)\rho$ that can not be treated systematically by Weinberg's "theorem".

In contrast, to implement EPTs for multiquark systems three ingredients are necessary — a wave equation (differential or integrodifferential), an interquark potential and effective quark masses:

A wave equation.
Since EPTs are not derived from more basic principles, the forms of the wave equations are not predetermined and can vary considerably. Even for two-quark systems there is a choice.

- *A non-relativistic Schrödinger equation.* This is suitable for heavy-quark mesons ($Q\bar{Q}$) with $m_Q \gg 1$ GeV and was the form used in Refs. [28, 29] for extracting the $Q\bar{Q}$-potentials. The best cases for this are the Bottomonium mesons such as the $\Upsilon(b\bar{b}, 9.5\text{GeV})$ since $m_b \approx 5$ GeV.
- *The Dirac equation.* Once the quarks are light, *i.e.* $m_q \ll 1$ GeV, relativistic effects become important and we, therefore, enter the realm of large/small wave function components, pair-creation and quantum field theory. This means that the use of the Dirac equation is less "clean" phenomenologically than the Schrödinger equation, since it deals explicitly with large/small wave function components but not with the related effect of pair-creation. In spite of this, for $Q\bar{q}$ systems such as the B-meson, where one quark is heavy, the Dirac equation is essentially a one-body equation and so the full complications of the two-body relativistic problem are avoided. There are many references where this one-body Dirac equation has been applied, *e.g.* [30, 31, 32, 33].
- *The Bethe–Salpeter equation.* When the two quarks are both light the correct relativistic scattering equation is the full Bethe-Salpeter equation. Unfortunately, direct use of this equation — for physically interesting cases — presents severe problems (see Ref. [34] for a recent discussion). Therefore, it is usually reduced from a four- to a three-dimensional scattering equation by inserting appropriate δ-functions of the energy. This can be carried out in several ways and leads to a number of different two-body equations that are covariant and satisfy relativistic unitarity. Examples are the equations of Blankenbecler–Sugar, Gross, Kadyshevsky, Thompson, Erkelenz–Holinde, ...see Chapter 6 in Ref. [35]. Depending on the problem, these alternatives have their separate advantages. For example, the Gross form — unlike those of Blankenbecler–

Sugar and Thompson — treats the two quarks asymmetrically and has the feature that it reduces to the Dirac equation when one of the quarks becomes infinitely heavy. This suggests that this form is perhaps more appropriate for describing the B-meson. On the other hand, when the quarks are equal in mass, as in the $J/\psi(c\bar{c})$ system, the Blankenbecler–Sugar or Thompson equations are probably preferable.

- *Multi-quark wave equations.* For multi-quark systems the choice of wave equation is very limited with the Resonating Group approach being the most usual — see the review by Oka and Yazaki in Chapter 6 of Ref. [36] and also more recently Refs. [37, 38]. This reduces the interaction between quark clusters ($q^3 + q^3$ for the NN-potential and $q\bar{q} + q\bar{q}$ for meson–meson interactions) to a *non-local* Schrödinger-like equation involving the relative distance between the clusters. This is achieved by integrating out the explicit quark degrees of freedom, which usually requires the introduction of gaussian radial factors in order to carry out the multiple integrals involved. At first sight such factors may seem to be unrealistic with exponential or Yukawa forms being more physical. However, it will be seen in Subsec. 5.9.3.2 that the one-body Dirac equation with a linear confining potential can lead to gaussian forms asymptotically. Unfortunately, the effective mass needed for the quarks is $\approx$ 300–400 MeV and so makes this non-relativistic approach somewhat questionable. On the other hand, the rôle of relativity in many-body systems is still an open question — a recent summary being given by Coester [39].

An interquark potential ($V_{Q\bar{Q}}$) — the second EPT ingredient. For two-quark systems the interquark potential is often taken to have the form suggested by the static limit of infinitely heavy quarks, namely

$$V_{Q\bar{Q}}(r) = -\frac{e}{r} + b_s r + c, \tag{5.6}$$

where the first term is that expected at short distances due to one-gluon exchange and the second term that expected from quark confinement — c simply being an additive constant. However, it should be emphasized that the form in Eq. 5.6 is, strictly speaking, only appropriate for the interaction between static quarks. At the present time, there seems to be no really convincing evidence for a significant one-gluon exchange interaction in systems where only light constituent quarks (up to ~400 MeV) are involved. This becomes evident in the N and Λ spectra. There a one-gluon interaction is unable to describe the empirical ordering of the positive and negative parity states. However, this ordering can be accomplished by Goldstone-boson exchange mechanisms [40], even though that model also has bad features — some states in the Λ and Δ spectra are poorly reproduced. The reason why the one-gluon exchange becomes ineffective with light quarks is because the use of minimal relativity — in, say, the Blankenbecler-Sugar equation for describing the $Q\bar{q}$ interaction — introduces relativistic square root factors. Indeed, if both quarks are light, then the effective one-gluon interaction is essentially flat and very weak for short distances. Since this damping of one-gluon exchange also enters in meson spectra, another mechanism is needed for the necessary short range attraction. A possible candidate for this is an effective interaction generated by instantons, which is usually expressed in terms of an attractive δ-function in r [41].

More details on $V_{Q\bar{Q}}(r)$ can be found in Chapter 3, which is devoted to this topic. The above is a two-body potential. However, for multi-quark systems there are strong indications that multi-quark potentials and/or potentials involving excited gluon states could also play a major rôle — see Secs. 5.5 and 5.7.

For light quarks and for nucleons in dense matter, relativistic effects enter and in some cases these can be expressed as corrections to the non-relativistic potential involved. An example of this is the so-called Relativistic Boost Correction shown to be important in high density nuclear matter [42].

Effective quark masses — the third ingredient for an EPT.
The masses of the quarks involved in this "Wave Equation + Potential" approach are not those of bare quarks *i.e.* not ≈ 10 MeV for the u and d quarks. They are essentially free parameters, which are often taken to be $\approx M_{\text{nucleon}}/3 \approx 300$ MeV. However, it has been suggested [43] that a more natural choice would be $\approx M_{\Delta}/3 \approx 400$ MeV.

So we see that the idea of EPTs covers an enormous number of theories, models and approaches that are frequently used in nuclear physics. The subject of the next section is to see how these EPT ideas can possibly be utilized in the understanding of QCD. The magnitude of this step should not be underestimated, since as the authors of Ref. [21] say:

"On the one hand, the concensus of the majority of the nuclear physics community holds that in nuclei

- nucleons are non-relativistic
- they interact via essentially two-body forces, with smaller contributions from many-body forces
- the two-nucleon interaction generally possesses a high degree of isospin symmetry
- external probes usually interact with mainly one nucleon at a time.

By contrast, in QCD

- the u, d and even s quarks are relativistic
- the interaction is manifestly multi-body, involving exchange of multiple gluons
- there is no obvious isospin symmetry
- external probes can, and often do, interact with many quarks at once.

It should not be surprising, then, that some new ideas are required to merge these two extraordinarily different bodies of theory." Others might simply say that this is a case of "Mission impossible".

5.2 What Is Meant by "A Bridge"?

The main theme of this article is the study of bridges between lattice QCD and nuclear physics. Unfortunately, what constitutes a possible bridge is rather subjective, since the basic idea is to compare some quantity that can be measured by lattice QCD with a "corresponding quantity" that arises in more conventional physics as the outcome of some EPT or is directly connected with experiment. Of course, the question instantly arises as to whether these two quantities are indeed comparable *i.e.* to what extent are we confident that we indeed have "corresponding quantities". In this chapter the main quantities to be related will be energies or radial correlations. This is probably best illustrated by the following simple example.

5.2.1 *A simple example of a bridge*

Since it is the main goal in this chapter, it may prove useful to the reader to first see a simple example of what is meant by "Bridges from LQCD to Nuclear Physics".

The results of any lattice QCD calculation are quantities expressed as dimensionless numbers. Therefore, to be able to make a connection with "real life", one — or more — of these numbers must be compared with its continuum counterpart that can actually be measured experimentally. This then sets the physical scale for lattice QCD.

5.2.1.1 *Setting the scale from the string tension*

For many years a quantity frequently used for this comparison was the string tension (b_s), which — as its name implies – is simply the energy/unit length of the flux-tube (*i.e.* string) connecting two quarks and appears in Eq. 5.6. Experimentally, estimates can be made of this string tension from the spectra of mesons and baryons with increasing orbital angular momentum (L) — a series of energies (E) that depend crucially on the string increasing in length. This can be carried out with varying degrees of sophistication. By simply plotting L versus E^2 this so-called Regge trajectory is found to be linear for both mesons and baryons — the slope (α) in each case being about 0.9 GeV^{-2} — see, for example, Figs. 7.33 and 7.34 in Ref. [44]. As shown in Ref. [45] using a simple classical model this slope is directly related to the string tension by the expression $\alpha \approx 1/(8b_s)$. This results in a value of $\sqrt{b_s} \approx 380$ MeV — a number that is somewhat smaller than the accepted value of ≈ 440 MeV, which is more in line with estimates from string models that give $\alpha \approx 1/(2\pi b_s)$ [46].

A less direct, but more precise, way to extract the string tension is to first find an effective quark–antiquark potential ($V_{Q\bar{Q}}$) that describes — *by way of a non-relativistic Schrödinger equation* — the above meson energy spectra. Naturally, for this non-relativistic approach to be realistic the mesons must be constructed from quarks that are much heavier than the proton. This, therefore, restricts the analysis to the Bottomonium $b\bar{b}$ mesons, where the b quark has a mass of about 5 GeV and possibly the $b\bar{c}$ and $c\bar{c}$ mesons, where the c quark has a mass of about 1.5 GeV. In fact, these spectra can be described by a *single* effective potential of the form given in Eq. 5.6 so that a value for the string tension of $\sqrt{b_s} \approx 440$ MeV results — equivalent to $b_s \approx 1$ GeV/fm or ≈ 5 fm^{-2} in other units. The potentials most frequently quoted are those of Richardson [28] and Cornell [29]. This value of b_s is now the experimental number to which the lattice estimate of $V_{Q\bar{Q}}(r)$ must be matched. Usually the latter is extracted by measuring a rectangular Wilson loop $W(l,t)$ of area a^2lt for two infinitely heavy quarks a distance $r = al$ apart on a lattice and propagating a Euclidean time at — a being the lattice spacing to be determined. Wilson loops will be discussed in more detail in Sec. 5.4.1.2. A key observation, first made by Wilson [47] in 1974, was that

$$W(l,t) \to \exp[-tV_{Q\bar{Q}}(l)] \quad \text{as} \quad t \to \infty. \tag{5.7}$$

Therefore, for sufficiently large l, $V_{Q\bar{Q}}(l) \to b'_s l$, where b'_s is the dimensionless counterpart to the experimental string tension b_s defined in Eq. 5.6. The two are then matched by way of the "bridging equation" $b'_s = a^2 b_s$ to give a. A typical number for b'_s is ≈ 0.05 giving $a \approx \sqrt{0.05/5} \approx 0.1$ fm.

5.2.1.2 *Sommer's prescription for setting the scale*

In order to set the scale, the above has compared the lattice result with experiment for the most simple of quantities — the string tension. However, the experimental data is mainly probing distances of $r \approx 0.2$ fm to $r \approx 1$ fm and *not* $r \to \infty$, since the rms–radii of the $b\bar{b}$ mesons cover the range from about 0.2 to 0.7 fm and the $c\bar{c}$ mesons the range from about 0.4 to 1 fm. Therefore, the experimental data encoded in the potential $V_{Q\bar{Q}}(l)$ is not optimal for studying the string tension. Furthermore, lattice calculations of the Wilson loop $W(l,t)$ require $t \gg l$, so that those $W(l,t)$ dominated by the string tension need to be evaluated for large values t. Unfortunately, as t increases the Signal/Noise ratio on $W(l,t)$ also increases — eventually making measurements for large $r = al$ meaningless. Therefore, on both the experimental and lattice sides there are problems for making a reliable estimate of a from the string tension.

In an attempt to overcome this problem, Sommer [48] proposed comparing the *potential* $V_{Q\bar{Q}}(l)$ as extracted from the lattice with that from experiment. However, to use the words of Sommer, "We must remember that the relationship between the static QCD potential and the effective potential used in phenomenology is *not* well understood." In spite of this, he suggests that the comparison be made at some value of $r = al$ in the optimal experimental range of $r \approx 0.5$ fm. Also at these values of r, $V_{Q\bar{Q}}(l)$ can be more reliably extracted on a lattice. In practice, it is the force, defined essentially as

$$F(l) = \frac{V_{Q\bar{Q}}(l) - V_{Q\bar{Q}}(l-a)}{a}, \tag{5.8}$$

that is compared through the expression

$$r^2 F(r)|_{r=R(c)} = c. \tag{5.9}$$

Sommer chose the dimensionless parameter $c = 1.65$, since — for the experimental potentials — this corresponds to a distance $R(1.65) \equiv R_0 \approx 0.5$ fm. Using the lattice forms of $F(r)$ that fit the lattice potential, Eq. 5.9 can be solved for r in lattice units a. Comparing this r with R_0 then gives a — see Ref. [48] for more technical details. However, it should be added that the value of c is somewhat uncertain with some authors [49] preferring $c = 2.44$, which corresponds to $R_0 \approx 0.66$ fm. This second choice of c gives values of a that are a few percent larger than before and also in better agreement with the string tension estimate.

The reason for this rather lengthy description for extracting the scale a is to show that "A Bridge from Lattice QCD to Nuclear Physics" has existed for many years. It should be added that another way of extracting a, when dealing with light quarks, is to use directly the mass of the ρ-meson as a corner stone and simply compare the experimental mass m_ρ with the outcome of the lattice QCD calculation — the dimensionless combination am_ρ [50]. However, m_ρ is known to be a sensitive indicator of scaling violations *i.e.* how the lattice results depend on a as $a \to 0$. It is, therefore, sometimes reserved for this purpose with the above method of Sommer being used to extract actual values of a [51]. The reason for choosing m_ρ and not m_π is because the π–meson, being so light, is more difficult to treat on a lattice.

5.2.2 *Are there bridges other than $V_{Q\bar{Q}}$?*

The above simple example showed how the $Q\bar{Q}$ potential $V_{Q\bar{Q}}$ could be related to its lattice QCD counterpart and serve as a means for extracting the lattice spacing a. The question then arises concerning the possibility of there being other quantities that could be compared. However, it must be noted that $V_{Q\bar{Q}}$ and the related force $F(r)$ are somewhat special and that it is still true what Sommer wrote in 1993: "As to today's knowledge, the force $F(r)$ between two static quarks is the quantity which can be

Table 5.1 Possible bridges between lattice QCD and nuclear physics

System	Quantity matched	Model	Refs.
$(Q\bar{Q})$	String Tension	Regge Trajectory	[45]
	$V_{Q\bar{Q}}$	Schrödinger Equation	[48]
$[(Q\bar{Q})(Q\bar{Q})]$	Energies	Matrix diagonalistion	[52, 53]
	Flux tube structure	Discretized String	[54]
		Dual Potential Model	[55]
$(Q\bar{q})$	Energies	Dirac Equation	[56, 57]
	Density distributions	Dirac Equation	[57]–[59]
$[(Q\bar{q})(Q\bar{q})]$	Energies	Variational	[60]
(QQq)	Energies		[61]
	Density distributions		[61]

calculated most precisely". There are several reasons for this and they should be kept in mind in the following discussion. Firstly, both the lattice and experimental determination of $V_{Q\bar{Q}}$ can be done with good statistical precision and, secondly, the two are what we think they are. In contrast, the string tension can only be extracted at values of r that are not necessarily sufficiently asymptotic and where there could be corrections from model dependent sub-leading terms. So for the purposes of setting a scale the use of $V_{Q\bar{Q}}$ is still the best.

Possibilities for bridges, in addition to the use of $V_{Q\bar{Q}}$, are listed in Table 5.1. This is essentially a "Table of Contents" for the rest of this chapter. The Lattice QCD $\leftrightarrow$ Nuclear Physics relationship changes as we go through this list. The first two rows for the $Q\bar{Q}$ static quark system have been discussed above. Here the rôles of the string tension and $V_{Q\bar{Q}}$ are to set a scale for QCD — a necessary step in order for lattice QCD to be compared with experiment *i.e.* the flow of information is Nuclear Physics $\rightarrow$ Lattice QCD. However, once the results of lattice QCD can be expressed reliably with physical dimensions then the information flow is completely Lattice QCD $\rightarrow$ Nuclear Physics. We can now consider the results of lattice

QCD on the same footing as experimental data — assuming that the $a \to 0$ limit is under control and that the quarks are sufficiently light as discussed in Subsec. 5.1.1.

It is lattice data that models must attempt to fit. Many of these models resort to the use of interquark potentials — often the above $V_{Q\bar{Q}}$ — in various forms of wave equation. Now the lattice QCD data will possibly be able to justify — *or rule out* — such models. At present these models are often simply mimicking techniques that have proven successful in Nuclear Physics. Hence my earlier statement that they are "Nuclear Physics–inspired" and not "QCD-inspired" as is often claimed.

Above I said that the results of lattice QCD can be considered on a similar footing as experimental data. However, when setting up models, in some ways lattice QCD data can sometimes be superior to experimental data, since it can be generated in "unphysical worlds".† Such worlds can have the following unphysical features that should, in some cases, also be inserted into the corresponding models to test the generality of these models:

(1) The real world of three coloured quarks (*i.e.* SU(3)) can be replace by one with two coloured quarks (*i.e.* SU(2)). Such a world is easier to deal with in lattice QCD — for example, there is essentially no distinction between quarks and antiquarks. Also the system corresponding to a baryon now consists of only two quarks. However, this is not simply an academic exercise, since, in practice, it is found that the ratio of many observables are similar in both SU(2) and SU(3) — but with a computer effort that is about an order of magnitude smaller. An example of this is the ratio $R = m_{\rm GB}/\sqrt{b_s}$, where $m_{\rm GB}$ is the glueball mass (see Chapter 2) and b_s the string tension. It is found for the glueball with the lowest mass (0^{++}) that $R \approx 3.5$ for both SU(2) and SU(3) — see Sec. 2.2.1 in Chapter 2 and Ref. [62]. In Ref. [63] this is extended to the general case of SU(N_C) for several glueball states and for the 0^{++} case results in $R(0^{++}) = 3.341(76) + 1.75/N_C^2$.

(2) The real world with 3 space coordinates and 1 time coordinate (3+1) can be replaced by one with 1 or 2 space coordinates and 1 time coordinate (1+1 and 2+1). On the lattice the latter are easier to study so that results with such high accuracy can be achieved

†In Sec. 3.1 of Chapter 3 these are referred to as "virtual worlds".

that there is little ambiguity in any final conclusions. Also the (2+1) world has interesting features in its own right and enables comparisons to be made between SU(2), SU(3), SU(4),..., SU(N_C) [64].

(3) In the real world space is isotropic, but this need not be so on a lattice, since the four axes can be treated differently by having unequal lattice spacing — in principle we could have $a_x \neq a_y \neq a_z \neq a_t$. However, the most common choice is $a_x = a_y = a_z \neq a_t$. This is appropriate for finite temperature systems [65], where the temperature is defined to be inversely proportional to the lattice size in the t-direction. For high temperatures this would mean a lattice that contained fewer steps in the t-direction and so lead to difficulties in extracting accurate correlation functions. However, if a_t is made smaller than the three spatial a's, then a given temperature is defined by more steps in the t-direction and better correlation estimates — see Ref. [12]. For studying high-momentum form factors such as $B \to K^*\gamma$, $B \to \pi l\nu$ or $B \to \rho l\nu$ it has been suggested that the 2+2 anisotropic lattice $a_x = a_y \neq a_z = a_t$ is more suitable [66]. Explicit anisotropic forms of the clover action — see Eq. 5.5 — and the pure gauge action in Eq. 5.17 can be found from Ref. [67]. More recent studies can be found in Refs. [68].

(4) In the real world the vacuum (sea) contains $q\bar{q}$-pairs, where the q are sea–quarks that can have any flavour u, d, s, c, b or t. These pairs are being continuously created and annihilated. Lattice QCD calculations that take this into account are said to involve **dynamical quarks**. In practice, only u, d and, possibly, s sea–quarks are included and these two cases are usually referred to as having $N_f = 2$ or 3. However, frequently this effect is neglected to give the so-called **quenched** approximation, which is numerically an order of magnitude less demanding on computer resources. In view of this last point, there has been much work attempting to show how realistic quenched results can be compared with their dynamical quark counterparts. The conclusion seems to be that, although no formal connection has been established between full QCD and the quenched approximation, the similarity of the results ($\approx 10\%$ differences) has led to the belief that the effects of quenching are generally small, so that quenched QCD provides a reasonable approximation to the full theory [69]. This has been taken one step

further in Ref. [70], where it is suggested how quenched results can be corrected in a systematic way to retrieve the corresponding full QCD prediction.
However, there are cases where the quenched versus unquenched comparison can lead to qualitative differences. In Ref. [71] the asymptotic potential between two quark clusters is calculated algebraically in a quenched approximation effective field theory and found to have a *pure exponential* decay and not the usual Yukawa form. But it is not clear that this qualitative difference for large intercluster distances leads to any overall quantitative differences. It must be remembered that, at these distances, calculations involving only quarks are expected to be incomplete with the introduction of, in particular, explicit pion fields being necessary. The same authors also study a **partially–quenched** effective field theory, where the masses of the sea–quarks are much larger than the valence quarks connected to the external sources [72]. This they suggest would help in the understanding of the NN–potential, when extrapolations are made of NN–lattice calculations to realistic quark masses.

(5) The real world of fixed quark masses can be replaced by one where the quark masses take on other values. In many cases, this is of necessity, since for light hadrons the use of realistic light quark masses of ≈ 10 MeV is computationally too heavy — see subsection 5.1.1. However, the variation of quark masses has interesting features in its own right. In particular, by carrying out lattice calculations with a series of light quark masses, we can extract the differential combination $J = m_{K^*} \frac{dM_V}{d(M_P^2)}$, where M_V and M_P are the corresponding vector and pseudoscalar meson masses and m_{K^*} is the mass of the K^*. This quantity J, which can be shown to be independent of a and the so-called hopping parameter that is related to the bare quark masses, serves as a check on the consistency of lattice QCD — see Refs. [73] and [74]. Such analyses can be performed by varying the valence– and sea–quark masses separately. In Subsection. 5.9.4 a similar argument is used for estimating the matter sum rule.
In Ref. [75] the authors have emphasized the importance of the quark-mass dependence of the nucleon-nucleon interaction by saying: "While the m_q-dependence of the nuclear force is unrelated

to present day observables, it is a fundamental aspect of nuclear physics, and in some sense serves as a benchmark for the development of a perturbative theory of nuclear forces. Having this behaviour under control will be essential to any bridge between lattice QCD simulations and nuclear physics in the near future."
In Ref. [76] the authors are more interested in the reverse situation, namely, how to extrapolate nuclear forces calculated in the chiral limit to larger pion masses pertinent for the extraction of NN-observables from lattice calculations.

(6) In the real world, hadron–hadron scattering is thought to be described directly in terms of quark-gluon physics at small interhadron distances, but at larger distances a description in terms of meson-exchange is expected to be more appropriate. Both of these limits must be included in a single model, if a direct comparison with *experiment* is to be made. However, lattice QCD can concentrate on just the small distance physics and generate "data" that is exact not only for that limit but, in principle, for larger interhadron distances — until the numerical signal becomes unmeasurable. In this way, models can be constructed and compared directly with the lattice data — ignoring the effect of explicit meson exchange that enters in the real world. However, it should be added that, in some cases, lattice calculations based purely on quarks and gluons seem to be able to generate effects that resemble meson exchange. An example of this is Ref. [77] discussed in Subsec. 5.10.1.

(7) In the real world, space is essentially infinite, whereas quantities calculated on a lattice are restricted to volumes L^3 with $L \approx 1 - 2$ fm often being comparable to the size of the object under study. This leads to results that could depend on L. Usually this is considered a negative feature and so lattices must be chosen sufficiently large to avoid this problem. However, it was shown by Lüscher in Refs. [78] that this volume dependence can be utilized to extract the interaction between hadrons. Recent examples of this idea consider the two-nucleon interaction [79, 80] and $\pi\pi$ scattering [81]. This approach is discussed in more detail in Chapter 4 Subsec. 4.1.5.

(8) In the real world, the systems encountered contain a number of valence-quarks and antiquarks. However, on a lattice, systems consisting of only gluons can be studied — with quarks only entering as sea-quarks in the case of dynamical quarks mentioned in item 4

above. This pure-glue world enables a cleaner study to be made of glueballs — see Sec. 2.2 of Chapter 2.

These possibilities of a different number of colours, spatial dimensions and quark masses greatly expand the scope of lattice QCD and give model builders much more data on which to test their models.

5.3 The Energies of Four Static Quarks ($QQ\bar{Q}\bar{Q}$)

5.3.1 *Quark descriptions of hadron–hadron interactions*

Much of particle and nuclear physics studies the interaction between hadrons. With increasing complexity in terms of the number of quarks thought to be involved, this ranges from meson–meson scattering up to heavy ion collisions. Clearly, any attempt to describe such processes at the quark level must begin with an understanding of meson–meson scattering. Unfortunately, even with this system there are several major complications preventing a direct comparison between theory and experiment. Firstly, the only mesons for which there is suitable experimental data are the pseudoscalars — the π, K and B, since beams of these can now be generated at the various π-, K- and B-factories [(π) PSI (Villigen) and TRIUMF (Vancouver); (K) DAPHNE (Frascati) and KEK (Tsukuba); (B) BaBaR at SLAC, Belle at KEK, Hera-B at DESY and CLEO III at Cornell]. Of course, having beams of mesons does not lead directly to obtaining data on meson–meson scattering. This can only be done indirectly, as a final state interaction, with the net result that essentially only $\pi\pi$ scattering data (and considerably less πK data) are at present available — and even those are very limited. This means that most theoretical attempts to understand meson–meson scattering concentrate on the $\pi\pi$ system. However, quark descriptions of this particular system are then complicated by the fact that the pion, being a Goldstone boson, does not have a quark structure as simple as a single $q\bar{q}$ configuration. Models of the pion (see, for example, Weise in Chapter 2 of Ref. [36]) suggest that it has large multiquark components *i.e.* $\phi_\pi = \sum_n a_n(q\bar{q})^n$. Furthermore, the total interaction between the two pions can not be only due to interquark interactions between the constituent quarks, since for large interpion distances it is expected that meson exchange — another multiquark mechanism — also plays a rôle *i.e.* $\pi\pi$-scattering involves much more than a discussion of the $(q\bar{q})(q\bar{q})$

system. Having said that, it should be added that these complications have not detered the construction of models for $\pi\pi$-scattering that are essentially nothing more than a $(q\bar{q})_\pi$ interacting with a $(q\bar{q})_\pi$ through an interquark potential of the form in Eq. 5.6. The references are too numerous to list here and, furthermore, they often involve physicists who are my friends. In my opinion, these models are, as yet, not justified. They are simply hoping that the success in treating multi-hadron systems in terms of two-body potentials will repeat itself.

5.3.2 *The rôle of lattice QCD*

To make a bridge between quark and hadron descriptions of, say, meson–meson scattering needs reliable experimental data. But, as said above, this is not available — and this is where lattice QCD enters. The latter is based on QCD, which is thought to be the exact theory of quark–gluon interactions, and its implementation on a lattice leads (in principle) to exact results — upto the lattice spacing, lattice finite size and quark mass reservations mentioned in Subsec. 5.1.1. Therefore, if we want to study, for example, $(q\bar{q})(q\bar{q})$ systems we simply calculate these on a lattice and we get *exact* results that can now be considered as "data". Model builders then try to understand these data in terms of $(q\bar{q})(q\bar{q})$ states. Such a procedure guarantees one of the necessary requirements of bridge building — the need to compare like-with-like. In this way, the lattice data generated for the $(q\bar{q})(q\bar{q})$ system can possibly be modelled with purely $(q\bar{q})(q\bar{q})$ configurations. In other words the conventional approach for model building

$$\text{Experimental data} \xrightarrow{1} \text{Hadron description} \xrightarrow{2} \text{Quark description}$$

is replaced by the alternative

$$\text{Lattice data} \xrightarrow{3} \text{Quark description} \xrightarrow{4} \text{Hadron description}$$

i.e. by concentrating on step 3 we avoid: a) At step 1 the shortage of experimental data; b) At step 2 the need to guess the hadron quark structure and how a model based on this structure matches on to models more appropriate at larger interhadron separation where meson exchange dominates. However, there are several problems when attempting to implement this second alternative:

- Step 4 is similar to step 2 each with their uncertainty in the physical hadron structure. However, now it is less serious, since step 3 enables a cleaner description to be made at the quark level. This is in contrast to the conventional approach, where the quark description is "shielded" from the experimental data by needing to go via the hadron description, which could well also involve explicit meson exchange.

- Step 3 is only feasible technically for a very restricted number of quark systems — four quarks being essentially the limit at present. However, there are a few six quark lattice calculations — mainly studies of the NN–interaction for static quarks (see Subsec. 4.5.2 in Chapter 4) and, in addition, attempts to determine whether or not the H dibaryon is bound. In Ref. [82] the indications are that such a ($uuddss$) bound state is ruled out. Also recently there have been multiquark lattice calculations [83] with $uudd\bar{s}$ quenched configurations in an attempt to describe the $\Theta^+(1540)$ seen in several experiments and thought to be the first observed pentaquark system [84].
- The best lattice calculations are carried out with dynamical quarks — the so-called unquenched formulation. There the possibility arises for the creation (and annihilation) of quark–antiquark pairs. This is in contrast to the quenched approximation where such pairs do not enter. This means that in the unquenched approximation the configurations included do not have a fixed number of quarks and antiquarks. In model building this effect is often ignored and only a fixed type of configuration is used, *e.g.* $(q\bar{q})(q\bar{q})$ for meson–meson interactions. Fortunately, it is found that in most cases of interest here the refinement of dynamical fermions does not lead to significant corrections to the quenched results. But this is only known in hindsight and should, if possible, always be checked.

Ideally, in the above example of meson–meson scattering, we would want to perform a lattice calculation with four light quarks *i.e.* where the quarks in $(q\bar{q})(q\bar{q})$ are u, d –quarks with masses of less than 10 MeV. Unfortunately, with present day computers, this is not yet possible. Therefore, the problem must be simplified — a process that can be done in several stages by making more and more of the quarks infinitely heavy *i.e.* $q \to Q$. This makes the lattice calculations easier and easier. Examples are as follows:

- $[(\mathbf{Q\bar{q}})(\mathbf{Q\bar{q}})]$ configurations. The energy of this system can be expressed as $V(R)$, where R is the distance between the two static quarks (Q). Ground and excited state energies in $V(R)$ can then be calculated in the Born-Oppenheimer approximation by assuming that the Q's have some definite finite mass. A specific example would be the $B-B$ system, where m_Q is the mass of the b–quark *i.e.* about 5 GeV. This will be discussed in more detail later in Sec. 5.10.

- $[(\mathbf{Q\bar{q}})(\mathbf{\bar{Q}q})]$ configurations. These are very similar to the above but with the added feature that we can now have the $q\bar{q}$ annihilating to give simply a $(Q\bar{Q})$ configuration. Such a process is a model for the string breaking mechanism $Q\bar{Q}(R) \to (Q\bar{q})(\bar{Q}q)$, where for some sufficiently large R it becomes energetically favourable for a $q\bar{q}$-pair to be created from the vacuum. This mechanism must occur in nature, but at present it has only been conclusively demonstrated in simplified versions of QCD. A specific example of this type of configuration would be the $B\bar{B}$-system — to be discussed later in Sec. 5.11.
- $[(\mathbf{Q\bar{Q}})(\mathbf{Q\bar{Q}})]$ configurations. These are the most simple four-quark configurations to deal with on the lattice. Unfortunately, they are also the most distant from appropriate experimental data — the nearest being $\Upsilon(b\bar{b}, 9.5\text{GeV})\Upsilon(b\bar{b}, 9.5\text{GeV})$ scattering. However, this class of experiments is still far in the future.

Each of these three possibilities will be discussed separately in the following sections.

The problem we are faced with is that the Lattice and Nuclear Physics approaches are extremes that concentrate on two opposite aspects. In the Nuclear Physics approach only the quark degrees of freedom are introduced with the gluon degrees of freedom entering only *implicitly* in the interquark potential. In contrast, with the $[(\mathbf{Q\bar{Q}})(\mathbf{Q\bar{Q}})]$ configurations — the ones studied here in most detail — the lattice calculations treat the quarks in the static quenched limit, where they play no dynamical rôle — all the effort being to deal explicitly with the gluon field.

5.4 $(Q\bar{Q})$ and $[(Q\bar{Q})(Q\bar{Q})]$ Configurations

The most important observable for $Q\bar{Q}$ states is the interquark potential $V_{Q\bar{Q}}$, which is the topic of Chapter 3 — see also Ref. [85] for a more detailed account. Here only those aspects of the $(Q\bar{Q})$–system will be mentioned that are relevant to the later $[(Q\bar{Q})(Q\bar{Q})]$ discussion — the main interest in this section.

When lattice calculations on four-quark systems were first attempted, it was not clear whether an acceptable signal could be achieved. Therefore, various simplifications were made in an attempt to maximize the possibility of success [52, 53, 86, 87, 88]. The most important of these was the use of quarks with only two colours *i.e.* SU(2). In addition, only configurations where the four quarks were at the corners of a square were studied, since it was hoped that this degenerate situation would lead to the maximum interaction — a feature that in fact turned out to be so and also one that had been assumed in the flip-flop model of Ref. [89].

5.4.1 *Lattice calculations with $(Q\bar{Q})$ configurations*

Schematically, in analogy to thermodynamics, the expectation value of some operator O can be obtained with an action S through the chain

$$\langle O\rangle = \frac{\int DUO(U)e^{-S(U)}}{\int DUe^{-S(U)}} \xrightarrow{1} \frac{\sum_i DU_iO(U_i)e^{-S(U_i)}}{\sum_i e^{-S(U_i)}} \xrightarrow{2} \frac{\sum_j^{N_M} O(U_j)}{N_M}, \tag{5.10}$$

where, at the first stage, the variables U are continuous. However, for QCD, even though we know the exact form of the action $S(U)$, the resulting integrals are singular. As discussed earlier, it is necessary to carry out step 1, in which the variables U are discretized into variables U_i that sit on the links (i) of a lattice. This now removes the singularities in a consistent manner and reduces the problem to a numerical approximation for a multiple integral. Unfortunately, there are very many U_i's. For example, in QCD the U_i are in fact colour matrices U_i^{ab} — defined by 8 independent real parameters. Furthermore, for a 10^4 space–time lattice there are about 4×10^4 links. In all this amounts to about 320000 integrations to be done. If each of these were approximated by 10 points, we would then be approximating the multiple integral with about 10^{320000} terms — not a feasible task. However, many of the link configurations lead to large values of $S(U_i)$ and so would have a negligible effect. These configurations can be avoided by "importance sampling", which essentially generates configurations with the probability given by the Boltzmann factor $\exp[-S(U_i)]$. This automatically encodes the feature that the ratio of the values of nearby links are exponential in form. Now only a relatively few configurations (often ~ 10's) need to be generated to get a good estimate of $\langle O \rangle$. This is depicted as step 2 in Eq. 5.10, where the Boltzmann factor has been replaced by the N_M configurations that tend to maximize this factor — see, for example, p.252 in Ref. [13].

The problem is, therefore, reduced to generating these "important" configurations and to finding an appropriate operator. These are the topics of the next two subsections.

5.4.1.1 *Generating lattice configurations*

Earlier, practitioners of lattice QCD generated their own lattice configurations. However, nowadays there are groups that specialize in this very time consuming computer task, *e.g.* the UKQCD group in Edinburgh. The rest of us are then able to simply perform expectation values of our operators in the knowledge that we are using configurations that are well tested and essentially independent of each other. This also has the added benefit that different groups use the same configurations, *i.e.* with the same lattice parameters, but with different operators. At times, this can lead to useful direct comparisons of the lattice results with experiment that do not have to be scaled in any way beforehand, *i.e.* if one group has extracted the lattice spacing for their particular problem, then the rest of the lattice community can use this same lattice spacing to convert all of their dimensionless lattice results directly into physical quantities.

In spite of this modern trend, it is probably useful to remind the reader of some of the problems that arise when generating configurations:

Equilibration
If a lattice simulation is ever started from scratch, then the first lattice must be simply "guessed". This could be "cold", where all the U_i are taken to be the same, or "hot", where the U_i are random numbers. To carry out the above "importance sampling" these lattices must first be equilibrated or, in the terminology of spins-on-a-lattice, thermalized. This is where clever programming enters to generate quickly configurations that are as independent of each other as is conveniently possible. For example, in our early work [86], where we generated all of our own configurations, the lattice was equilibrated by the heat bath method.

Updating
Once the lattice is equilibrated, measurements of the operator O can begin.

However, the lattice must be continuously updated after each measurement to ensure these measurements are made on lattices that are, as far as possible and convenient, independent of each other. In Eq. 5.10 the number of such lattices is referred to as N_M. As an example, in Ref. [86] — after equilibration — the lattice was further updated by a combination of three **over–relaxation** sweeps, each of which can change the configuration a lot but leave the energy unchanged. This is followed by one **heat bath** sweep, which changes the configuration less but can change the energy as is necessary for ergodicity. Then a measurement is made of the appropriate correlation functions. In that particular calculation, the lattice was updated 6400 times with 1600 measurements being made. The latter were divided into 20 blocks of 80 measurements for convenience in carrying out an error analysis. In later work [59] 40 updates were made between measurements to reduce possible correlations between successive measurements. However, it should be added that, in our experience, there has never been any sign of a significant correlation due to insufficient updating between measurements, when dealing with the types of problem we have been studying. This is usually checked by calculating the autocorrelation between blocks of measurements. In the above, I mention spins-on-a-lattice — the classical example being the Ising model — in order to remind the reader that lattice QCD and the thermodynamics of spins-on-a-lattice have very much in common. Over the years, many ideas were first developed in the spin case before attempting to implement them in the more complicated case of lattice QCD.

5.4.1.2 *Appropriate operators on a lattice*

Most of the operators evaluated on a lattice take the form of correlations between different Euclidean times. The most simple example is the Wilson loop involved in extracting $V_{Q\bar{Q}}$ in Eqs. 5.6 and 5.7. Consider a Q and a $\bar{Q}$ are at the lattice sites $\mathbf{x}$ and $\mathbf{x}+\mathbf{l}$ — with the notation that $\mathbf{l}$ is the lower case letter corresponding to the upper case L, which is reserved for the spatial size of the lattice. A $Q\bar{Q}(\mathbf{l})$-state, is then constructed from a sequence of connected lattice links U_i, at a fixed Euclidean time t_1, as

$$\Psi(Q\bar{Q},\, \mathbf{l}) = \phi^a_Q(\mathbf{x})U^{ab}U^{bc}U^{cd}\ldots U^{za}\bar{\phi}^a_{\bar{Q}}(\mathbf{x}+\mathbf{l}). \tag{5.11}$$

Here the colour indices $a, b, c, \ldots$ on successive links are coupled to give overall a gauge invariant chain of U's. Also the Q and $\bar{Q}$ have the same colour to ensure the meson is a colour singlet. To be more specific, if the Q and $\bar{Q}$ are a distance of two links apart *i.e.* they are at lattice sites $(x, 0, 0, t_1)$ and $(x+2, 0, 0, t_1)$, then one possibility is the direct path

$$\Psi(Q\bar{Q},\ t_1) = \\ \phi^a_Q(x,\ t_1)U^{ab}(x, x+1; t_1)U^{ba}(x+1, x+2; t_1)\bar{\phi}^a_{\bar{Q}}(x+2,\ t_1), \quad (5.12)$$

where the two U-links are simply along the x-axis. Other choices of less direct paths between the Q and $\bar{Q}$ are possible and, indeed, necessary if excited states are of interest. To construct a Wilson loop a similar wave function is written down at time $t_2 = t_1 + t$ as

$$\Psi(Q\bar{Q},\ t_2) = \\ \phi^{a'}_Q(x,\ t_2)U^{a'b'}(x, x+1; t_2)U^{b'a'}(x+1, x+2; t_2)\bar{\phi}^{a'}_{\bar{Q}}(x+2,\ t_2). \quad (5.13)$$

This gives the two horizontal (wavy) sides of the Wilson loop in Fig. 5.1. To complete the loop we need to insert the propagators of the Q and $\bar{Q}$

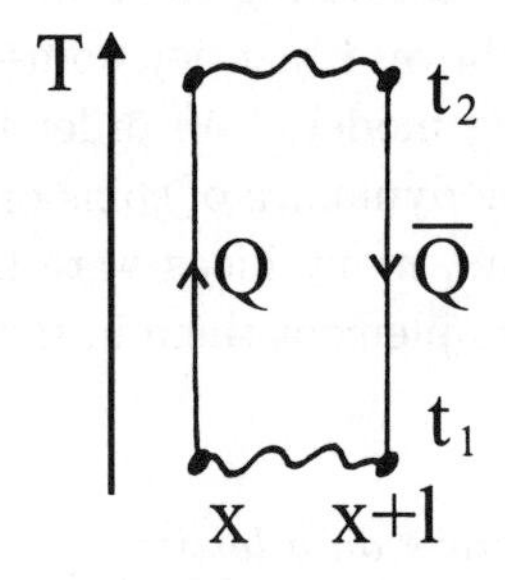

Fig. 5.1 The Wilson loop for a $Q\bar{Q}$ state $\Psi[Q(\mathrm{x})\bar{Q}(\mathrm{x}+1)]$ propagating from Euclidean time t_1 to t_2.

from t_1 to t_2. For static quarks these are, for $t_2 - t_1 = 2$,

$$\phi^a_Q(x; t_1)U^{ab'}(x; t_1, t_1+1)U^{b'a'}(x; t_1+1, t_1+2)\bar{\phi}^{a'}_Q(x; t_2)$$

and

$$\phi^a_{\bar{Q}}(x+2; t_1)U^{ab'}(x+2; t_1, t_1+1)U^{b'a'}(x+2; t_1+1, t_1+2)\bar{\phi}^{a'}_{\bar{Q}}(x+2; t_2),$$

where here the U's are all in the T direction and, without loss of generality, can be arranged by gauge fixing to be simply *unity.* The Wilson loop of

Eq.5.7 then reduces to the overlap

$$W(x,x+2;t_1,t_2)=\delta^{aa'}U^{ab}(x,x+1,t_1)U^{ba}(x+1,x+2,t_1)\times \left[U^{a'b'}(x,x+1,t_2)U^{b'a'}(x+1,x+2,t_2)\right]^{\dagger}. \quad (5.14)$$

This is simply a number, which in principle $\to \exp[-tV_{Q\bar{Q}}(l=2)]$ as $t\to\infty$. The emergence of an exponential factor from this overlap should not be surprising, since the essence of the importance sampling in Subsec. 5.4.1 was to encode the Boltzmann factor in Eq. 5.10 into the values of the links. Of course, a single overlap $W(x,x+2;t_1,t_2)$ would not be very informative. Only when — for all three orientations x,y,z — this is averaged over the whole L^3T lattice and the N_M different lattices does one expect a reasonable numerical signal for $W(2,t)$ to emerge. In general, this has the form

$$W(l,t)=\frac{1}{N_M L^3 T}\sum_{N_M}\sum_{l_0=x_0,y_0,z_0}^{L}\sum_{t_1}^{T}W(l_0,l_0+l;t_1,t_1+t)$$

$$\to \exp[-tV_{Q\bar{Q}}(l)] \quad \text{as} \quad t\to\infty, \quad (5.15)$$

where the average over the different lattices includes:

(1) All values of x in the range $0\le x\le L$,
(2) All three spatial directions x,y,z
(3) All values of t_1 in the range $0\le t_1\le T$

and it is repeated for many as–independent–as–possible lattices N_M.

Having introduced the notion of a lattice link U_i representing the gluon field, we can combine four links to form a closed loop called an elementary plaquette $U_\Box$. For example, a plaquette in the xy–plane would be constructed from links $U(x\to x':y\to y')$ as

$$U_\Box^{xy}=U^{ad}(x\to x+1:y)U^{dc}(x+1:y\to y+1)\times U^{cb}(x+1\to x:y+1)U^{ba}(x:y+1\to y). \quad (5.16)$$

Similarly there can be plaquettes in the $xz,\ldots,zt$–planes. It was in terms of these plaquettes that Wilson [47] first expressed the discretized form of the QCD action mentioned in Sec. 5.1.1, which for SU(N_C) has the form

$$S_W=\frac{2N_C}{g^2}\sum_\Box[1-\frac{1}{N_C}\,\mathrm{Tr}\,U_\Box]. \quad (5.17)$$

Usually the basic coupling (g) is expressed in terms of $\beta = 2N_C/g^2$.

5.4.1.3 *Fuzzing*

The above Wilson loops were constructed from the basic lattice links $U^0_\mu(n)$, where — using a slight change of notation for convenience — the earlier colour indices $a, b, \ldots$ are omitted and replaced by an index $j = 0$, n is a lattice site and μ is either the x, y or z direction. These basic $(j = 0)$ links can now be generalised by "fuzzing, blocking or smearing" [90, 91] by means of which the basic link is supplemented by a combination of neighbouring links. Here I will concentrate on the the fuzzing option, but blocking and smearing are similar. Fuzzing is depicted in Fig. 5.2.

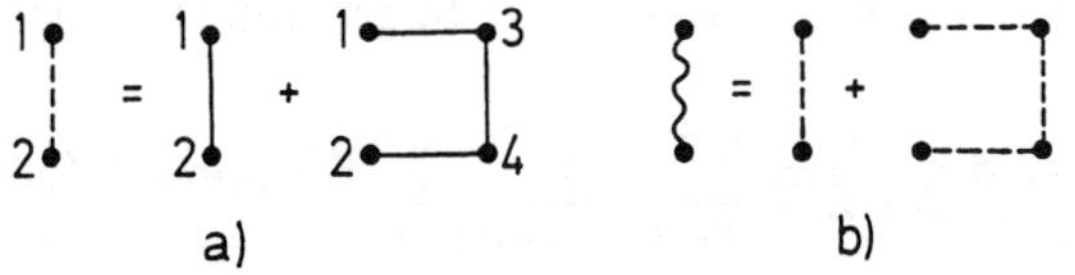

Fig. 5.2 Fuzzing. In a) the solid line 12-link is replaced by the dashed line 12-link. In b) this dashed line 12-link is itself now replaced by the wavy line 12-link.

This illustrates the replacement

$$U^0_\mu(n) \to U^1_\mu(n) = A^1_n \left[cU^0_\mu(n) + \sum_{\substack{\pm\nu\neq\mu \\ \nu\neq 4}} U^0_\nu(n) U^0_\mu(n+\bar{\nu}) U^{0\dagger}_\nu(n+\bar{\mu}) \right] \quad (5.18)$$

to give links U^1 constructed from the basic links U^0 by a single fuzzing $j = 1$. This is followed by

$$U^1_\mu(n) \to U^2_\mu(n) = A^2_n \left[cU^1_\mu(n) + \sum_{\substack{\pm\nu\neq\mu \\ \nu\neq 4}} U^1_\nu(n) U^1_\mu(n+\bar{\nu}) U^{1\dagger}_\nu(n+\bar{\mu}) \right] \quad (5.19)$$

to give links U^2 constructed from the U^1 by a second fuzzing $j = 2$ and so on. In general

$$U^{m-1}_\mu(n) \to U^m_\mu(n) =$$
$$A^m_n \left[cU^{m-1}_\mu(n) + \sum_{\substack{\pm\nu\neq\mu \\ \nu\neq 4}} U^{m-1}_\nu(n) U^{m-1}_\mu(n+\bar{\nu}) U^{m-1\,\dagger}_\nu(n+\bar{\mu}) \right]. \quad (5.20)$$

Here the A_n^j are normalisation factors chosen to project the $U_\mu^j(n)$ into $SU(N_C)$ and c is, in principle, a free parameter. However, experience when measuring energies [90, 91] has shown for SU(2) that $c = 4$ is a suitable value for the present class of problems. The value of c could be optimized, but it is found that results are never crucially dependent on the precise value of c. Another degree of freedom is the amount of fuzzing (m). For correlations over large distances the greater the fuzzing the better the efficiency of the calculation in the sense that the state $\Psi(Q\bar{Q})$ in Eq. 5.11, generated by connecting together a series of fuzzed links, is expected to have a greater overlap with the true ground state wave function. In some cases this has been carried to great lengths *e.g.* in Ref. [92] m =110 fuzzing iterations were used, since there the emphasis was on quark separations upto 24 lattice spacings.

In the calculation of the interquark potential in Eq. 5.15, the fuzzing procedure plays the second rôle of generating different paths (P_i) between quarks, for use in the variational approach of Ref. [90, 91]. In this notation the basic path from Eq. 5.11 is $P_0 = U^{0,ab}U^{0,bc}U^{0,cd}.....U^{0,za}$ and the fuzzed paths $P_i = U^{i,ab}U^{i,bc}U^{i,cd}.....U^{i,za}$. The variational approach then yields the correlation matrix between paths with different fuzzing levels i, j

$$W_{ij}^t =< P_i|\bar{T}^t|P_j > . \tag{5.21}$$

Here $\bar{T} = \exp(-a\bar{H})$ is the transfer matrix for a single time step a, with the basic QCD Hamiltonian $\bar{H}$, the $P_{i,j}$ are paths constructed as products of fuzzed basic links and, as before, t is the number of steps in the imaginary time direction. As shown in Ref. [90, 91], a trial wave function $\psi = \sum_i a_i|P_i >$ leads to the eigenvalue equation

$$W_{ij}^t a_j^t = \lambda^{(t)} W_{ij}^{t-1} a_j^t. \tag{5.22}$$

For a single path this reduces to

$$\lambda^{(t)} = \frac{W_{11}^t}{W_{11}^{t-1}} = \exp(-aV_{Q\bar{Q}}), \tag{5.23}$$

where $V_{Q\bar{Q}}$ is the potential of the quark system being studied. Unfortunately, in this single path case *i.e.* with only P_0 as in the previous subsection, t needs to be large and this can lead to unacceptable error bars on the value of $V_{Q\bar{Q}}$ extracted. However, if — in addition — a few fuzzed paths are included, it is found that t need only be small ($t \approx 5$) to get a good convergence to $V_{Q\bar{Q}}$ with small error bars.

A further very important advantage of fuzzing is that not only can the lowest eigenvalue be extracted but also higher ones, since a matrix diagonalisation is involved. These higher states correspond to excitations of the gluon field. However, with the direct paths from the Q to $\bar{Q}$ in Eq. 5.12, these excitations are purely S-wave. To generate non-S-wave gluonic excitations combinations of indirect paths are needed — see Chapter 2.

The above fuzzing is an attempt to improve the overlap of the lattice wavefunction with the true ground state wavefunction and so only involves spatial links (N.B. $\nu \neq 4$ in the summations in Eq. 5.18 *i.e.* the staples in Fig. 5.2 are purely spatial). However, a similar procedure can be applied to a time-like link [93]. In the notation of Eq. 5.18, the basic time-like link $U^0_{\mu=0}(n)$ is now replaced by $W^0_{\mu=0}(n)$ — the average of the six staples in the planes $(\pm x, T)$, $(\pm y, T)$ and $(\pm z, T)$. The benefit gained from this is a reduction in the Noise/Signal ratio $R_{\rm NS}$ for configurations involving a static quark. Without this time-like fuzzing, it has been observed for B-meson correlations, with time extent x_0, that $R_{\rm NS} \propto \exp(x_0 \Delta E)$, where $\Delta E = E_0 - m_\pi$ with E_0 being the ground state energy of the B-meson [94]. This leads to very noisy signals as x_0 increases. However, using $W^0_{\mu=0}(n)$ results in almost an *order of magnitude* reduction in $R_{\rm NS}$ for $x_0 \approx 1.5$ fm. This is shown in Fig. 5.3.

An even greater reduction in R_{NS} can be achieved, if the single staple fuzzing in Fig. 5.2 is replaced by the hypercubic blocking of Ref. [95]. There each level of fuzzing is described by three equations (and so needing three free parameters c_i) similar to Eq. 5.18. But these three equations mix links that are connected to the original link only *i.e.* this "fuzzing" only involves links in the hypercube defined by the original link. This procedure yields a fuzzed link $W_{\rm HYP}$ that is much more local than three fuzzing levels from Eq. 5.18 and so preserves short distance spatial structure. Furthermore, when this is applied to a time-like link, then it reduces $R_{\rm NS}$ even further than the use of $W^0_{\mu=0}(n)$. The use of hypercubic links is still in its infancy and we should expect to see much more of this development in the future — see Refs. [96]. This is also shown in Fig. 5.3.

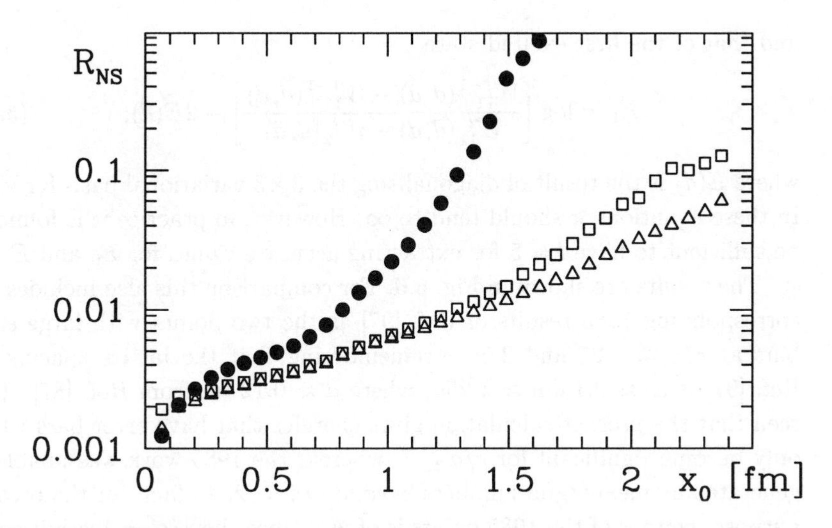

Fig. 5.3 Noise to signal ratio ($R_{\rm NS}$) of a $B-$meson correlation function as a function of the time extent x_0. The solid dots are when using simply $U^0_{\mu=0}(n)$. The open squares are for $W^0_{\mu=0}(n)$ and the open triangles for $W_{\rm HYP}$ [93].

5.4.2 *Lattice calculations with* $[(Q\bar{Q})(Q\bar{Q})]$ *configurations*

The previous section outlines the techniques for extracting the two-quark potential. This is now generalised to the $[(Q\bar{Q})(Q\bar{Q})]$ case, where the quarks are on the corners of a rectangle with sides of length d and r. During the Monte Carlo simulation, the correlations $W_{ij}(r)$ in Eq. 5.21 — appropriate for extracting the two-quark potential $V_{Q\bar{Q}}(r)$ — and the correlations $W_{ij}(d,r)$ for the four-quark potential $V_4(d,r)$ are evaluated at the same time. As an example, in Ref. [86] for $W_{ij}(r)$ three paths were generated by the three different fuzzing levels m =12, 16 and 20. On the other hand, for the $W_{i'j'}(d,r)$ only the level 20 was kept, with the variational basis $(i'j')$ being the two configurations A and B in Figs. 5.4. These lead to the Wilson loops in Fig. 5.5. In the case of squares (*i.e.* $r = d$), Eq. 5.22 gives the potential energy of the ground state as

$$\begin{aligned} E_0 =& aV_4(d,d) = aE_4(d,d) - 2E(d) \\ =& \log\left[\frac{W^{t-1}_{11}(d,d) + W^{t-1}_{12}(d,d)}{W^t_{11}(d,d) + W^t_{12}(d,d)}\right] - 2E(d) \end{aligned} \tag{5.24}$$

and that of the first excited state

$$E_1 = \log\left[\frac{W_{11}^{t-1}(d,d) - W_{12}^{t-1}(d,d)}{W_{11}^{t}(d,d) - W_{12}^{t}(d,d)}\right] - 2E(d), \qquad (5.25)$$

where $E(d)$ is the result of diagonalising the 3×3 variational basis for $V_{Q\bar{Q}}$. In these equations, t should tend to ∞. However, in practice it is found to be sufficient to have $t \approx 5$ for extracting accurate values for E_0 and E_1.

The results are shown in Fig. 5.6. For comparison this also includes the corresponding 1985 results of Ref. [97] as the two points with large error bars at $r/a \approx 1.25$ and 2.5 — remembering that the lattice spacing for Ref. [97] is $a' \approx 0.15$fm $\approx 1.25a$, where $a \approx 0.12$ fm from Ref. [87]. It is seen that the present calculation gives energies that have error bars which only become significant for $r/a \geq 7$, whereas the 1985 work was unable to generate any meaningful numbers beyond $r/a' = 2$. In fact, for the present purpose, neither of the 1985 points is of use, since the $r/a' = 1$ result could

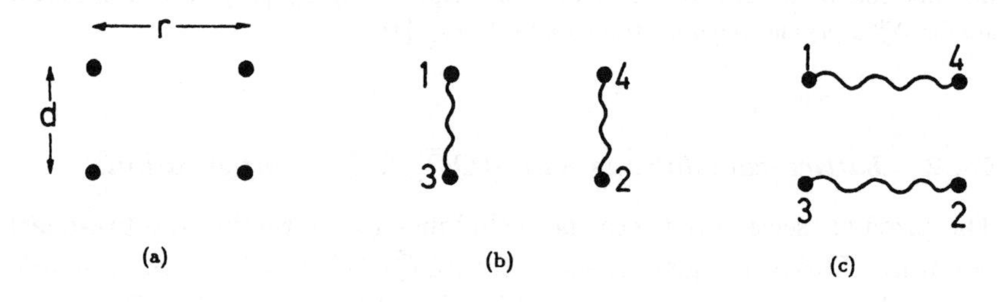

Fig. 5.4 a) Four quarks in a rectangle of sides r and d:
The two partitions b) $A = [Q_1\bar{Q}_3][Q_2\bar{Q}_4]$ and c) $B = [Q_1\bar{Q}_4][Q_2\bar{Q}_3]$

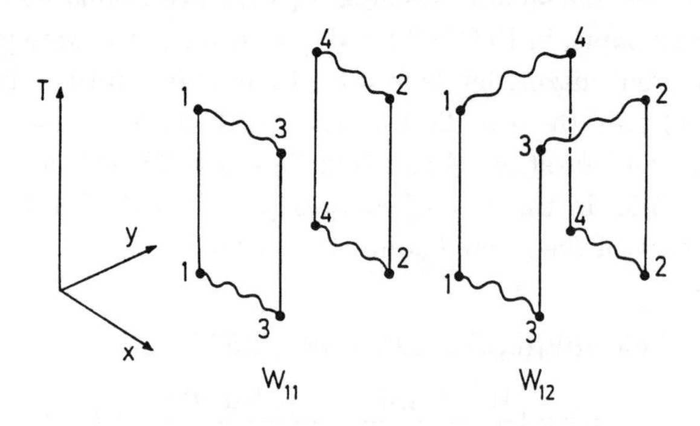

Fig. 5.5 The 4-Q Wilson loops W_{11} and W_{12}.

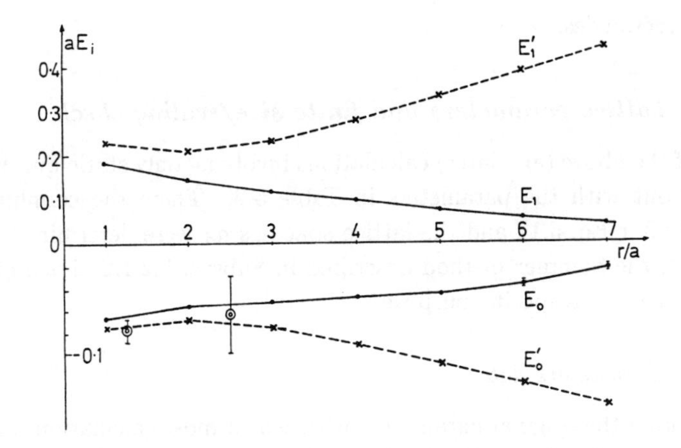

Fig. 5.6 The comparison of the lattice data $E_{0,1}$ (continuous line) with the $f = 1$ model results $E'_{0,1}$ (dashed line) in Subsec. 5.5.1 — see Ref. [87]. Also the two lattice data points from Ref. [97] are included.

well suffer from lattice artefacts — a problem afflicting all calculations of configurations involving a single lattice spacing — and the $r/a' = 2$ result has error bars that are too large for the analysis in Subsec. 5.5 to be carried out. However, the following should be added in defence of Ref. [97]. Firstly, their calculation was for SU(3) and so it was, at least, an order of magnitude more demanding on CPU time — especially in 1985. Secondly, it should be remembered that the energies of interest ($E_{0,1}$) are very small compared with the total four-quark energy E_4 in Eq. 5.24. For example, at $r = d = 4a$ the value of $aE_0 = -0.050(1)$ is obtained from the difference between aE_4 and $2aE(d)$, which is 1.505–1.555 *i.e.* the error quoted on aE_0 is less than 0.1% of aE_4.

One of the most outstanding features of the results is that for r *equals* d the value of aE_0 decreases smoothly in magnitude from –0.07 for $r = d = a$ to –0.04 for $r = d = 6a$. However, for the few cases where $r \neq d$ the value of E_0 is *an order of magnitude smaller* than the adjacent $r = d$ cases *e.g.* $E_0(2,3) \approx -0.006$, whereas $E_0(2,2)$ and $E_0(3,3)$ are ≈ -0.055. This result is reminiscent of the conclusion found with the flux-tube model [98] and the ansatz made in the flip-flop model of Refs. [89]. In both of these models the interaction between the two separate two-quark partitions of Fig. 5.4 is very small (in fact zero in the flip-flop model) except when the unperturbed energies of the two partitions is the same *i.e.* at $r = d$ in the case of rectangles.

5.4.3 *Lattice parameters and finite size/scaling check*

Most of the above (and later) calculations involving only static quarks were carried out with the parameters in Table 5.2. There the coupling β is defined after Eq. 5.17 and the lattice spacings a_n were determined by, for example, the Sommer method described in Subsec. 5.2.1.2. Each of these three sets serve a specific purpose.

5.4.3.1 *Benchmark data*

Set 1 shows the original parameters with which most calculations are performed (*e.g.* for the results in Fig. 5.6) and against which the following results with Sets 2 and 3 are compared. These parameters are chosen so that — unlike Sets 2 and 3 — the problems of computer memory and CPU time do not prevent generating many "independent" measurements to ensure good error estimates.

5.4.3.2 *Finite size effect*

With Set 2 the effect of the finite lattice size is checked. However, in the present case it was found that the results for the size of squares considered (*i.e.* up to 7×7) were unchanged within error bars — see Table 2 in Ref. [88]. This type of check is important for large squares because of the spatial periodic boundary conditions at 0 and L. Two quarks can

Table 5.2 Typical SU(2) lattice parameters used in $[(Q\bar{Q})(Q\bar{Q})]$ calculations. Much of the notation is explained in the text with M/G being the number of Measurements per Geometry.

Set	$L^3 \times T$	β	a_n (fm)	M/G	Refs.
1	$16^3 \times 32$	2.4	≈ 0.12	720	[86]–[88]
2	$24^3 \times 32$	2.4	≈ 0.12	160	[88]
3	$24^3 \times 32$	2.5	≈ 0.082	660	[88]

be connected by two paths — a direct path on the lattice (*e.g.* that in Eq. 5.12) or an indirect path that *encircles* the boundary. If the two quarks are separated by $r > L/2$, then the indirect path is of length $L - r < r$ and so is potentially more important than the direct path — the only one of the two that is explicitly treated. In the present case with $L = 16$, such problems should only begin to occur seriously with 8×8 squares. For this larger lattice, fewer configurations per square were generated, since the storage and computer time increased by about a factor of $\approx (24/16)^3 \approx 3$.

5.4.3.3 *Smaller lattice spacing*

In most cases the possible importance of finite size effects can be seen beforehand by using simple arguments. We know, for example, that the length r of the direct path should be smaller than the length $L - r$ of the indirect path. However, the effect as $a \to 0$ is not obvious and needs checking. In order to isolate this scaling effect from the finite size effect, it is convenient to use lattices that have approximately the same *physical* size. Therefore, since the spatial volume for Set 1 is $(16 \times 0.12)^3 \approx 7$ fm^3, the value of a_3 for a 24^3 lattice should be $\approx 7^{1/3}/24 \approx 0.08$ fm — a lattice spacing that corresponds to $\beta \approx 2.5$. For the results to show **scaling**, the physical energies (*i.e.* in, say, MeV) at the same values of r (in fm) should be the same. In this case, it means

$$E[\text{Set 1 (or 2)},\ r/a_1] \approx E[\text{Set 3},\ r/a_3], \tag{5.26}$$

where the a_nE(Set n, r/a_n) are the *dimensionless* numbers given by the lattice calculations and the r/a_n are the number of lattice links between the quarks. Of course, in general only r/a_1 or r/a_3 is an integer, so that the comparison in Eq. 5.26 must be done by interpolation. This approximate equality is sufficiently well satisfied in the present problem (see Table 3 in Ref. [88]). In Ref. [49] a more complete test of the $a \to 0$ limit is carried out with the addition of the non-rectangular geometries to be discussed in Sec. 5.6. There the β values (and lattice sizes) used were β= 2.35, 2.4 ($16^3 \times 32$), 2.45 ($20^3 \times 32$), 2.5 ($24^3 \times 32$) and 2.55 ($26^3 \times 32$) and led to 4-quark energies that were essentially independent of a over this range of β *i.e.* scaling was achieved for $\beta \geq 2.35$.

It should be added that the above procedure is called *scaling* in contrast to *asymptotic scaling.* The latter relates results from different values of $a(\beta)$ by *perturbative* arguments and is not expected to apply at the comparatively large values of a in Table 5.2 — see Ref. [99].

5.5 Potential Model Description of the Lattice Data

The assumption often made by those who create models for multi-quark systems is that these systems can be treated in terms of two-body potentials. This is the Nuclear Physics inspired approach that is very successful for, say, multi-nucleon systems, where *three-* and *four-*body forces are small. Of course, the fact that the latter were small was at first simply an assumption. However, later this was found to be justified by Weinberg [23] using various low-energy theorems that force the $\pi N \to \pi N$ interaction — the main mechanism for multi-nucleon forces — to essentially vanish in nuclei. At present, there seems to be no such simplifying feature for multi-quark interactions. In the following, I will first show the consequences of using — in the four-quark system — simply two-quark potentials unmodified by the presence of the other two quarks. Then this model will be improved by including also a direct four-quark interaction. These are examples of what was referred to as Effective Potential Theories (EPTs) in Subsec. 5.1.2.2. However, since the quarks are now static, there is no kinetic energy and so the question of which quark wave equation to use does not arise. Similarly, the effective quark masses do not enter, since the quantities of interest are the *binding energies* between the two 2-quark clusters. Therefore, of the three ingredients usually needed for an EPT of Subsec. 5.1.2.2 only freedom with the interquark potential remains. This is both a weakness, by being a model that is unrealistic and not comparable with any experimental data — except for possible future $\Upsilon(b\bar{b}, 9.5\text{GeV})\Upsilon(b\bar{b}, 9.5\text{GeV})$ scattering — and a strength, in that only the potential ingredient plays a rôle.

5.5.1 *Unmodified two-body approach*

Once the quark–quark potential $V_{Q\bar{Q}}$ is known and, if it is assumed to be the *only* interaction between the quarks, then the energy of a multi-quark system can be readily calculated — provided the wave function for that system is expressed in terms of a sufficient number of basis states. For the present situation, the most obvious choices for such states are A and B in Figs. 5.4 b) and c).

In this extreme two-body approach, since the presence of the gluon fields have been explicitly removed — their only effect now being in the colour indices of the quarks — the states A and B form a complete but non-orthogonal basis. This implicitly assumes in Figs. 5.4 the quark (Q), antiquark ($\bar{Q}$) assignment $A = [Q_1\bar{Q}_3][Q_2\bar{Q}_4]$. In SU(2) the other quark assignment $[Q_1Q_3][\bar{Q}_2\bar{Q}_4]$ is numerically equivalent and so leads to nothing new. In this approach, states are excluded in which the gluon fields are excited. However, later this restriction is relaxed in an extension to the model in Sec. 5.7 and Appendix A.3. The energies (E_i') of this static four-quark system can be extracted from the eigenvalues of the Hamiltonian

$$(\mathbf{V} - \lambda_i \mathbf{N})\, \Psi_i = 0, \tag{5.27}$$

where the normalisation matrix

$$\mathbf{N} = \begin{pmatrix} 1 & 1/2 \\ 1/2 & 1 \end{pmatrix} \tag{5.28}$$

and the potential energy matrix

$$\mathbf{V} = \begin{pmatrix} v_{13} + v_{24} & V_{AB} \\ V_{BA} & v_{14} + v_{23} \end{pmatrix}. \tag{5.29}$$

Several points need explaining in these equations.

(1) The off-diagonal matrix element $N_{12} = \langle A|B\rangle = 1/2$ shows the non-orthogonality of the A, B basis. In general, for SU(N_C) this becomes $N_{12} = 1/N_C$ and arises simply by recoupling the colour components of the quark terms *i.e.*

$$|1_{1\bar{4}}1_{2\bar{3}}\rangle = \frac{1}{N_C}|1_{1\bar{3}}1_{2\bar{4}}\rangle + \frac{\sqrt{N_C^2 - 1}}{N_C}|A_{1\bar{3}}A_{2\bar{4}}\rangle, \tag{5.30}$$

where $|1_{i\bar{j}}\rangle$ and $|A_{i\bar{j}}\rangle$ are the SU(N_C) singlet and adjoint representations.

At this stage, the lack of orthogonality could have been avoided by simply using the basis $A \pm B$. However, later it will be seen that the A, B basis is in fact more convenient and suggestive, when the gluon fields are reintroduced in a more explicit manner.

(2) The interquark potential $V_{Q\bar{Q}}(ij) = v_{ij}$ in Eq. 5.6 has been extracted as the potential energy between a single static quark (Q)

and a single static antiquark $(\bar{Q})$. In order to evaluate general multiquark potential energy matrix elements, a further assumption is needed concerning the colour structure of v_{ij}. Here the usual identification

$$V_{ij} = -\frac{1}{3}\tau_i.\tau_j v_{ij} \tag{5.31}$$

will be made, where the τ_i are the Pauli spin matrices appropriate for SU(2). This choice ensures, for a colour singlet meson-like state $[ij]^0$, that $\langle [ij]^0|\tau_i.\tau_j|[ij]^0\rangle = -3$ and

$$\langle [ij]^0|V_{ij}|[ij]^0\rangle = v_{ij}. \tag{5.32}$$

Strictly speaking, the form in Eq. 5.31 is only true in the weak coupling limit of one-gluon exchange, since this has replaced the local gauge invariance — ensured by the series of U-links of Eq. 5.11 connecting the two quarks — by the global gauge invariance reflected by the $\tau_i.\tau_j$ factor.

(3) With the choice of V_{ij} in Eq. 5.31, the off-diagonal potential matrix element becomes

$$\begin{aligned}\langle A|V|B\rangle =& V_{AB} = V_{BA} \\ =& \frac{1}{2}\left(v_{13} + v_{24} + v_{14} + v_{23} - v_{12} - v_{34}\right).\end{aligned} \tag{5.33}$$

Since the following discussion only involves quark configurations in the rectangular geometries of Figs. 5.4, it is convenient to use the notation

$$v_{13} = v_{24} = v_d, \quad v_{14} = v_{23} = v_r \text{ and } v_{12} = v_{34} = v_x, \tag{5.34}$$

where the suffix x refers to the diagonals of the rectangles in Fig. 5.4.

Even though the form of Eq. 5.33 is derived in the one-gluon exchange limit, it is now assumed that a more realistic model emerges if the v_{ij} are taken to be the complete potential of Eq. 5.6 and not just the one-gluon exchange component. This clearly has the correct form when the distance between the two two-quark clusters of Fig. 5.4 are far apart, since in this case the only interactions are those *within* the separate clusters — due to the cancellation in Eq. 5.33 of v_{12}, v_{34} with either v_{13}, v_{24} or v_{14}, v_{23}.

The 2×2 Hamiltonian of Eq. 5.27 is easily diagonalised to give the eigenvalues $\lambda_{0,1}$. Since it is the binding energy E'_i of the four-quark system that is of interest, and also that was extracted in Subsec. 5.4.2 from the Monte Carlo simulation, the internal energy of the meson-like state with the lowest energy (*i.e.* $2v_d$ for $d \le r$) is now subtracted from the λ_i to give

$$E_i' = \lambda_i - 2v_d \tag{5.35}$$

in analogy with the lattice expressions in Eqs. 5.24 and 5.25.

Therefore, in this simplest version of the two-body approach, the E_i''s should correspond to the E_i's from the Monte Carlo simulation — a comparison which is made in Fig. 5.6. Since the values of E_0 for squares (*i.e.* $r = d$) are much larger than those for neighbouring rectangles, only the results for squares are shown in Fig. 5.6. In these cases

$$E_0' = -\frac{2}{3}(v_x - v_d) \text{ and } E_1' = 2(v_x - v_d) = -3E_0', \tag{5.36}$$

In addition, the corresponding normalised wave functions are of the form $\psi(E_0') = \frac{1}{\sqrt{3}}|A + B\rangle$ and $\psi(E_1') = |A - B\rangle$.

Two comments should be made on these results:

- For the smallest squares and rectangles the agreement between E_i and E_i' is best. This is reasonable, since it is expected that at small interquark distances perturbation theory is adequate *i.e.* the lowest order gluonic effects are already incorporated correctly into the interquark potential.
- As the squares and rectangles get larger the differences between the E_i and the E_i' grow until E_0' is more than three times E_0, and E_1' more than seven times E_1, for the largest squares $d \approx 7a \approx 0.8$ fm. Since E_0' is too attractive and E_1' too repulsive, this suggests that the off-diagonal matrix element V_{AB} in Eq. 5.33 is too large.

The main conclusion to be drawn from the above is that the two-body potential of Eq. 5.31 does *not* give the potential energy of the four-quark system — the indication being that the off-diagonal potential energy V_{AB} is too large. This is nothing more than the well known van der Waals effect of Ref. [100] — see Oka and Yazaki in Chapter 6 of Ref. [36].

5.5.2 *The effect of multiquark interactions*

In the above model it is assumed that all of the gluonic effects are incorporated into the two-body potential v_{ij}. However, this is clearly an oversimplification that is at best only applicable in situations where perturbation theory holds, namely at short distances, as already noted in the discussion of Fig. 5.6. In more realistic models, the QCD coupling is sufficiently strong to constrain the gluon field into flux-tubes connecting the quarks in a given meson — as visualised by the wavy lines between the quarks in states A and B in Figs. 5.4. Therefore, the overlap of states A and B, *i.e.* $N_{12} = \langle A|B\rangle$, is not simply the colour recoupling factor of 1/2, but should also involve the lack of overlap of the gluon fields in states A and B. This can be incorporated by introducing an entity f, which simply multiplies the original N_{12}, and which is an unknown function of the position coordinates of the four quarks. With this interpretation of f as a gluon field overlap factor, it is also necessary to multiply the off-diagonal potential matrix element V_{AB} of Eq. 5.33 by the *same* factor f. This factor must be the same in $N_{12} = f/2$ and $V_{12} = fV_{AB}$, since otherwise the binding energies $E'_{0,1}$ would be dependent on the self-energy term c in the form of v_{ij} given in Eq. 5.6 — which would be unphysical. The $E'_{0,1}$ are now extracted from Eq. 5.35 after diagonalising

$$[\mathbf{V}(f) - \lambda_i(f)\mathbf{N}(f)]\,\Psi_i = 0 \tag{5.37}$$

with

$$\mathbf{N}(f) = \begin{pmatrix} 1 & f/2 \\ f/2 & 1 \end{pmatrix} \text{ and } \mathbf{V}(f) = \begin{pmatrix} v_{13} + v_{24} & fV_{AB} \\ fV_{BA} & v_{14} + v_{23} \end{pmatrix}. \tag{5.38}$$

The two equations (5.37) and (5.38) are the basis of the following analysis. They give a procedure for extending the model of Eqs. 5.27 – 5.29, which was justified in the weak coupling limit, into the domain beyond one-gluon exchange. The off-diagonal potential matrix element performs this extension in two ways. Firstly, even though V_{AB} in Eq. 5.33 still has the same algebraic structure in terms of the v_{ij} as dictated by the one-gluon exchange limit, the v_{ij}'s themselves are the full two-quark potential of Eq. 5.6. Secondly, in the off-diagonal correlation W_{12} of Fig. 5.5, the one-gluon exchange model suggests the presence of the overall multiplicative factor f due to gluon exchange within the *initial* and *final* states A and B at euclidean times $T = 0$ and ∞. In this interpretation, the terms in V_{AB} arise during the period of propagation between $T = 0$ and ∞.

The strategy is now to adjust f to get an exact fit to E_0 or E_1. In the case of squares, since

$$E'_0 = \frac{-f}{1+f/2}(v_x - v_d) \text{ and } E'_1 = \frac{f}{1-f/2}(v_x - v_d), \tag{5.39}$$

the appropriate expressions are

$$f(E_0) = \frac{E_0}{v_d - v_x - E_0/2} \text{ and } f(E_1) = \frac{E_1}{-v_d + v_x + E_1/2} \tag{5.40}$$

for fitting E_0 and E_1 respectively — the results being shown in Fig. 5.7. For general rectangles the corresponding equations are somewhat more complicated.

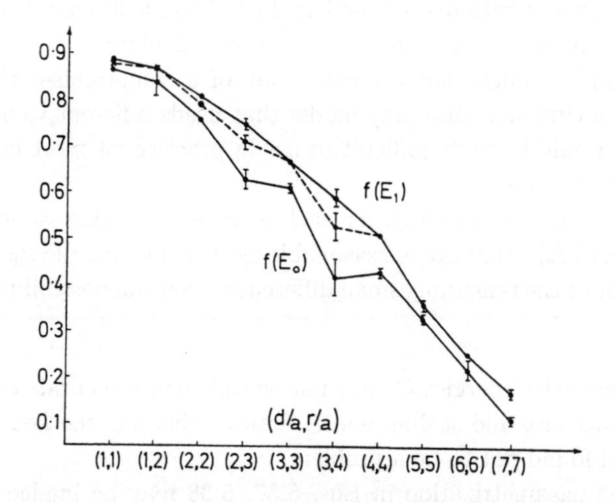

Fig. 5.7 The values of $f(E_0)$, $f(E_1)$ and $\bar{f}$ (from Subsec. 5.5.3).

Two points should be noted from this figure:

- All values of f are less than unity as is expected from the interpretation of f as a gluon field overlap factor. In addition, this idea is supported by the fact that the values of f decrease as the quarks get further apart.
- The values of $f(E_0)$ and $f(E_1)$ are rather similar, which suggests that a compromise value of f would give a reasonable description of both E_0 and E_1 — as will be seen below.

5.5.3 *A compromise for the overlap factor f*

The previous subsection suggests that for each geometry a *single* value of f could give a reasonable description of both E_0 and E_1. Here one such possibility is given by finding that value of $\bar{f}$ which minimizes the expression

$$D(\bar{f}) = \left(\frac{E_0 - E'_0(\bar{f})}{\Delta E_0}\right)^2 + \left(\frac{E_1 - E'_1(\bar{f})}{\Delta E_1}\right)^2, \tag{5.41}$$

where the ΔE_i are the errors quoted for the E_i from the lattice calculation. The result is shown by the dashed line in Fig. 5.7.

It is seen that indeed a single value of $f = \bar{f}$ suffices to explain reasonably well both energies. This is a non-trivial observation, since it indicates that the parametrization suggested in Eqs. 5.37, 5.38 contains the most important features of the more precise lattice calculation.

It should be added that the extraction of a compromise value of f is not simply a curiosity, since any model that needs different values of f for E_0 and E_1 would be more difficult to use in practice for more complicated multi-quark systems.

At this point, even though for each geometry a single value of $f = \bar{f}$ gives values of $E'_{0,1}$ that are in reasonable agreement with the $E_{0,1}$, it might be asked about the remaining small differences. Several possibilities are now open:

- The lattice energies E_i may not be sufficiently accurate due to finite lattice size and scaling uncertainties. This was checked in Ref.[88] and found not to be a problem.
- The parametrization in Eqs. 5.37, 5.38 may be inadequate. One possibility would be to combine the notion of a gluon overlap factor f with a generalized form for the two-quark potential in Eq. 5.6, since this could introduce more free parameters.
- Any model based only on states A and B in Fig. 5.4 is incomplete and other states are necessary in addition. This point will be discussed further in Sec. 5.7.
- The step from Eq. 5.27 to Eq. 5.37 was motivated by one gluon exchange. However, this could possibly be extended by performing a *two-gluon* exchange calculation to see what new terms arise and to then be guided by this in making an improved parametrization — the topic of the next subsection.

5.5.4 *The effect of two-gluon exchange*

In Ref. [101] it was noted that the two-body $f = 1$ models discussed in Subsec. 5.5.1 correspond to lowest order perturbation theory in the quark–gluon coupling *i.e.* to order $\alpha = g^2/4\pi$ in the notation of Eq. 5.17. This was extended in Ref. [102], where a perturbative calculation to fourth order in the quark–gluon coupling (*i.e.* to $O(\alpha^2)$) was made for the potential of the $QQ\bar{Q}\bar{Q}$ system. This was performed in the general case of colour SU(N). Considering the quarks to be at points R_i define

$$V_A = V(R_{13}) + V(R_{24}),\ V_B = V(R_{14}) + V(R_{23}),\ V_C = V(R_{12}) + V(R_{34}),$$

where $V(R_{pq})$ is the two-body potential between the different quark and antiquark pairs a distance R_{pq} apart. For the two possible energy eigenstates, diagonalization yields the following two potentials correct to $O(\alpha^2)$:

$$V_0 = \frac{(N^2-2)\,(V_A+V_B) + 2V_C - N\sqrt{N^2\,(V_A-V_B)^2 + 4\,(V_A-V_C)(V_B-V_C)}}{2\,(N^2-1)}$$

$$V_1 = \frac{(N^2-2)\,(V_A+V_B) + 2V_C + N\sqrt{N^2\,(V_A-V_B)^2 + 4\,(V_A-V_C)(V_B-V_C)}}{2\,(N^2-1)}. \tag{5.42}$$

These potentials are *exactly equal* to those given by the naive ($f = 1$) two-body model in Eqs. 5.27 – 5.29. This fact that a straightforward two-body model is correct also to next-to-leading order in the quark–gluon coupling may be surprising in light of the non-abelian nature of QCD. However, Ref. [102] does go on to show that this two-body model fails at $O(\alpha^3)$ as three- and four-body forces appear due to the onset of three-gluon vertex effects. In general, their nature seems to be complicated, but for some geometries simplifications are possible; *e.g.* for the four quarks on the corners of a regular tetrahedron there will be no contribution from quark self–interactions to four-body forces to $O(\alpha^3)$.

The overall conclusion from Ref. [102] is as follows: "Looking at the Monte Carlo lattice calculations for the $QQ\bar{Q}\bar{Q}$-system in Refs. [87, 88], it is observed that for small interquark distances of a few lattice spacings (with $a \approx 0.12$ fm) the $f = 1$ two-body model gives a reasonable approximation in the sense that the four-quark potentials calculated from Eq. 5.42 using the Monte Carlo two-body potentials are comparable to the four-quark potentials from the lattice simulation. The agreement improves the smaller the distances get. By comparing the perturbative (*i.e.* $1/R$) and non-perturbative (*i.e.* linear) part in the usual parametrization of the $Q\bar{Q}$-potential in Eq. 5.6, one would expect to start entering the perturbative regime when distances get down to about two lattice spacings. However, at that stage the approximation provided by the two-body model is already very good. The fact that the two-body model is correct to fourth order in perturbation theory certainly suggests that it should be a reasonable approximation in the perturbative domain. This result supports the belief that the results of the lattice simulations for small enough distances indeed are correlated to continuum perturbation theory, and thus that continuum physics is extracted from the Monte Carlo calculations."

So far in this section various models have been proposed in an attempt to understand the results of the Monte Carlo simulations of lattice QCD. The main outcome — summarised in Fig. 5.7 — is the emergence of a function $\bar{f}$ that depends on the coordinates of the four quarks involved. This shows that the usual models based on purely two-quark interactions need to be modified considerably — essentially by the factor $\bar{f}$, which becomes $\ll 1$ for large interquark distances. This observation is in itself of much interest, but at this stage it is not clear how the effect can be incorporated into more realistic situations in which the quarks are not so restricted in their geometry. It is the purpose of the next section to tackle this problem by first studying how $\bar{f}$ can be parametrized.

5.5.5 *Parametrizations of the gluon-field overlap factor f*

In Subsecs. 5.5.2, 5.5.3, models were introduced in an attempt to understand the ground state binding energy (E_0) and excited state energy (E_1) emerging from a Monte Carlo simulation, in which four quarks were at the corners of a rectangle. These models are summarised by Eqs. 5.37, 5.38. For each quark configuration, both of the energies $E_{0,1}$ are described in terms of a function f of the four quark positions — Eq. 5.39. As it stands, this is not particularly useful, when wishing to extend these ideas to systems containing more than four quarks, unless f can be parametrized in some sensible and convenient manner. In the literature, several parametrizations have been suggested. In Refs. [101, 103], motivated by strong coupling arguments, the phenomenological form is taken to be

$$f_1 = \exp[-\alpha b_s S], \tag{5.43}$$

where $b_s = 0.0736$ is the string tension in the interquark potential of Eq. 5.6 and S is the minimal area of the surface bounded by the straight lines connecting the quarks and antiquarks. The other form, the one proposed in [104], is

$$f_2 = \exp\big[-\frac{kb_s}{6}\sum_{i<j} r_{ij}^2\big] \tag{5.44}$$

i.e. the cut-down is governed by the average of the six links present in a $Q^2\bar{Q}^2$ system. In Eqs. 5.43, 5.44 the α and k are at present free parameters to be determined later. Both of these parametrizations of f accommodate the following two extreme models.

- Weak coupling, which has $f = 1$ when all $r_{ij} = 0$. This $f = 1$ limit was assumed to apply for all values of r_{ij} in Subsec. 5.5.1.
- Strong coupling, which has $f = 0$ when any $r_{ij} \to \infty$. In this limit the flux-tubes between the quarks in the separate mesons — seen in Fig. 5.4 — are then very narrow and straight. In this case the flux-tube overlap of configurations A and B tends to zero in the limit that any $r_{ij} \to \infty$.

With squares ($r = d$), for which the most accurate values of f exist, k equals $3\alpha/4$. One measure of how meaningful these parametrizations really are, is given by extracting α and k for each quark configuration. The hope would then be that, for squares, α (and therefore k) would be independent of the separate configurations. Only the few points for non-square rectangles ($r \neq d$) would be able to distinguish between f_1 and f_2. The outcome is depicted in Fig. 5.8 for the values of $k(E_0), k(E_1)$ and $\bar{k}$ corresponding to the values of $f(E_0), f(E_1)$ and $\bar{f}$ in Fig. 5.7. This shows that $k(E_0)$ for the squares appears to decrease slowly from about 0.7 to about 0.5 as the sizes of the squares increase from (2×2) to (7×7), whereas for non-squares $k(E_0)$ appears to be stable at about 0.7±0.1. On the other hand, $k(E_1)$ and $\bar{k}$ decrease somewhat less and also the square and non-square values are consistent with each other.

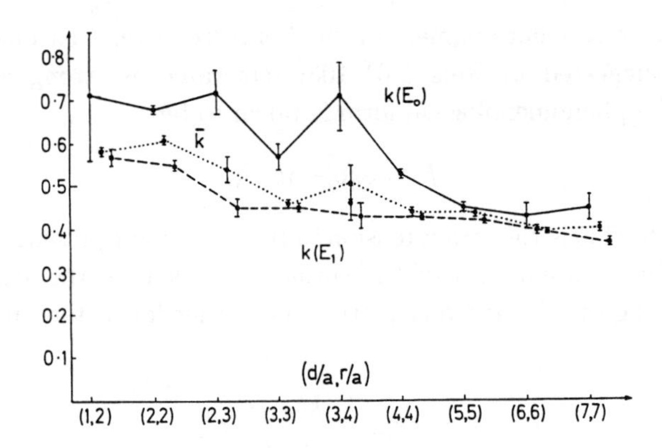

Fig. 5.8 The values of $k(E_0), k(E_1)$ and $\bar{k}$ corresponding to the values of $f(E_0), f(E_1)$ and $\bar{f}$ in Fig. 5.7.

The indications from this are that the *single* parameter $\bar{k} \approx 0.5(1)$ is a suitable compromise value that results in a reasonable fit to both E_0 and E_1 for a series of square and near-square geometries.

5.5.5.1 *A reason for* $f_1 = \exp[-\alpha b_s S]$

In the limit of large l, Eq. 5.7 becomes

$$W(l,t) \to \exp[-tb'_s l] = \exp[-b_s S] \quad \text{as} \quad t \to \infty, \tag{5.45}$$

where S is the *minimal* space–time area enclosed by the loop in Fig. 5.1. For this $Q\bar{Q}$ case, the meaning of S is clear — it is simply a^2lt. But for the $[Q\bar{Q}][Q\bar{Q}]$ case in Fig. 5.5 the form of the appropriate minimal space–time area is less clear. However, in the extreme strong coupling limit the area S is the one produced by the *minimum* number of elementary plaquettes needed to tile the enclosed area in question. Any fluctuations about this space-time surface would need more plaquettes and so be higher order in the strong coupling model. In this limit, the diagonal loops W_{ii} in Fig. 5.5 are again simply two Wilson loops each tiled separately by the minimum number of plaquettes. However, the off-diagonal loops W_{ij} are more complicated. A model for this was suggested in Ref. [105] and developed in Refs. [103, 106, 107]. In the notation of Fig. 5.9 the Euclidean Green's function for a $Q_A, \bar{Q}_B, Q_C, \bar{Q}_D$ system can be thought of as a 2×2 matrix for a two channel problem with the $(A\bar{B})(C\bar{D})$ and $(A\bar{D})(C\bar{B})$ configurations. The transition potential between these two configurations may then

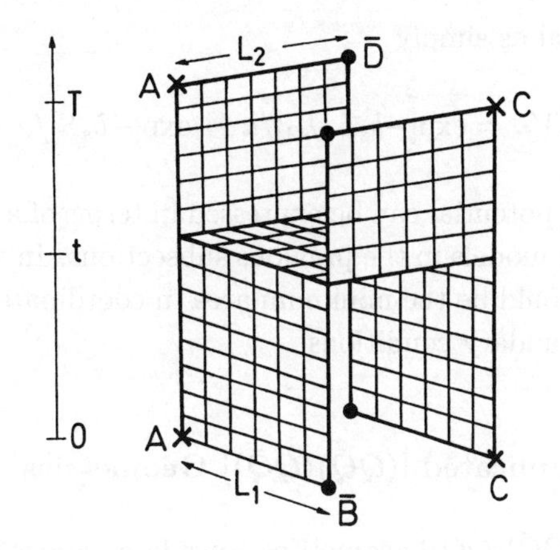

Fig. 5.9 Tiling the transition between the states $[(Q_A\bar{Q}_B)(Q_C\bar{Q}_D)]$ and $[(Q_A\bar{Q}_D)(Q_C\bar{Q}_B)]$.

be extracted from the expression for the Wilson loop of this system,

$$\mathbf{W}(T) = \begin{pmatrix} \exp[-2b_sL_1T] & \epsilon \\ \epsilon & \exp[-2b_sL_2T] \end{pmatrix}, \tag{5.46}$$

where L_1 and L_2 are the minimum flux tube lengths in the basis states and b_s is the string tension of Subsec. 5.2.1.1. The diagonal terms are each simply the product of two propagators — $\exp[-b_sL_1T]$ for W_{11} and $\exp[-b_sL_2T]$ for W_{22}. In a language more familiar in nuclear physics, these are simply the Green's functions

$$W_{11} = G_{A\bar{B}}(T)G_{C\bar{D}}(T) \text{ and } W_{22} = G_{A\bar{D}}(T)G_{C\bar{B}}(T). \tag{5.47}$$

The mixing term

$$\epsilon = \sum_t \exp[-2bL_2(T-t)]\exp[-bL_1L_2]\exp[-2bL_1t] \tag{5.48}$$

corresponds to tiling of the off-diagonal loops W_{ij}. This now resembles the standard expression for the lowest order transition of a Green's function

$$G_{fi}(T) = \int dtG_f(T-t)V_{fi}G_i(t), \tag{5.49}$$

where here i, f denote the initial and final channels $[(Q_A\bar{Q}_B)(Q_C\bar{Q}_D)]$ and $[(Q_A\bar{Q}_D)(Q_C\bar{Q}_B)]$. Remembering that $\sum_t \to a^{-1}\int dt$ we can identify the transition potential as simply

$$V_{fi} = \exp[-b_s L_1 L_2]/a \to \exp[-b_s S]/a \qquad (5.50)$$

i.e. the transition potential can be expressed in terms of a *spatial* area — as anticipated by the models in the previous subsections. In more complicated geometries this should be the minimum area in coordinate space associated with the given boundary conditions.

5.6 More Complicated $[(Q\bar{Q})(Q\bar{Q})]$ Geometries

So far the only $[(Q\bar{Q})(Q\bar{Q})]$ geometries considered above were squares and rectangles. This suggested that the strongest interaction between two separate two-quark clusters occurs when the clusters are degenerate in energy — namely for square geometries compared with rectangles. To test this more, in Ref. [52] the six different geometries in Fig. 5.10 were studied. Since this confirmed that configurations degenerate in energy always gave the largest binding — not just in the square versus rectangle case — in Ref. [49] the study concentrated only on those configurations that were near degenerate.

In Ref. [53] the case of tetrahedral configurations was considered in some detail, since in SU(2) this has the *three* degenerate partitions seen in Fig. 5.10(T) when $d = r$. Since tetrahedral and linear configurations have certain interesting features, they will be discussed separately below.

However, real life is an average over all spatial configurations, so why should we be so interested in such special geometries for four static quarks? Firstly, it must be remembered that model (bridge) builders should consider all spatial possibilities — a failure with one configuration indicating that the proposed model is faulty. Also, in some ways the tetrahedral and linear configurations have simplicities and symmetries not present in other cases, which could make "Lattice QCD $\leftrightarrow$ Model" comparisons easier.

5.6.1 *Tetrahedral configurations on a lattice*

Since tetrahedral configurations are so symmetrical, at first sight there is no reason to consider only two of the three possible cluster partions $A+B$, $B+C$ or $A+C$. In this case, using the notation that the suffices $1,2,3$ are any combination of A,B,C that forms a basis, then the appro-

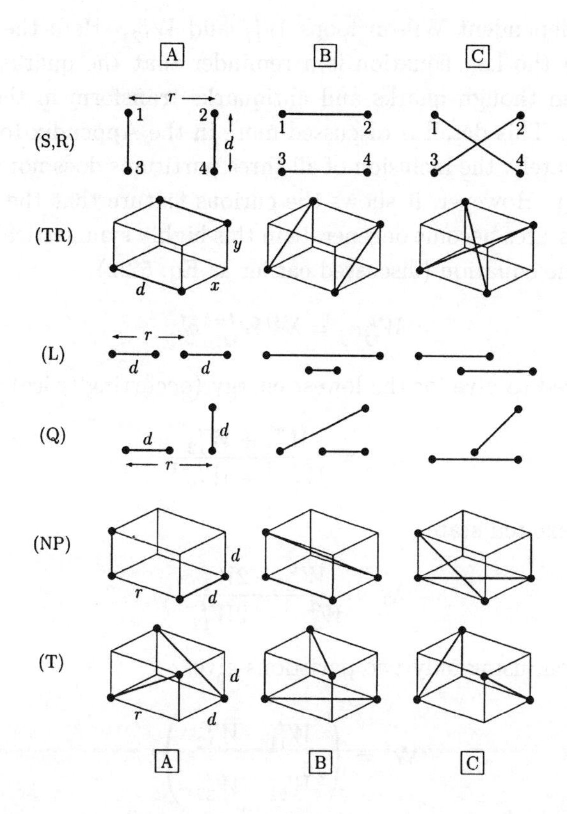

Fig. 5.10 Six four-quark geometries a) Squares (S) and Rectangles (R), b) Tilted rectangles (TR), c) Linear (L), d) Quadrilateral (Q), e) Non-Planar (NP) and f) Tetrahedra (T). Their energies are calculated using Lattice QCD in Refs. [49, 52, 53, 86, 87, 88] and analysed in Ref. [108].

priate Wilson loop matrix is [53]

$$\mathbf{W}^t = \begin{pmatrix} W_{11}^t & W_{12}^t & W_{13}^t \\ W_{21}^t & W_{22}^t & W_{23}^t \\ W_{31}^t & W_{32}^t & W_{33}^t \end{pmatrix}. \tag{5.51}$$

For regular tetrahedra, the general symmetries $W_{11}^t = W_{22}^t = W_{33}^t$, $W_{21}^t = W_{12}^t$, $W_{31}^t = W_{13}^t$, $W_{32}^t = W_{23}^t$ are expected and, in addition, there are the equalities $W_{13}^t = W_{12}^t$ and $W_{23}^t = -W_{13}^t$. Therefore, in all, there are only *two* independent Wilson loops W_{11}^t and W_{12}^t. Here the minus sign appearing in the last equation is a reminder that the quarks are in fact fermions even though quarks and antiquarks transform in the same way under SU(2). This detail is discussed more in the Appendix to Ref. [52].

Unfortunately, the inclusion of all three partitions does not lead to even more binding. However, it shows the curious feature that the ground and first excited states become degenerate in this highly symmetrical limit, since the eigenvalue equation (discussed earlier as Eq. 5.22)

$$W_{ij}^t a_j^t = \lambda^{(t)} W_{ij}^{t-1} a_j^t, \tag{5.52}$$

is easily solved to give for the lowest energy (occurring twice)

$$\lambda_{1,2} = \frac{W_{11}^t + W_{12}^t}{W_{11}^{t-1} + W_{12}^{t-1}} \tag{5.53}$$

and for the excited state

$$\lambda_3 = \frac{W_{11}^t - 2W_{12}^t}{W_{11}^{t-1} - 2W_{12}^{t-1}}. \tag{5.54}$$

In comparison, using only two partitions gives

$$\mathbf{W}^t = \begin{pmatrix} W_{11}^t & W_{12}^t \\ W_{21}^t & W_{22}^t \end{pmatrix}, \tag{5.55}$$

where not only is the general symmetry $W_{12}^t = W_{21}^t$ expected but also for a regular tetrahedron $W_{11}^t = W_{22}^t$. In this case Eq. 5.52 is again easily solved to give for the lowest energy

$$\lambda_1 = \frac{W_{11}^t + W_{12}^t}{W_{11}^{t-1} + W_{12}^{t-1}} \tag{5.56}$$

i.e. the **same** as λ_1 in Eq. 5.53. However, for the energy of the first excited state

$$\lambda_2 = \frac{W_{11}^t - W_{12}^t}{W_{11}^{t-1} - W_{12}^{t-1}}, \tag{5.57}$$

which is quite different to the complete result in Eqs. 5.53, 5.54. So the degeneracy is easily explained as a feature of the more complete 3×3 lattice QCD simulation. The effect is depicted in Fig. 5.11 as $d \to r$. This also shows that, whereas ground state energies can be quite stable, those of excited states are more model dependent.

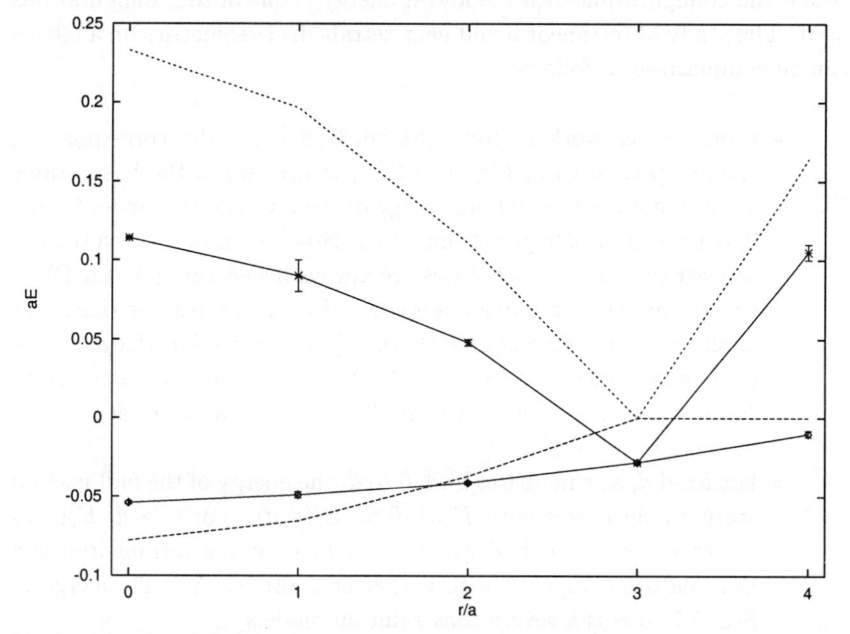

Fig. 5.11 The binding energies – in units of the lattice spacing – of the four-quark states for the tetrahedral geometry of Fig. 5.10 (T) for $d/a = 3$ and $r/a = 0, 1, 2, 3, 4$.
Solid lines show lattice results: $\diamond$ – the ground-state binding energy E_0.
$\times$ – the first excited state energy E_1.
Dashed and dotted lines show model results with $f = 1$ from Eqs. 5.27 – 5.29, for E_0 and E_1 respectively — see Ref [53].

The dominance of the cluster interaction by degenerate configurations was carried to the extreme in the so-called flip-flop model of Ref. [89], which makes the ansatz that the 4-quark interaction occurs *only* in the case of exact degeneracy. However, the model is developed for a purely *linear* interquark potential *i.e.* with only the b_s term in Eq. 5.6, which means that the interaction only occurs when — in the notation of Fig. 5.4 — the spatial distances $r_{13} + r_{24}$ and $r_{14} + r_{23}$ are equal.

In several of these geometries it is not clear what is the best "natural" partition into two-quark clusters. Therefore, in the lattice simulation all three possibilities A, B and C in Fig. 5.10 should be taken into account. However, in some cases it is found that one (or more) of the combinations

$A + B$ or $A + C$ or $B + C$ is sufficient to give for both E_0 and E_1 similar results to the complete $A + B + C$ simulation. This is particularly true if the configuration with the lowest energy is one of the configurations used. The study of tetrahedral and near-tetrahedral geometries on a lattice can be summarized as follows:

- From earlier work in Refs. [52, 86, 87, 88], for the corresponding squares [*i.e.* $(d, 0)$ in Fig. 5.10 (T)], where two of the basis states are degenerate, the binding energy of the lowest state ranges from –0.07 to –0.05 as d/a goes from 1 to 5. However, now — even though at least two of the basis states are always degenerate (A and B) — the ground state binding energy $[E_0(d,r)]$ is always less than that of the corresponding square $[E_0(d,0)]$. For a fixed d, $|E_0(d,r)|$ decreases as r increases from $r = 0$. Nothing interesting happens to E_0 at $r = d$, the point at which all the basis states are degenerate in energy.
- For fixed d, as r increases from 0 to d, the energy of the first excited state E_1 decreases until $E_1(d,d) = E_0(d,d)$. For $r > d$, $E_1(d,r)$ increases again. This degeneracy of $E_{0,1}$ for the tetrahedron is a new feature compared with earlier geometries. As will emerge in Sec. 5.7, this is a severe constraint on models.
- The choice of which 2×2 basis to use depends on the particular geometry, because one of these two basis states must have the lowest unperturbed energy. Since A and B are degenerate in energy, for a given d this amounts to using $A + B$ for $r \leq d$. On the other hand, for $r > d$ it is necessary to use $B + C$, since C now has the lowest unperturbed energy.
- Except for the tetrahedra, the values of E_1 are essentially the same in the 2×2 and 3×3 bases.
- The values of $E_2(d,r)$ are always much higher than $E_1(d,r)$. However, as will be discussed in the Sec. 5.7, this second excited state is dominated by excitations of the gluon field and so is outside the scope of the models introduced in that section.

The above is for colour SU(2). However, in the real world of SU(3) the state $C = [Q_1Q_2][\bar{Q}_3\bar{Q}_4]$ cannot appear asymptotically as two clusters — see Eq. A.3. Even so this SU(2) lattice data should be understandable in terms of models that are similar to those of SU(3)— see Subsec. 5.2.2.

5.6.2 *QCD in two dimensions (1+1)*

The above analyses unavoidably involve numerical results extracted from Monte Carlo simulations on a 4-dimensional lattice. However, it is possible to study colinear colour sources in a simple approximation for which exact theoretical results are known [88]. This is QCD in two dimensions (QCD2) – the one spatial direction allowing colinear (and only colinear) configurations. For quenched QCD2, the spectrum is known *exactly* even on a lattice. This can be summarised as the requirement that each link is in a representation of the colour group [*i.e.* SU(2) here]. Therefore, the links can be in the singlet ground state (J=0), or they can be excited to a fundamental representation (J=1/2) or an adjoint representation (J=1) and so on. The energy per unit length for such links is given by $b_J = \frac{4}{3}KJ(J+1)$. For 4 colinear quarks, the lowest state [A in Fig. 5.10 (L)] has energy $2Kd$, while the first excited state has, in the middle, an adjoint link of length $(r-d)$ resulting in an energy $2Kd + 8K(r-d)/3$. This exact result applies at any β-value, *i.e.* strong coupling or weak coupling.

A comparison with the above f-mixing model of Eqs. 5.37, 5.38 shows that there is agreement with the exact QCD2 results provided $f = 1$ — independent of which combination of states is used for the analysis A+B, A+C or B+C. This is also true in the more general case when $d_{13} \neq d_{24}$. If the interpretation of f being a gluon-field overlap factor is correct, then it is easy to understand that here f must be unity, since with only one spatial direction the colour flux must overlap fully. Earlier, the main motivation for the f-mixing model had been from weak coupling arguments, so that this agreement between the mixing model and QCD2 does suggest that the model is sensible even at large β. This adds support to the claim that it may be a useful phenomenological tool when the extension to $f \neq 1$ is made.

5.7 Extensions of the 2×2 f-Model

In Sec. 5.5 a model for describing 4-quark interactions was developed as a 2×2 matrix equation for the two heavy-quark states $A = [Q_1\bar{Q}_3][Q_2\bar{Q}_4]$ and $B = [Q_1\bar{Q}_4][Q_2\bar{Q}_3]$ in Fig. 5.4 — see Eqs. 5.37 and 5.38. There the only geometries studied were those in which the quarks were at the corners of squares or rectangles. However, when the four quarks are in more general geometries, such as those in Fig. 5.10, the choice of which two partitions out of the possible three to use is less clear *i.e.* $C = [Q_1Q_2][\bar{Q}_3\bar{Q}_4]$ should be included. This is completely analogous to the problem that arose in Subsec. 5.6.1 when deciding which configurations to use in the corresponding lattice calculations.

In Appendix A and in Ref. [109] the model of Sec. 5.5 is extended from being a 2×2 matrix equation for states A, B into a 3×3 version, where all 3 basic partitions A, B, C are included. Finally a 6×6 extension is developed, in which interquark excited states are introduced to give three additional basis states A^*, B^*, C^*.

Much of the discussion will concentrate on the problems that arise in trying to model the energies of regular tetrahedral configurations. This might be considered a minor point to worry about, since such a configuration is very special. However, the philosophy is that, if *any* configuration cannot be fitted, then the model fails, since then there is no reason to expect other configurations not checked explicitly to be fitted.

A summary of the main points that emerge from Appendix A are as follows:

1) The $2 \times 2 \longrightarrow 3 \times 3$ extension, to some extent, clarifies the reason why the earlier 2×2 f-model was, in many cases, quite successful. Also it shows that an understanding of the tetrahedron spectrum — in particular the degenerate ground state — requires a generalisation of the two-quark approach.

2) The step $3 \times 3 \longrightarrow 6 \times 6$ has the very positive feature for tetrahedra of giving a ground state binding energy that initially *increases* with the size of the tetrahedron — a result also seen in the corresponding lattice data. This arises naturally, since the energies of the additional states A^*, B^*, C^* are each excited by an energy of π/R with respect to the A, B, C states in the 3×3 model. Here R is the interquark distance for that pair of quarks containing the excited gluon field. Therefore, as the four-quark configuration gets larger spatially, the energies of the A^*, B^*, C^* states approach from above the energies of the A, B, C states. The subsequent mixing between the two sets of states then manifests itself as an additional overall attraction that also grows with the spatial size.

3) An even more interesting conclusion from the 6×6 extension is that it partially justifies the $f = 1$ model of Subsec. 5.5.1. In Appendix A it is seen that, by fitting simultaneously the lattice energies for the geometries in Fig. 5.10, the 6×6 model shows two features:

- Outside the range where perturbation theory holds (*i.e.* beyond about 0.2 fm) the binding is dominated by the A^*, B^*, C^* configurations.
- The overlap factor between the A^*, B^*, C^* states (corresponding to f in Eq. 5.38 for the A, B, C states) is essentially *unity*.

We, therefore, come to the following scenario for the four-quark interaction: At the shortest distances, up to about 0.2 fm, perturbation theory is reasonable with the binding being given mainly by the A, B, C states interacting simply through the two-quark potentials with little effect from four-quark potentials. However, for intermediate ranges, from about 0.2 to 0.5 fm, the four-quark potentials act in such a way as to reduce the effect of the A, B, C states so that the binding is dominated by the A^*, B^*, C^* states, which now interact amongst themselves again simply through the two-quark potentials *with little effect from four-quark potentials.* This suggests that models involving only two-quark potentials could be justified — *provided excited gluon states (such as A^*, B^*, C^*) are included on the same footing as the standard states A, B, C.* The above result that excited states play an important rôle in the overall binding is reminiscent of the nucleon-nucleon interaction, where nucleon excitations — especially the $\Delta(1236)$ — are also responsible for a sizeable part of the attraction.

5.8 Heavy-Light Mesons ($Q\bar{q}$)

In the previous sections only static (*i.e.* infinitely heavy) quarks with two colours were discussed. Even though this is a far cry from the real world of finite mass quarks with three colours, it resulted in several interesting conclusions. However, these were of a somewhat academic nature useful for creating models, but could not be compared directly with experimental data. In this section a compromise situation of two quarks is studied where one of the quarks is still static but with the other being light. This is essentially the "hydrogen atom" of quark physics and is expected to be a reasonable representation of the heavy-light B-mesons. This also means that in the development of the Effective Potential Theories of Subsec. 5.1.2.2, it is possible that the *one-body* Schrödinger or Dirac equation is applicable. If this proves to be so, then it will be a further reason for studying B-mesons in addition to the more basic ones discussed below.

It should be added that the hydrogen atom analogy also partially holds for the interaction — the coulomb potential $\propto e^2/r$ of the hydrogen atom versus the one-gluon exchange $\propto \alpha/r$ in the heavy-light meson. However, as mentioned in Subsec. 5.1.2.2, for light quarks there are indications that this $1/r$ attraction gets damped by form factor effects and that much of the needed attraction could arise from a short ranged instanton–generated interaction. Also, it must not be forgotten that beyond $r \approx$0.2–0.3 fm the interaction in Eq. 5.6 becomes dominated by the linear confining potential.

From Eq. 5.2, for lattice calculations with a heavy quark Q, we had the condition that a should satisfy $a \ll m_Q^{-1}$. At present, this rules out direct lattice calculations with b quarks of mass 5 GeV, since they would require $a \ll 0.04$ fm. This prompted the weaker condition $a \sim m_Q^{-1}$ in Eq. 5.3. One way to partially avoid this problem is to perform lattice calculations for lighter quarks with $m_Q \sim 2$ GeV, which do not require such a fine mesh, and then to *extrapolate* the results to the b quark mass. However, if the results from a *static*–light system are also included in the analysis we end up having to *interpolate* to $m_Q \sim 5$ GeV. Since heavy quarks with such masses are essentially non-relativistic, the appropriate interpolation parameter is $1/m_Q$, so that the actual interpolation is between $1/m_Q = 0$ to 0.5 GeV^{-1}. This is usually a more reliable procedure [110].

5.8.1 *Bottom (B)-mesons*

Bottom mesons are the bound $q\bar{q}$ states of a $\bar{b}$-antiquark of mass $\approx$5 GeV and a lighter quark. These are listed in Table 5.3. This century has opened with there being renewed interest in B-physics — see the Bottom meson summary in the Particle Data listings of Ref. [111]. There are new generations of B-meson experiments at BaBaR (SLAC), Belle, CLEO III and Hera-B. These machines have started to accumulate B-mesons and the long-awaited B-factory era has begun. The hope is that these experiments will deliver the fundamental constants of the Standard Model and also improve our understanding of CP violation. However, having seen that B-mesons are important objects, it must be confessed that their study by lattice QCD is very incomplete. This will be the topic of the present section, where we concentrate on their energies, and in Sec. 5.9, where we extract density distributions.

Table 5.3 Properties of the Bottom mesons [111]. The state marked with * is from Ref. [112] and those marked with ? do not, at the time of writing, have $I(J^P)$ confirmed.

Meson	$q\bar{b}$	$I(J^P)$	nL	Mass(GeV)
B^+	$u\bar{b}$	$\frac{1}{2}(0^-)$	$1S$	5.2790(5)
B^0	$d\bar{b}$	$\frac{1}{2}(0^-)$	$1S$	5.2794(5)
B'^0	$d\bar{b}$	$\frac{1}{2}(0^-)$	$2S$	5.859(15)*
B^{*+}	$u\bar{b}$	$\frac{1}{2}(1^-)$	$1S$	5.325(6)
B^{**}_J	$u\bar{b}$	$\frac{1}{2}(0,1,2^+)$	$1P$	5.698(8) ?
B^0_s	$s\bar{b}$	$0(0^-)$	$1S$	5.3696(24)
B^*_s	$s\bar{b}$	$0(1^-)$	$1S$	5.4166(35) ?
B^{**}_{Js}	$s\bar{b}$	$0(0,1,2^+)$	$1P$	5.853(15) ?
B^+_c	$c\bar{b}$	$0(0^-)$	$1S$	6.4(4)

5.8.2 *Lattice parameters*

In the past few years there have been several detailed measurements of B-meson excited state energies. Some of the parameters used in these studies are given in Table 5.4. The quenched calculations of Refs. [56, 58] are very similar to each other with the latter having a somewhat larger lattice. The parameter C_{SW} has been tuned for the clover improved action mentioned earlier [5, 6]. The hopping parameter κ essentially determines the mass m_q of the light quark as being slightly smaller than the accepted value of the strange quark mass *i.e.* $m_{\bar{q}} = 0.91(2)m_{\rm s}$. The ratio of the pseudoscalar to vector masses (M_{PS}/M_V) — *i.e.* the ratio of the "π"- and ρ-meson masses generated by those particular configurations — is another measure of m_q. Since this ratio is much larger than the experimental value of 0.18, we again see that $m_q \gg m_{u,d}$. Refs. [59, 114] are expected to be a distinct improvement, since these are unquenched calculations with smaller lattice spacings. Now $m_{\bar{q}} = 1.28,\ 1.12m_{\rm s}$ respectively. However, in practice, it is found that the energy splittings of the excited states are only

Table 5.4 Lattice parameters. The notation is explained in the text. Refs. [56, 58] are in the quenched approximation and Refs. [59, 114] are unquenched.

Ref.	$L^3 \times T$	β	C_{SW}	κ	a (fm)	M_{PS}/M_V
[56]	$12^3 \times 24$	5.7	1.57	0.14077	≈ 0.17	0.65
[58]	$16^3 \times 24$	5.7	1.57	0.14077	≈ 0.17	0.65
[59]	$16^3 \times 24$	5.2	1.76	0.1395	≈ 0.14	0.72
[114]	$16^3 \times 32$	5.2	2.02	0.1350	≈ 0.11	0.70

weakly dependent on $m_{\bar{q}}$ for the range of values used here.

5.8.3 *Maximal Variance Reduction (MVR)*

One of the reasons why the energies of $Q\bar{q}$-states can now be calculated reliably is not only the improvement in computer capabilities, but also by developments in formalism. It has been demonstrated that light-quark propagators can be constructed in an efficient way using the so-called Maximal Variance Reduction (MVR) method. Since this has been explained in detail elsewhere, for example Subsec. 4.5.3 of Chapter 4 and in Ref. [56], the emphasis here will be mainly on the differences that arise when estimating on a lattice the two- and three-point correlation functions C_2, C_3 needed for measuring spatial charge and matter densities (C_3) in addition to the energies (C_2). In the MVR method the inverse of a positive definite matrix A is expressed in the form of a Monte Carlo integration

$$A^{-1}_{ji} = \frac{1}{Z} \int D\phi \; \phi^*_i \phi_j \; \exp(-\frac{1}{2}\phi^* A \phi), \tag{5.58}$$

where the scalar fields ϕ are pseudofermions located on lattice sites i, j. For a given gauge configuration on this lattice, N independent samples of the ϕ fields can be constructed by Monte Carlo techniques resulting in a stochastic estimate of A^{-1}_{ji} as an average of these N samples *i.e.* $A^{-1}_{ji} = \langle \phi^*_i \phi_j \rangle$. The N samples of the ϕ fields can be calculated separately and stored for use in any problem involving light quarks with the same gauge configurations. In practice, $N \approx 25$ is found to be sufficient.

In LQCD the matrix of interest is the Wilson–Dirac matrix $Q = 1 - \kappa M$, where M is a discretized form of the Dirac operator $(\not{\partial} + m)$ and is the mechanism for "hopping " the quarks from one site to another. However, Q is not positive definite for those values of the hopping parameter κ that are of interest. Therefore, we must deal with $A = Q^\dagger Q$, which is positive definite. Since M contains only nearest neighbour interactions, A — with at most next-to-nearest neighbour interactions — is still sufficiently local for effective updating schemes to be implemented. In this case the light-quark propagator from site i to site j is expressed as

$$G_q = G_{ji} = Q^{-1}_{ji} = \langle (Q_{ik}\phi_k)^* \phi_j \rangle = \langle \psi_i^* \phi_j \rangle. \tag{5.59}$$

This is the key element in the following formalism. The Wilson–Dirac matrix also leads to an alternative form for the above light-quark propagator from site i to near site j

$$G'_q = G'_{ji} = \gamma_5 \langle (Q_{jk}\phi_k) \phi_i^* \rangle \gamma_5 = \gamma_5 \langle \psi_j \phi_i^* \rangle \gamma_5. \tag{5.60}$$

In practice both forms are used, since — for the *same* correlation — they lead to independent measurements, which can then be averaged to improve the overall statistical error. Later, it will be essential to use at some lattice sites operators that are *purely local.* This then restricts us to using at such sites only the ϕ fields that are located on single lattice sites. In contrast the ψ_i fields, defined above as $Q_{ik}\phi_k$, are not purely local, since they contain ϕ fields on next-to-nearest neighbour sites.

In the above, the term "Maximal Variance Reduction" comes from the technique applied to reduce the statistical noise in Eq. 5.59. The lattice is divided into two boxes ($0 < t < T/2$ and $T/2 < t < T$) whose boundary is kept fixed. Variance of the pseudofermionic fields is then reduced by numerically solving the equation of motion inside each box. This allows the variance of propagators from one box to the other to be greatly reduced. However, in the case of a three-point correlation in Subsec. 5.9 two propagators are needed and this is best treated by choosing one of the points to be on the boundary of the boxes while the other two are inside their own boxes. Furthermore, the field at the boundary must be local to avoid the two propagators interfering with each other. This means that only the ϕ fields should be used on the boundary and there they can couple to the charge, matter or any other one-body operator. For the points in the boxes, the temporal distances from the boundary should be approximately equal to give the propagators a similar degree of statistical variance.

5.8.4 *Energies of heavy-light mesons ($Q\bar{q}$)*

5.8.4.1 *Two-point correlation functions C_2*

The basic entities for measuring energies are the two-point correlation functions C_2. These are depicted in Fig. 5.12 and are seen to be constructed from essentially two quantities — the heavy–quark (static–quark) propagator G_Q and the light quark propagator G_q. As discussed in detail in

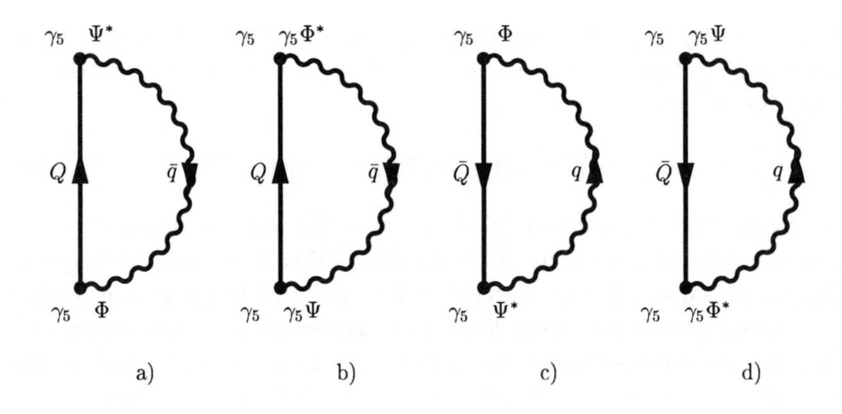

Fig. 5.12 The four contributions to the two-point correlation function C_2.

Ref. [58], when the heavy quark propagates from site $(\mathbf{x},\ t)$ to site $(\mathbf{x},\ t{+}T)$, G_Q can be expressed as

$$G_Q(\mathbf{x},\ t\ ;\ \mathbf{x},\ t+T) = \frac{1}{2}(1+\gamma_4)U^Q(\mathbf{x},t,T), \tag{5.61}$$

where $U^Q(\mathbf{x},t,T) = \prod_{i=0}^{T-1} U_4(\mathbf{x},t+i)$ is the gauge link product in the time direction. On the other hand, as the light-quark propagates from site i to site j, it can be schematically expressed as one of the two alternatives (G_q or G'_q) in Eqs. 5.59 and 5.60. Knowing G_Q and G_q, the general form of a two-point correlation can be constructed from a heavy quark propagating from site $(\mathbf{x},\ t)$ to site $(\mathbf{x}',\ t+T)$ and a light quark propagating from site

$(\mathbf{x}',\ t+T)$ back to site $(\mathbf{x},\ t)$ as

$$\begin{aligned} C_2(T) =& \mathrm{Tr}\langle \Gamma^\dagger\, G_Q(\mathbf{x},t;\mathbf{x}',t+T)\,\Gamma\, G_q(\mathbf{x}',t+T;\mathbf{x},t)\rangle \\ =& 2\langle \mathrm{Re}\left[U^Q[\psi^*(\mathbf{x},t+T)\phi(\mathbf{x},t)+\phi^*(\mathbf{x},t+T)\psi(\mathbf{x},t)]\right]\rangle. \end{aligned} \quad (5.62)$$

Here Γ is the spin structure of the heavy quark – light quark vertices at t and $t+T$. In this case $\Gamma=\gamma_5$, since we are only interested in pseudoscalar mesons such as the B-meson. For clarity, the Dirac indices have been omitted. The four contributions to C_2 are depicted in Fig. 5.12. Here the a) term uses the light quark propagator G_q in Eq. 5.59 and term b) the alternative G'_q in Eq. 5.60 — the two terms in Eq. 5.62. Terms c) and d) are the corresponding ones for a heavy antiquark ($\bar{Q}$). It is necessary to include the $\bar{Q}$-terms to ensure C_2 is real. It would be sufficient to use only a)+c) or b)+d), since both combinations correspond to measuring C_2. However, since these two combinations are independent measurements of the *same* correlation, keeping both improves the statistics on the final measurement.

The above has been written down for a single type of gauge field. The correlations can now be greatly improved by fuzzing as discussed in Subsec. 5.4.1.3. This makes the two-point correlation function into the fuzzing matrix $C_{2,ij}$. Since i,j usually take on 2 or 3 values, this means that S-wave *excited* state energies and properties can now be studied in addition to those of the ground state.

5.8.4.2 *Analysis of C_2 to extract energies*

There are several ways of analysing the correlation functions C_2 in order to extract the quantities of interest *i.e.* energies. For a review of these methods see Ref. [113] — with more details using the present notation being found in Ref. [58].

The actual analysis gives not only the energies (m_α) but also the eigenvectors ($\mathbf{v}$) for the states of the $Q\bar{q}$-system. These values of m_α and $\mathbf{v}^\alpha$ are later fixed when analysing the three-point correlation data C_3 to give the charge and matter densities $x^{\alpha\beta}(r)$ in Sec. 5.9. Each element $C_{2,ij}(T)$ is fitted with the form

$$C_{2,ij}(T)\approx \tilde{C}_{2,ij}(T)=\sum_{\alpha=1}^{M_2} v_i^\alpha \exp(-m_\alpha T)v_j^\alpha, \quad (5.63)$$

where M_2 is the number of eigenvalues to be extracted (usually 3 or 4) and m_1 is the ground state energy of the heavy-light meson. The values of m_α

and $v^\alpha_{i,j}$ are then determined by minimizing the difference between the C_2 data from the lattice and the parametric form $\tilde{C}_2$. The function actually minimized is the usual

$$\chi^2 = \sum_{i,j} \sum_{T_{2,\min}}^{T_{2,\max}} \left[\frac{C_{2,ij}(T) - \tilde{C}_{2,ij}(T)}{\Delta C_{2,ij}(T)} \right]^2 , \tag{5.64}$$

where $\Delta C_{2,ij}(T)$ is the statistical error on $C_{2,ij}(T)$ and $T_{2,\min}$, $T_{2,\max}$ are the minimum and maximum values of T_2 used in the fit. The latter depend on the lattice size and the future use to which the m_α and $v^\alpha_{i,j}$ are destined. A typical outcome is depicted in Fig. 5.13 from Ref. [114]. This is for a

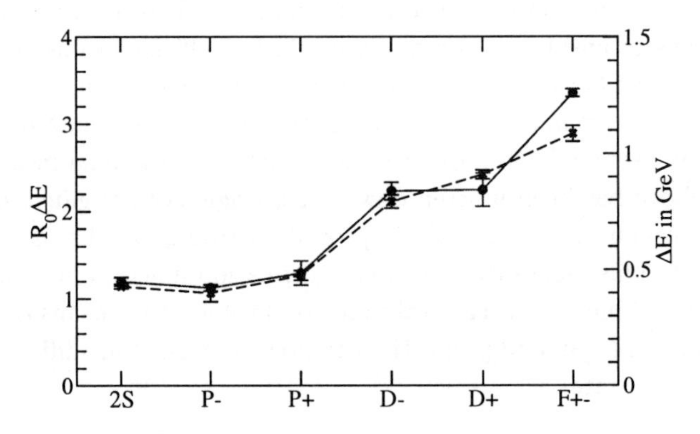

Fig. 5.13 The energies of $Q\bar{q}$ states from Ref. [114]. The solid line uses dynamical fermions and the dashed line is a corresponding quenched calculation. The energies of the L-wave excited states and the S-wave radial excited state are relative to the ground state (1S) — see text for notation. These energies are given both in terms of GeV — right axis — and in the more usual Sommer units of $R_0 \approx 0.5$ fm defined in Subsec. 5.2.1.2 — left axis.

dynamical fermion calculation and a corresponding quenched calculation on a $16^3 \times 24$ lattice. In both cases, for numerical reasons, the light quark has a mass that is approximately that of a strange quark. Since the heavy quark is static, the energies can be labelled by $L_\pm$, where the coupling of the light quark spin to the orbital angular momentum gives $j = L \pm 1/2$. The total angular momentum (J) is then obtained by coupling j to the heavy quark spin giving $J = j \pm 1/2$. However, since the heavy quark spin interaction can be neglected, the latter two states are degenerate in energy *i.e.* the P_- state will have $J^P = 0^+, 1^+$ and the P_+ state will have $J^P = 1^+, 2^+$ *etc.* The D-waves show the interesting feature that there appears to be *little or no spin-orbit splitting* between the D_- and D_+ states — contrary to some expectations [115, 116] that there should be an inversion of the level ordering (with L_+ lighter than L_-) at larger L or for radial excitations. This has important implications for phenomenological interpretations of the data. We return to this in Subsec. 5.9.3.2. For F-waves, only the energy from a spin independent mixed operator $F_\pm$ is shown. The latter is expected to approximately correspond to the usual spin-average of the F_- and F_+ states.

5.9 Charge and Matter Distributions of Heavy-Light Mesons ($Q\bar{q}$)

In many cases, when phenomenological models are constructed to describe lattice data, the emphasis is on fitting the energies. However, there are other observables that can be measured on a lattice. Of potential value when constructing models are lattice data for radial distributions of various quantities such as the charge and matter densities, which can be measured using three-point correlation functions C_3. Here we concentrate on the radial distributions of the $\bar{q}$ in the $Q\bar{q}$ system, whereas in Ref. [117, 118] the much more ambitious task of measuring radial distributions in the π, ρ, N and Δ is tackled. The reasons why distributions in few-quark systems have received much less attention are two-fold. Firstly, unlike energies, these distributions are not directly observable, but only arise in integrated forms such as sum rules, form factors, transition rates *etc.* Secondly, as will be seen later, their measurement on a lattice is more difficult and less accurate than the corresponding energies. In spite of this, it is of interest to extract lattice estimates of various spatial distributions.

5.9.1 *Three-point correlation functions C_3*

When the light-quark field is probed by an operator $\Theta(\mathbf{r})$ at $t = 0$ as the heavy quark propagates from $t = -t_2$ to $t = t_1$, the result is the three-point correlation function depicted in Fig. 5.14. This involves two light-quark propagators — one (G'_q) going from $t = -t_2$ to $t = 0$ and a second (G_q)

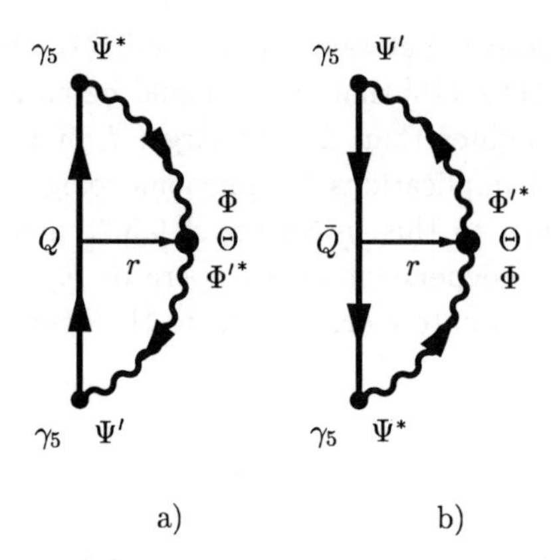

Fig. 5.14 The two contributions to the three-point correlation function C_3.

going from $t = 0$ to $t = t_1$ and has the form

$$C_3(-t_2,\ t_1,\ \mathbf{r}) = \\ \mathrm{Tr}\langle \Gamma^\dagger G_Q(\mathbf{x}, -t_2; \mathbf{x}, t_1)\Gamma G_q(\mathbf{x}, t_1; \mathbf{x}+\mathbf{r}, 0)\ \Theta(\mathbf{r}) G'_q(\mathbf{x}+\mathbf{r}, 0; \mathbf{x}, -t_2)\rangle, \tag{5.65}$$

which can be expressed in terms of the pseudofermion fields $\phi(\mathbf{x}, t)$ and $\psi(\mathbf{x}, t)$ — similar to that for C_2 in Eq. 5.62. Here the vertex $\Theta = \gamma_4$ for the charge (vector) distribution and 1 for the matter (scalar) density. Again the $\bar{Q}$-term in Fig. 5.14 ensures that C_3 is real, but now there are only two terms compared with the four for C_2 in Fig. 5.12. This is because the fields connected to the probe Θ must be local, since the purpose of the probe is to measure the charge or matter distribution at a definite point $\mathbf{r}$. Therefore, only those light quark propagators that involve the local basic field ϕ at $\mathbf{r}$ can be used, since the ψ field contains contributions from ϕ fields at next-to-nearest neighbour sites and so is *non-local.* This must also be kept in mind when fuzzing is introduced to give a matrix $C_{3,ij}(-t_2,\ t_1,\ \mathbf{r})$. Here the fuzzing indices i, j refer to the various fuzzing options of the ψ's at the $Q\bar{q}$ vertices. As with the energies extracted from C_2, the fuzzing permits the measurement of excited state distributions.

5.9.2 *Analysis of C_3*

The analysis of the three-point correlation functions $C_3(\Theta, T = t_1 + t_2, \mathbf{r})$ is performed using a generalisation of the one for C_2 in Eq. 5.63, namely, fitting $C_{3,ij}(\Theta, T, \mathbf{r})$ with the parametric form

$$\tilde{C}_{3,ij}(\Theta, T, \mathbf{r}) = \sum_{\alpha=1}^{M_3} \sum_{\beta=1}^{M_3} v_i^\alpha \exp[-m_\alpha t_1] x^{\alpha\beta}(r) \exp[-m_\beta(T - t_1)] v_j^\beta . \quad (5.66)$$

The m_α and $\mathbf{v}$-vectors are those obtained by minimizing the C_2 in Eq. 5.63 and, for each value of r, the $x^{\alpha\beta}(r)$ are varied to ensure a good fit to $C_{3,ij}(\Theta, T, \mathbf{r})$ by the model expression $\tilde{C}_{3,ij}(\Theta, T, \mathbf{r})$.

Two forms of $x^{\alpha\beta}(r)$ have been used:

(1) A non-separable (NS) form, where each $x^{\alpha\beta}(r)$ is treated as a single entity.
(2) A separable (S) form $x^{\alpha\beta}(r) = y_\alpha(r) y_\beta(r)$. It is seen from Eq. 5.65 that this appears to be a more natural parametrization, since it contains the product of two light-quark propagators G_q. Also the $y_\alpha(r)$, to some extent, resemble a wave function for the state α, since its square yields a distribution.

The outcome is shown in Fig. 5.15 for the ground state and in Fig. 5.16 for the excited states. Several points are of interest in Fig. 5.15:

- At small values of r the two densities are comparable *i.e.* $x_C^{11} \approx x_M^{11}$.
- As r increases from zero the matter density drops off faster than the charge density. A similar difference has also been seen in Ref. [118], where the authors measure these densities for the π, ρ, N and Δ on a $16^3 \times 32$ lattice with $\beta = 6.0$ for both quenched and unquenched configurations.
- The densities calculated with the quenched approximation in Ref. [58] are the same, within error bars, as those for the full dynamical quark calculation of Ref. [59]. However, as will be discussed in Subsec. 5.9.4, the matter *sum rule* does seem to differ.
- The densities do not have a smooth variation with r, but, as will be shown in Subsec. 5.9.3, many of the kinks can be understood in terms of latticized forms of standard Yukawa, exponential or gaussian functions.

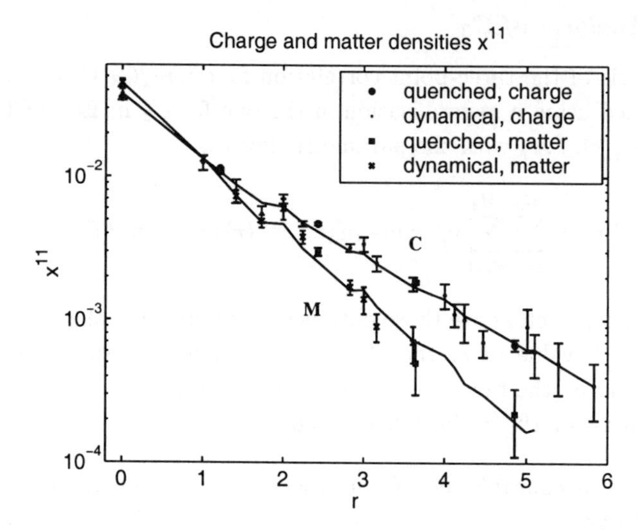

Fig. 5.15 The ground state charge (C) and matter (M) densities $[x^{11}(r)]$ as a function of r/a from Ref. [59]. The lines shows a fit to these densities with a sum of two lattice exponential functions — see Subsec. 5.9.3. The scaled quenched results of Ref. [58] are also shown by filled circles and squares.

In Fig. 5.16 the excited state results are shown for different types of analysis. The label "sep" refers to the separable assumption for x^{ij} and "non-sep" to the non-separable assumption. The two numbers associated with each set of data refer to $T_{2,\text{min}}$ and $T_{3,\text{min}}$ in Eq. 5.64 and the corresponding equation for Eq. 5.66. In this figure the main point of interest is the appearance of nodes. The presence of these nodes is very clear and also the number of nodes is as expected. The first excited state has a single node at about 0.3 fm in the charge case and about 0.2 fm for the matter. The second excited state seems to have two nodes with one being at about 0.1 fm and a second at about 0.4 fm for the charge and 0.3 fm for the matter. It will be a challenge for phenomenological models to explain these data.

When discussing the use of the separable form $x^{\alpha\beta}(r) = y_\alpha(r)y_\beta(r)$, it was stated that $y_\alpha(r)$ can possibly be interpreted as a wave function for the state α. However, there are other radial distributions associated with the $Q\bar{q}$ system that can also be interpreted as wave functions. These are the Bethe–Salpeter wave functions $w_\alpha(r)$ discussed in Ref. [56]. They are extracted by assuming the hadronic operators $C_{\alpha\alpha}(r_1,\ r_2,\ T)$ to be of the form $w_\alpha(r_1)w_\alpha(r_2)\exp(-m_\alpha T)$, where the sink and source operators

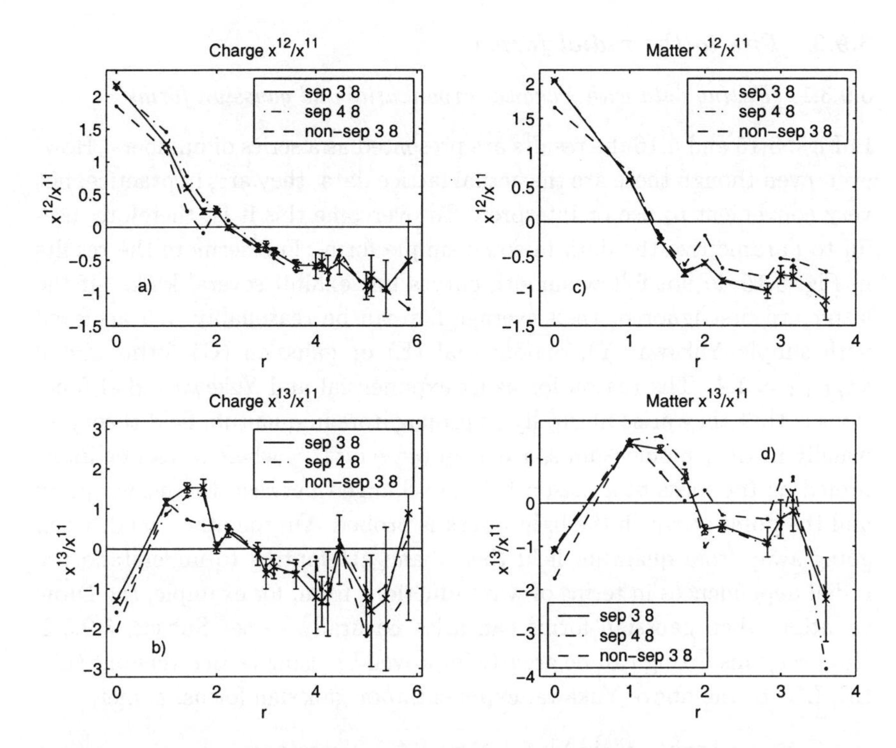

Fig. 5.16 a) and b): The ratios $y_2(r) = x^{12}/x^{11}$ and $y_3(r) = x^{13}/x^{11}$ for the charge distribution. c) and d): These ratios for the matter distribution — see Ref. [59].

are of spatial size r_1 and r_2. Qualitatively, the forms of $y_1(r)$ and $y_2(r)$ plotted in Fig. 5.16 and the corresponding ones for $w_1(r)$ and $w_2(r)$ are found to be similar. However, even though they do bear some similarities, it should be added that there are several reasons why these two types of wave function should *not* agree in detail with each other. In particular, the $[w_\alpha(r)]^2$ cannot be identified as a charge or matter distribution. In addition, they are found by using an explicit fuzzed path between Q and $\bar{q}$ and so are dependent on the fuzzing prescription, whereas the $y_\alpha(r)$ are defined in a path-independent way.

The above considers only S-wave distributions for both ground and excited states. The extension to other partial waves is now in progress [61]. Preliminary results for the P_- state indicate that there the distributions are qualitatively of the form expected from a Dirac equation description *i.e.* the charge and matter distributions are *not* zero for $r = 0$.

5.9.3 *Fits to the radial forms*

5.9.3.1 *Fitting data with Yukawa, exponential and gaussian forms*

In Figs. 5.15 and 5.16 the results are presented as a series of numbers. However, even though these are the actual lattice data, they are, in practice, not very convenient to use or interpret. To overcome this it is, therefore, useful to parametrize the data in some simple form. Furthermore the results in Fig. 5.15 do not follow smooth curves but exhibit several kinks. If the latter are first ignored, then average fits can be reasonably well achieved with simple Yukawa (Y), exponential (E) or gaussian (G) forms giving $\chi^2/n_{\text{dof}} \approx 1.4$. The reason for using exponential and Yukawa radial functions is that they arise naturally as propagators in quantum field theory — usually in their momentum space form $(q^2+m^2)^{-1}$, where m can be interpreted as the mass of a meson being exchanged between the heavy quark and the point at which the light quark is probed. On the other hand, when going away from quantum field theory and attempting to understand the radial dependences in terms of wave functions from, for example, the Dirac equation, then gaussian forms can arise naturally — see Subsec. 5.9.3.2. However, this fitting can be greatly improved by using lattice versions (LY, LE, LG) of the above Yukawa, exponential or gaussian forms, namely

$$\left[\frac{\exp(-r/r^{\text{LY}})}{r}\right]_{\text{LY}} = \frac{\pi}{aL^3}\sum_{\mathbf{q}}\frac{\cos(\mathbf{r}.\mathbf{q})}{D+0.25[a/r^{\text{LY}}]^2}, \tag{5.67}$$

$$[\exp(-r/r^{\text{LE}})]_{\text{LE}} = \frac{\pi a}{2r^{\text{LE}}L^3}\sum_{\mathbf{q}}\frac{\cos(\mathbf{r}.\mathbf{q})}{[D+0.25(a/r^{\text{LE}})^2]^2}, \tag{5.68}$$

$$[\exp[-(r/r^{\text{LG}})^2]]_{\text{LG}} = \left[\frac{r^{\text{LG}}\sqrt{\pi}}{aL}\right]^3\sum_{\mathbf{q}}\cos(\mathbf{r}.\mathbf{q})\exp[-(r^{\text{LG}}/a)^2 D]. \tag{5.69}$$

Here L is the lattice size along one axis and $D=\sum_{i=1}^{3}\sin^2(aq_i/2)$, where $aq_i = 0, \frac{2\pi}{L}, \ldots, \frac{2\pi(L-1)}{L}$.

These are able to give much of the kink structure in the data — as is seen in Fig. 5.17, where in each case two Yukawas, exponentials or gaussians are used to give the fits 2LY, 2LE and 2LG with $\chi^2/n_{\text{dof}} \approx 1$. All three of these forms are equally acceptable [59]. In the 2LE and 2LY fits it is of interest to express the range parameters r^{LY} and r^{LE} in terms of the

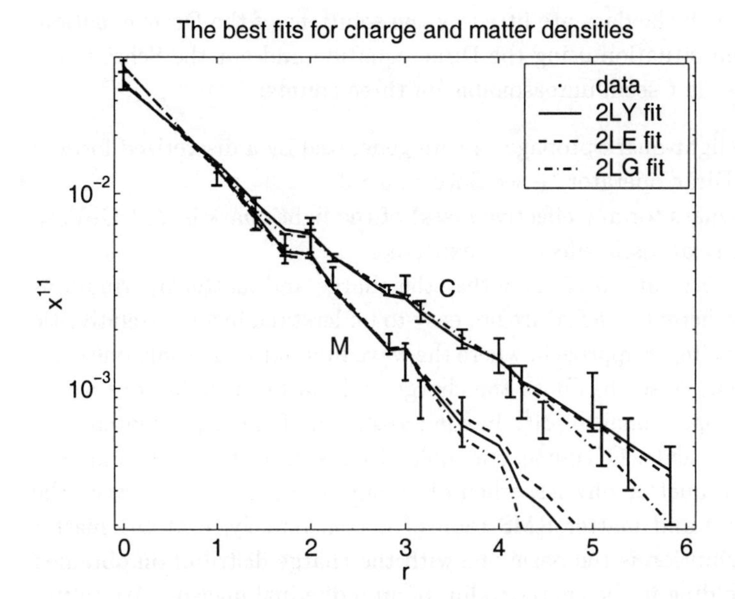

Fig. 5.17 Fits to the lattice data in Fig. 5.15 with lattice exponential (2LE), Yukawa (2LY) and gaussian (2LG) forms from Ref. [59].

mass ($1/r^i$) of the particle producing this range. Both types of fit find that the longest range corresponds to the exchange of a vector meson of mass $m_0^v \approx 1$ GeV for the charge density and a scalar meson of mass $m_0^s \approx 1.5$ GeV for the matter density. These masses of the mesons have been extracted in a rather indirect manner. However, in the literature there have been direct calculations of the energies of these $q\bar{q}$ states using the same lattice parameters and lattice size as those employed here. In Ref. [119] they got $m_0^v = 1.11(1)$ GeV and in Ref. [120] $m_0^s = 1.66(10)$ GeV — numbers consistent with the above indirect estimates.

It should be added that the expression LY in Eq. 5.67 is often used in the Coulomb limit $r^{\text{LY}} \to \infty$ for discretizing the $1/r$ term of $V_{Q\bar{Q}}(r)$ in Eq. 5.6. However, in that limit the condition $\mathbf{q} \neq 0$ is necessary.

5.9.3.2 *Fitting $Q\bar{q}$ data with the Dirac equation*

The above fits are purely phenomenological with there being no connection between the forms fitting the charge and matter distributions. However, there is an alternative approach in the spirit of an Effective Potential Theory

(EPT), in which the data are fitted by the solutions of the Dirac equation. In the present situation using the Dirac equation and not the Schrödinger equation does not seem unreasonable on three counts:

(1) The light-quark propagators are generated by a discretized form of the Dirac operator — see Subsec. 5.8.3.
(2) The mass (or any effective mass) of the light quark is $\ll 1$ GeV, so that relativistic effects are expected.
(3) Figs. 5.15 and 5.17 show that the charge and matter distributions are different — a feature not easy to understand in a non-relativistic Schrödinger approach, where the wave functions have only one component ϕ, so that both the charge and matter distributions would be proportional to $|\phi|^2$. In comparison the Dirac equation has two components. Of course, one could also resort to the argument familiar in nuclear physics, when discussing the difference between the charge and matter RMS radii of nuclei, namely, that the matter distribution is the basic one with the charge distribution obtained by folding in the charge radius of an individual nucleon. We return to this in the next subsection.

In the notation that G and F are the large and small components of the solution to the Dirac equation then the charge (C) and matter (M) distributions can be expressed as

$$x_C^{\alpha\beta}(r) = G_\alpha(r)G_\beta(r)+F_\alpha(r)F_\beta(r) \text{ and } x_M^{\alpha\beta}(r) = G_\alpha(r)G_\beta(r)-F_\alpha(r)F_\beta(r)$$

respectively. Attempts are now underway [57] to study to what extent the above distributions can indeed be interpreted by these two relationships.

Since we are now in the realm of EPTs, we need the three ingredients discussed in Subsec. 5.1.2.2 — a wave equation (now the Dirac equation), a potential $V_{Q\bar{q}}$ and an effective quark mass (a free parameter). The main problem is the form of $V_{Q\bar{q}}$, since it is not at all clear whether the "natural" generalisation of the form appropriate for static quarks in Eq. 5.6 to

$$V_{Q\bar{q}}(r) = -\frac{e}{r} + b_s r \gamma_4, \tag{5.70}$$

is correct. Here the first term is treated as a four-vector simply by the analogy between one-gluon exchange and one-photon exchange. However, if this term for large r is in fact the so-called Lüscher term [122], a vibrating string correction of $-\pi/(12r)$ to the leading string term $b_s r$, then it should be a *scalar*. On the other hand, there are theoretical reasons that partially justify the use of the Dirac equation for heavy-light mesons with the form for the potential in Eq. 5.70. For example, in Ref. [121] the authors derive a Dirac equation for a heavy-light meson by starting from the QCD Lagrangian and taking into account both perturbative and nonperturbative effects. The Coulomb-like effect is treated rigorously and the confining potential heuristically. The outcome is that the confining potential is a scalar and the Coulomb part is the fourth component of a 4-vector. However, the lack of any significant D-wave spin-orbit splitting, as is seen in Fig. 5.13, does suggest that the confining potential can not be purely scalar. This follows from the simple argument that, in a heavy(static)-light quark system, a central potential of the form $V(r) = a/r + br$ should give rise to a spin-orbit potential

$$V_{SO}(r) = -\frac{1}{4m^2 r}\frac{dV}{dr} = \frac{1}{4m^2}\left(\frac{a}{r^3} - \frac{b}{r}\right). \tag{5.71}$$

Such a potential would lead to inversion of the level ordering (with L_+ lighter than L_-) at larger L or for radial excitations and this is *not* seen in Fig. 5.13. The authors of Ref. [32], in fact, give arguments why the interquark confining potential should have the form

$$V(r) = a/r + br(1 + \gamma_4)/2 \tag{5.72}$$

leading to simply

$$V_{SO}(r) = \frac{1}{4m^2}\frac{a}{r^3}, \tag{5.73}$$

which would rapidly vanish at larger L. In Ref. [123] the electromagnetic structure functions of a heavy(static)-light quark system interacting via the potential in Eq. 5.72 are calculated. This ambiguity between the vector *vs* scalar structure of $V_{Q\bar{q}}$ is an ongoing argument — see Ref. [32] for a list of references on this controversy. Also, as mentioned in Subsec. 5.1.2.2, form factor effects reduce the rôle of any one-gluon exchange and that this loss of vector-like attraction could be replaced by a very short ranged instanton–generated *scalar* interaction.

An interesting property of the solutions to the Dirac equation is that, for a linearly rising potential, both the large (G) and small (F) components decay asymptotically as *gaussians*. This follows from the observation that the coupled Dirac equations for large r can be written as

$$-m(r)G(r) = -F'(r) \quad \text{and} \quad m(r)F(r) = G'(r), \tag{5.74}$$

where $m(r) = m + cr \to cr$ as $r \to \infty$ giving asymptotically the simple harmonic oscillator equation $G'' + (cr)^2 G = 0$. This was the reason for considering not only Yukawa and exponential but also gaussian forms in the last subsection.

5.9.3.3 *Fitting $Q\bar{q}$ data with the Schrödinger equation*

In the previous subsection three reasons were given for attempting to fit the lattice data with the Dirac equation. However, two of these reasons are far from compelling — only the fact that the effective light-quark mass being $\ll 1$ GeV seems unavoidable. The comparison with the discretized Dirac operator used in the lattice simulation, as mentioned in Subsec. 5.8.3, is nothing more than an analogy without theoretical basis. Secondly, the fact that the charge distribution has a longer range than that for the matter is reminiscent of the difference between the charge and matter radii of nuclei. There this is usually expressed as

$$\langle r^2 \rangle_{\text{charge}} = \langle r^2 \rangle_{\text{matter}} + \langle r^2 \rangle_{\text{proton}}, \tag{5.75}$$

relating the Mean Square Radius of the nuclear charge distribution with that for the matter distribution and the proton charge radius. This suggests that there is only *one* basic distribution — that of the matter — with the charge distribution arising from a finite size correction to the charge of the light-quark. This should not be surprising since the effective quark mass ($m_{q,\ \text{effective}}$) is quite different (larger) compared with that used in the lattice calculation, which is about that of the strange quark. Here we are now saying that, in addition to a mass renormalisation, the light-quark develops a *charge form factor*. A direct application of Eq. 5.75 on the fits to the data in Fig. 5.17 results in $\bar{q}$ RMS charge radii of 0.51(4), 0.49(7) and 0.35(3) fm for the Yukawa, exponential and gaussian fits respectively. These are sizes consistent with $m_{q,\ \text{effective}} \sim 500$ MeV (*i.e.* $1/m_{q,\ \text{effective}} \approx 0.4$ fm) and are surprisingly large being 2-4 lattice spacings. Also $\bar{q}$ charge form factors of this size qualitatively explain why in Fig. 5.16 the node in the x^{12}/x^{11} charge distribution is at a larger value of r than that for the matter.

Since the problem has been reduced to only one basic distribution, we are now able to use an Effective Potential Theory (EPT) based on a non-relativistic Schrödinger equation. As before, the other two ingredients for an EPT are $m_{q,\text{ effective}}$ (a free parameter) and the potential $V_{Q\bar{q}}$ presumably based on the form appropriate for static quarks in Eq. 5.6.

In Ref. [60], when analysing the $Q^2\bar{q}^2$ lattice data of Ref. [77], this strategy was used first to extract a value of $m_{q,\text{ effective}}$ by fitting the spin-averaged energies of the $Q\bar{q}$ system — see Fig. 5.18. The outcome was

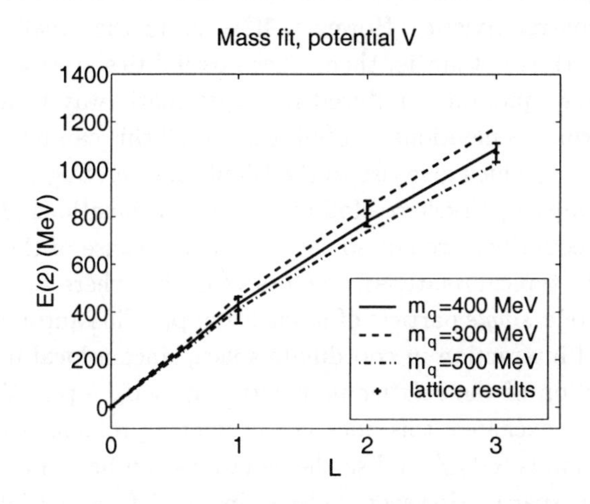

Fig. 5.18 Fits to the spin-averaged energies in Fig. 5.13 for a series of values for m_q.

$m_{q,\text{ effective}} \approx 400$ MeV — a value consistent with the above size estimate. We return to this in Subsec. 5.10.2. The above approach has also been carried out in Ref. [117], where nucleon charge correlations are measured and then fitted with a non-relativistic Schrödinger equation. First the authors extract two entities $C(r_\Delta)$ and $C(r_Y)$, which they interpret as the nucleon wave functions in either the coordinate $r_\Delta = (r_{12} + r_{23} + r_{13})$ — the so-called Δ-Ansatz — or the coordinate $r_Y = \text{Min}(r_{1\epsilon} + r_{2\epsilon} + r_{3\epsilon})$, where ϵ is the junction at which the three flux tubes meet. They then go on to fit this data with Airy functions that decay as $\exp(-cr^{3/2})$. These are the wave functions expected from a non-relativistic Schrödinger equation when using linearly rising potentials $V(r_\Delta) \propto r_\Delta$ or $V(r_Y) \propto r_Y$. In this way, fitting the $C(r_\Delta, r_Y)$ data seemed to slightly favour the Δ-Ansatz. However, it should be added that this conclusion is far from being universally accepted. For example, in Ref. [124] the authors say: "In particular the Δ-shape configuration debated in the literature is shown to be impossible and the well-known Y-shaped baryon is the only possibility." This latter result is supported in Refs. [125, 126].

5.9.3.4 *Fitting $Q\bar{q}$ data with semirelativistic equations*

The equations considered in the above subsections are two extremes with the Dirac equation being fully relativistic and the Schrödinger equation completely non-relativistic. However, if — as in the "real" B-meson — the heavy quark is not static, then other possibilities arise when the basic Bethe–Salpeter equation is reduced in a systematic way to a Lippmann–Schwinger form. As mentioned in Subsec. 5.1.2.2 this can be carried out in a variety of ways, which give rise to the Blankenbecler–Sugar, Gross, Kadyshevsky, Thompson, Erkelenz–Holinde and other equations [35]. Unfortunately, these equations are not so easy to treat because of the unavoidable presence of the typical relativistic factors $\sqrt{m_i/E_i}$, where $E_i = \sqrt{m_i^2 + \mathbf{p_i}^2}$ is the energy of a *single* particle of momentum $\mathbf{p_i}$. This automatically leads to a non-local interaction in coordinate space, since a local interaction requires a function of the *relative* momentum $\mathbf{q} = \mathbf{p_a} - \mathbf{p_b}$. When dealing with one-gluon exchange this does not present a problem, since there the basic interaction is $\propto 1/q^2$ and so the equations can be formulated directly in momentum space. However, as seen in Eq. 5.6, a crucial part of the interquark interaction is the confining term $b_s r$ — an interaction that can not be conveniently treated directly in momentum space. In the literature there are several attempts to fit directly the meagre experimental B-meson data with these forms. For example, in Refs. [126] and [127] the Gross and Blankenbecler–Sugar equations, respectively, are employed to interpret the B-meson spectrum and a series of available transition rates.

Some of the complications that arise when dealing with the above equations can be partially overcome by using instantaneous interactions. Shortly after formulating the Bethe-Salpeter (B-S) equation in 1951 [128], Salpeter in 1952 [129] replaced the interaction $G(q^2 = \mathbf{q}^2 - q_4^2)$ in the B-S equation by its three dimensional counterpart $G(\mathbf{q}^2)$. This resulted in an equation for two particles a and b, which could be written in the centre–of–mass system as

$$[E - H_a(\mathbf{p}) - H_b(\mathbf{p})]\phi(\mathbf{p}) = P\int d^3q\gamma_4^a G(\mathbf{q})\gamma_4^b\phi(\mathbf{p}+\mathbf{q}), \tag{5.76}$$

where $H_i(\mathbf{p}) = m_i\beta^i + \boldsymbol{p}.\boldsymbol{\alpha}^i$ with β and $\boldsymbol{\alpha}^i$ being the usual Dirac matrices.

The operator $P = \Lambda_+^a\Lambda_+^b - \Lambda_-^a\Lambda_-^b$ is a combination of the projection operators $\Lambda_\pm^a = [E_a(p) \pm H_a(\mathbf{p})]/E_a(p)$, where $E_a(p) = \sqrt{m_a^2 + \mathbf{p}^2}$. It should be added that Eq. 5.76 differs from an earlier one by Breit in 1929 [130] by the presence of the P factor. In the nonrelativistic limit both equations reduce to the same form, but in general the Breit equation has a more limited applicability than Eq. 5.76 — as discussed in Ref. [129].

In Refs. [131], Eq. 5.76 is further simplified by first considering only positive-energy solutions *i.e.* omit the $\Lambda_-^a\Lambda_-^b$ terms in P. This results in the reduced Salpeter equation

$$[E - \sqrt{m_a^2 + \mathbf{p}^2} - \sqrt{m_b^2 + \mathbf{p}^2}\,]\phi(\mathbf{p}) = \int d^3q \Lambda_+^a \gamma_4^a G(\mathbf{q}) \gamma_4^b \Lambda_+^b \phi(\mathbf{p} + \mathbf{q}). \tag{5.77}$$

Then the formalism is further restricted to the positive energy components to give the semirelativistic spinless–Salpeter equation

$$[\sqrt{m_a^2 + \mathbf{p}^2} + \sqrt{m_b^2 + \mathbf{p}^2} + V(x)]\psi = E\psi, \tag{5.78}$$

where $V(x)$ is the Fourier transform of the $G(\mathbf{q})$. Finally the authors of Refs. [131] go one more step by studying the equal mass case

$$[\beta\sqrt{m^2 + \mathbf{p}^2} + V(x)]\psi = E\psi, \tag{5.79}$$

where, instead of fixing β at 2, they show that $\beta > 0$ can simulate the effect of several particles all of mass m. They interpret this equation as "the generalization of the nonrelativistic Schrödinger Hamiltonian towards relativistic kinematics" and study algebraically its properties for a variety of forms for $V(x)$. The reason why these authors go through explicitly the simplification of Eq. 5.76 is to show that Eq. 5.79 — the obvious extension of the Schrödinger equation — can indeed be derived in a systematic manner and is not just an educated guess.

5.9.4 *Sum rules*

In addition to measuring $C_3(\mathbf{r})$ for various values of $\mathbf{r}$, the correlation where $\mathbf{r}$ is *summed* over the whole lattice is also obtained. This leads to the charge sum rule as discussed in Ref. [59]. For the charge distribution, the outcome is that $\sum_\mathbf{r} x^{11}(\mathbf{r}) = X^{11}$ is $\approx 1.3(1)$, which is consistent with the earlier quenched result [58]. The fact that X^{11} is not unity, as expected in the continuum limit, can be qualitatively understood by introducing a renormalisation factor of $\approx 1/1.3 \approx 0.8$ into the basic γ_4 vertex used to measure the charge density. Such a factor of this magnitude is reasonable as shown in Ref. [132]. It is also reassuring that the $X^{\alpha\beta}$ with $\alpha \neq \beta$ are, in general, consistent with zero — as expected in the continuum limit.

The matter sum rule has a somewhat wider spread of values with 0.9(1) being a reasonable compromise — a number that is about twice the estimate of 0.4(1) for the quenched calculation of Ref. [58]. Perhaps this is an indication that, unlike the corresponding matter radial distributions in Figure 5.15, the quenched and unquenched results, due to the effect of disconnected processes, can differ even with the present sea–quark masses of about that of the strange quark. Certainly differences should appear with very light sea–quarks, since then the contribution from the disconnected processes, that only enter for the matter distributions, can become significant. However, we do not have the data to cross check with Refs. [74], which advocate the existence of such differences for the matter sum rule.

The fact that the matter sum rule is considerably less than that for charge can be qualitatively understood by employing data from different hopping parameters (κ) and using the identity

$$X^{11} = \frac{d(am_1)}{d\kappa^{-1}}, \tag{5.80}$$

where am_1 is the ground state energy and κ the hopping parameter — see Subsec. 5.8.2 and Ref. [74]. When the m_1's correspond to the cases where the light quark is of about one and two strange quark masses, Refs. [74] and [56] give $X^{11} \approx 0.34(8)$ and 0.31(6) respectively — consistent with the present value of 0.4(1). These values are also consistent with the following simple estimate: If the $Q\bar{q}$-meson mass (am_1) is taken to be simply the sum of the quark masses *i.e.* $am_Q + am_q$ and $\kappa^{-1} = 8 + 2am_q$, then Eq. 5.80 gives

$$X^{11} = \frac{d(am_Q + am_q)}{d(8 + 2am_q)} = 0.5. \tag{5.81}$$

Another reason for expecting $X^{11}_M < X^{11}_C$ also follows from a potential approach using the Dirac equation as in Subsec. 5.9.3.2. This results in $X^{11}_C \sim G^2_1 + F^2_1$ and $X^{11}_M \sim G^2_1 - F^2_1$. Here G^2_1 and F^2_1 are integrals of the large and small components of the solution to the Dirac equation.

5.10 The $B - B$ System as a $[(Q\bar{q})(Q\bar{q})]$ Configuration

In Secs. 5.8 and 5.9 the energies and some radial distributions of a *single* heavy-light meson were studied. In this section the interaction between two such mesons is extracted using lattice QCD and the outcome is fitted with an extension of the f-model developed earlier for the $[(Q\bar{Q})(Q\bar{Q})]$ system in Secs. 5.5 and 5.7. However, it should be added that the study of $[(Q\bar{q})(Q\bar{q})]$ configurations is much less academic than their $[(Q\bar{Q})(Q\bar{Q})]$ counterparts. Many years ago simple multi-quark systems have been proposed to exist as bound states [133, 134, 135]. Also four quarks forming colour singlets or as bound states of two mesons are candidates for particles lying close to meson–antimeson thresholds, such as $a_0(980)$, $f_0(980)$ $(K\bar{K})$, $f_0(1500)$, $f_2(1500)$ $(\omega\omega,\ \rho\rho)$, $f_J(1710)$ $(K^*\bar{K}^*)$, $\psi(4040)$ $(D^*\bar{D}^*)$, $\Upsilon(10580)$ $(B^*\bar{B}^*)$ [111]. Systems involving b quarks are particularly interesting since they should be more easily bound provided the potential is attractive, since the repulsive kinetic energy of the quarks is smaller, while the attractive two-body potential remains the same. In so-called deuson models [136] the long-range potential between two mesons comes from one-pion exchange, suggesting that meson–meson systems are significantly less bound than meson–antimeson systems. Other models used for realistic four-quark systems include string-flip potential models (see Ref. [138] for a review), bag models [134], and a model-independent approach [139]. In fact four-quark states with two heavy quarks have been predicted to be stable [140]. Most models give stability for systems where the heavy quarks have the b mass, but long range forces might push the required heavy-to-light mass ratio down sufficiently so that $cc\bar{q}\bar{q}$ states would be bound as well.

5.10.1 *Lattice calculation of the* $[(Q\bar{q})(Q\bar{q})]$ *system*

The interaction between two $(Q\bar{q})$ states is depicted in Fig. 5.19 as a sum of two terms — the uncrossed and crossed diagrams. In the latter diagram the $\bar{q}$ hops from one Q to the other. Exploratory studies of two-meson systems have been made for the cross diagram only for SU(3) colour [141] and for both diagrams in Refs. [142], [143] for SU(2), SU(3) colour respectively. This topic is also discussed in Sec. 2.4 of Chapter 2 and Subsec. 4.4.3 of Chapter 4.

In Refs. [77, 144] quenched lattices are used with SU(3) colour and static

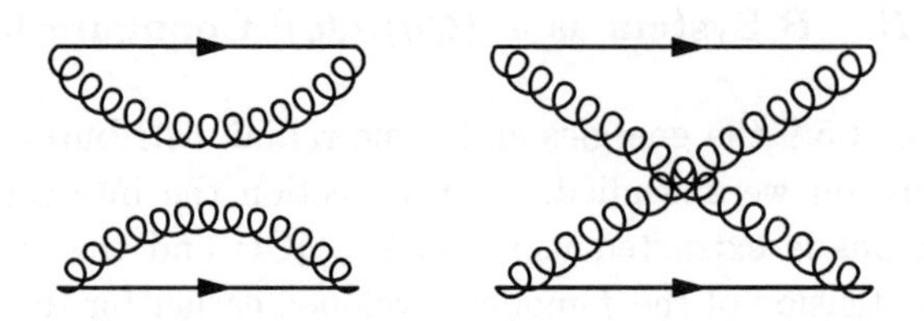

Fig. 5.19 The interaction between two $(Q\bar{q})$ states — the so-called uncrossed and crossed diagrams. The solid lines represent Q's and the wavy ones $\bar{q}$'s.
NB Compared with earlier figures such as Figs. 5.12 and 5.14 the Euclidean time direction is now horizontal.

heavy quarks with light quarks of approximately the strange quark mass. Also, when the two Q's are at the same point (*i.e.* $R = 0$) and by using the SU(3) colour relationship $3 \otimes 3 = \bar{3} \oplus 6$, then the two Q's behave like a $\bar{Q}$. This equivalence implies that the $I_q = 1$, $S_q = 1$ state will have the same light quark structure as the Σ_b baryon and the $I_q = 0$, $S_q = 0$ state as that of the Λ_b baryon. The other two allowed states at $R = 0$ correspond to a static sextet source. When the results for $R = 0$ are compared with the known spectrum of *baryons* with one heavy quark (Λ_b and Σ_b), good agreement is found. Note that this link to baryons at small separation R *cannot* be explored using SU(2) of colour.

Below, the mass of one quark in each meson is taken to be very heavy — the prototype being the B-meson. The static limit is then the leading term in the heavy quark effective theory for a heavy quark of zero velocity and there will be corrections of higher orders in $1/m_Q$, where m_Q is the heavy quark mass. In the limit of a static heavy quark, the heavy quark spin is uncoupled since the relevant magnetic moment vanishes which implies that the pseudoscalar B-meson and the vector B^*-meson will be degenerate. This is a reasonable approximation since they are split by only 46 MeV experimentally, which is less than 1% of the mass of the mesons — see Table 5.3. Since it is often convenient to treat these two mesonic states as if they were degenerate, here they are described collectively as the $\mathcal{B}$-meson. Because of the insensitivity to the heavy quark spin, it is then appropriate to classify these degenerate $\mathcal{B}$-meson states by the light quark spin. The system of two heavy-light mesons at spatial separation R will be referred to as the $\mathcal{BB}$ system. With both heavy-light mesons static, this $\mathcal{BB}$ system is then described by the two independent spin states of the two light quarks in the two mesons *i.e.* $S_q = 0$ and 1. Thus there are four possible states and it is necessary to classify the interaction in terms of these spin states.

This situation is very similar to that of the hydrogen molecule in the Born–Oppenheimer approximation — with, however, the additional possibility that the two "electrons" can have different properties. Another similarity is with the potential between quarks which has a central component and also scalar and tensor spin-dependent contributions.

Each $\mathcal{B}$-meson will have a light quark *flavour* assignment. For the $\mathcal{B}\mathcal{B}$ system, it will be appropriate to classify these states according to their symmetry under interchange of the light quark flavours. For identical flavours (*e.g.* ss or uu), we have symmetry under interchange, whereas for non-identical flavours (*e.g.* su or du), we may have either symmetry or antisymmetry. For two light quarks, it is convenient to classify the states according to isospin as $I_q = 1$ (with uu, $ud + du$ and dd) or $I_q = 0$ (with $ud - du$). To ensure overall symmetry of the wave function under interchange and, assuming symmetry for spatial interchange, the flavour, total light quark spin (S_q) and total heavy quark spin (S_b) must be combined to achieve this. Thus in the limit of an isotropic spatial wave function *i.e.* $L = 0$, there will be four different ground state levels of the $\mathcal{B}\mathcal{B}$ system labelled by $(I_q,\ S_q)$ in the following discussion. These are shown in Table 4.2 of Chapter 4.

To check for possible finite size effects, the numerical analysis was carried out on quenched lattices of sizes $12^3 \times 24$ and $16^3 \times 24$, at $\beta = 5.7$, corresponding to $a \approx 0.18$ fm. The bare mass of the light quark was near that of the strange mass and light quark propagators were generated using the Maximal Variance Reduction method in Subsec. 5.8.3. This enabled measurements of the strength of the interaction to be made out to separations of $R \approx 8$, which corresponds roughly to 1.4 fm.

As seen in Figs. 2.5, 2.6 of Chapter 2 and Figs. 4.21 of Chapter 4, attraction between two $\mathcal{B}$ mesons is found at small values of R for (I_q, S_q)=(0,0) and (1,1) and at moderate R ($\sim$0.5 fm) for (1,0) and (0,1). For very heavy quarks, this will imply binding of the $\mathcal{B}\mathcal{B}$ molecules with these quantum numbers and $L = 0$ — see Sec. 2.4 of Chapter 2 for more details.

It is also possible to extract from the lattice data quantities that can be identified as π- and ρ-exchange between the two $\mathcal{B}$-mesons. This is shown in Fig. 5.20. For the π-exchange at large R the potential is expected to be of the form

$$V(R) = \vec{\tau}_1.\vec{\tau}_2 \frac{g^2 M^2}{4\pi f_\pi^2} \frac{e^{-MR}}{R}, \tag{5.82}$$

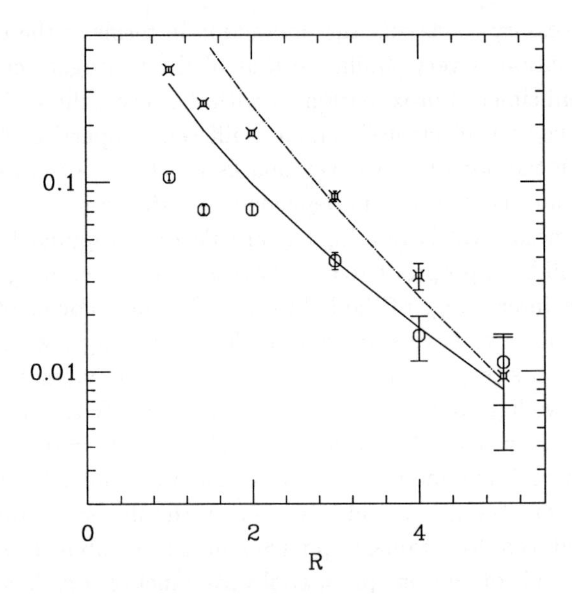

Fig. 5.20 The ratio of the crossed-diagram contributions to the spin-averaged uncrossed contribution to the $\mathcal{BB}$ correlation. The meson exchange expressions, $\exp(-MR)/R$, are compared with these results — using pion exchange with $M = 580$ MeV (continuous line) and rho exchange with $M = 890$ MeV (dash-dotted line). Note that the pion exchange expression is normalised as described in the text, whereas the rho exchange contribution has an *ad hoc* normalisation. Here R is in units of $a \approx 0.18$ fm [77].

where g/f_π is the pion coupling to quarks [136]. From the lattice studies of $BB^*\pi$ coupling [145] the value of $g = 0.42(8)$ is predicted and f_π is the pion decay constant (132 MeV). Because the comparison with the lattice results is for light quarks, the pion mass is taken to be $Ma = 0.53$ *i.e.* a "pion" of mass ≈ 580 MeV — a value appropriate for this lattice. Therefore, in Fig. 5.20 the solid curve being compared with that data containing one-pion-exchange is from Eq. 5.82 and is seen to have the correct features — giving support, at these large values of R, for the deuson model of Ref. [136]. However, the corresponding ρ-exchange comparison with the dash-dotted line in Fig. 5.20 is less informative, since the normalisation is *ad hoc.*

Even though the agreement with one-pion-exchange is seen to be very good, the authors of Ref. [77] are quick to point out that their result is "better than should be expected" and give arguments why "This implies that we should not take our estimate of the magnitude of one pion exchange as more than a rough guide at the R-values we are able to measure." It is possible that here we are seeing an effect reminiscent of the Chiral Bag Model described by Myhrer in Chapter 4 of Ref. [36]. In such models there are pion fields outside some radius $r_b \approx 0.5$ fm and only quarks inside r_b. Calculations of one-pion-exchange potentials (OPEP) between, say, two nucleons then naturally give the usual form of OPEP for $r > r_b$. However, for some smaller values of r the interaction seems to be simply a continuation of this "usual form of OPEP" *i.e.* there is a precocious onset of OPEP in a region where there are no pions. Perhaps it is this that is being seen in Ref. [77] and Fig. 5.20. It should be added that in some studies of interacting clusters, the OPEP is expected to emerge with an *exponential* dependence and not the Yukawa form in Eq. 5.82. This has been demonstrated in Ref. [71], where the authors show that this unconventional form of OPEP is due to the use of the *quenched* approximation. To avoid this problem with the long range part of the interaction, in a later paper [137] these authors study the $\Lambda_Q\Lambda_Q$ potential, which does not contain a one-pion or one-eta contribution.

5.10.2 *Extension of the f-model to the* $[(Q\bar{q})(Q\bar{q})]$ *system*

In Sec. 5.5 a model was developed for understanding the lattice energies of four static quarks $Q(\mathbf{r_1})Q(\mathbf{r_2})\bar{Q}(\mathbf{r_3})\bar{Q}(\mathbf{r_4})$ in terms of two-quark potentials and is summarised by Eqs. 5.37, 5.38. This f-model, although very simple, contains the same basic assumptions made in the more elaborate many-body models that incorporate kinetic energy *e.g.* the Resonating Group Method described by Oka and Yazaki in Chapter 6 of Ref. [36]. It is, therefore, reasonable that this simplified f-model can to some extent check the validity of these more elaborate counterparts. In Ref. [60] the model in Sec. 5.5 was extended as below to study the interaction between two $Q\bar{q}$ mesons. This resulted in a non-relativistic Schrödinger-like equation

$$|\mathbf{K}'(R) + \mathbf{V}'(R) - E(4,R)\mathbf{N}'(R)|\psi = 0 \qquad (5.83)$$

— for more details see Appendix B.

So far this model has only been developed for spin independent potentials, which means that it should only be compared with the spin-averaged results of Ref. [77] shown in Fig. 4.21 in Chapter 4. The outcome from Refs. [60, 146] is shown in Fig. 5.21, where it is seen by the solid line that the use of only two-quark potentials (*i.e.* in the weak coupling limit $f = 1$ or $k_f = 0$) results in a considerable overbinding at $R = 0.18$ fm. The dashed line shows the effect of using a form factor with $k_f = 0.6$. Admittedly this

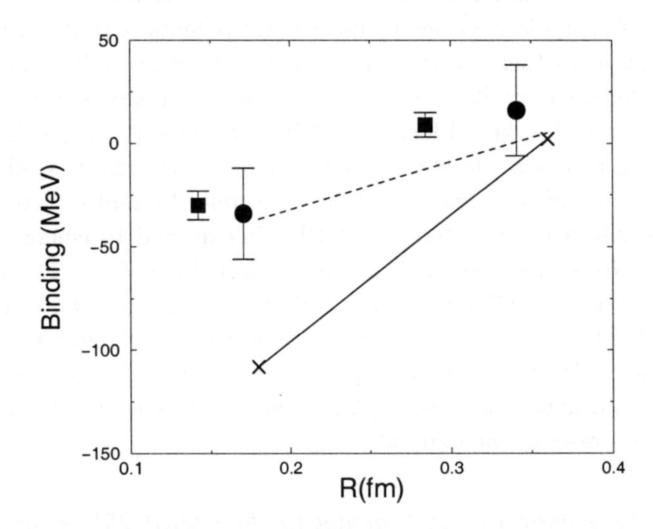

Fig. 5.21 Comparison between the spin independent part of the $Q^2\bar{q}^2$ binding energies calculated on a lattice [77] (solid circles – quenched approximation with $a = 0.170$ fm) [147] (solid squares – with dynamical fermions and $a = 0.142$ fm). The crosses, with the solid line to guide the eye, show the model in the weak coupling limit ($k_f = 0$). The dotted line shows the result with k_f =0.6 and m_q=400 MeV. The dynamical fermion data was not used in any fit. They are simply included to show that it is qualitatively consistent with the quenched data but with considerably smaller error bars.

is less convincing than the earlier four static quark case, since the conclusion depends essentially on only the two data points corresponding to the two Q's being 1 and 2 lattice spacings apart. However, the fact that k_f is consistent with $\bar{k} \approx 0.5(1)$ — see Fig. 5.8 — the corresponding parameter needed in Subsec. 5.5.5 for $[(Q\bar{Q})(Q\bar{Q})]$ configurations in squares or near-squares lends support to the general approach of this model.

In Refs. [148] BB-scattering has been treated in the weak coupling limit of $k_f = 0$ — a limit that appears to be ruled out by the comparison in Fig. 5.21. However, it is possible that this limit can, to some extent, be salvaged if the model is extended by including states with excited glue — as in Sec. 5.7 and Appendix A.3 for the $[(Q\bar{Q})(Q\bar{Q})]$ system.

5.11 The $B - \bar{B}$ System as a $[(Q\bar{q})(\bar{Q}q)]$ Configuration

The B-factories discussed in Subsec. 5.3.1 are not able to study directly BB reactions. However, the related $B\bar{B}$ system is accessible as a final state in the decay of the $\Upsilon(4S, 10580\text{ MeV})$, whose main branching ($\geq 96\%$) is into $B\bar{B}$.

At first sight it may be thought that the $[(Q\bar{q})(\bar{Q}q)]$ and $[(Q\bar{q})(Q\bar{q})]$ configurations have similar properties. However, this is not so, since now the q and $\bar{q}$ can annihilate each other. This means that there is a coupling between $(Q\bar{Q})$ and $[(Q\bar{q})(\bar{Q}q)]$ states *i.e.* the string (flux) connecting the Q and $\bar{Q}$ in the $(Q\bar{Q})$ state can break into two mesons $(Q\bar{q})$ and $(\bar{Q}q)$. This becomes very clear if the mechanism for the interaction between two $(Q\bar{q})$ states, shown in the Fig. 5.19, is compared with the corresponding ones for the present $[(Q\bar{q})(\bar{Q}q)]$ case on the first row of Fig. 5.22 — only the uncrossed contributions look similar. The diagram b) represents the two step process involving the annihilation and creation of a $q\bar{q}$ pair

$$[(Q\bar{q})(\bar{Q}q)] \longrightarrow (Q\bar{Q}) \longrightarrow [(Q\bar{q})(\bar{Q}q)].$$

This breaking of a long flux tube between two static quarks into a quark–antiquark pair is one of the most fundamental phenomena in QCD. Because of its highly non-perturbative nature it has defied analytical calculation, while its large scale, *e.g.* when compared to the sizes of composite particles

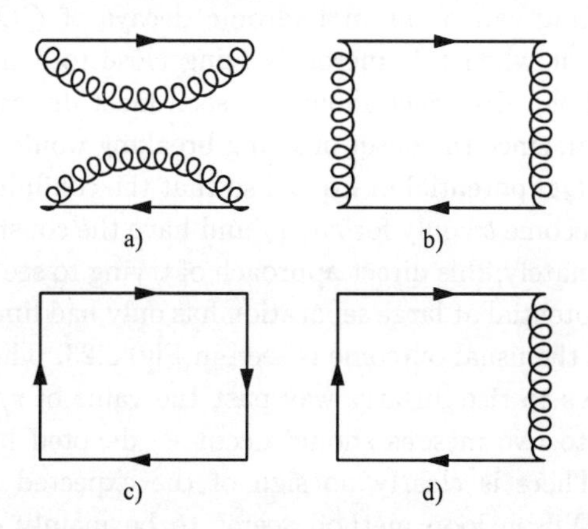

Fig. 5.22 The $[(Q\bar{q})(\bar{Q}q)]$ interaction and string breaking.
The a) uncrossed and b) crossed $[(Q\bar{q})(\bar{Q}q)] \to [(Q\bar{q})(\bar{Q}q)]$ contributions to the interaction; c) The $(Q\bar{Q}) \to (Q\bar{Q})$ Wilson loop and d) the $[(Q\bar{q})(\bar{Q}q)] \to (Q\bar{Q})$ off-diagonal correlation.

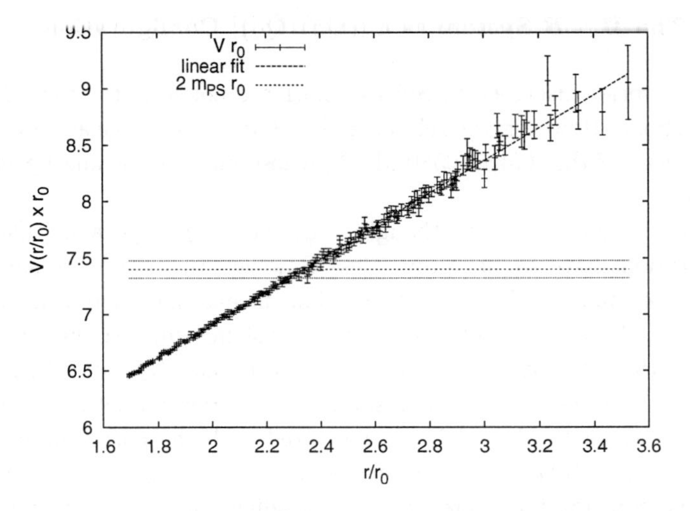

Fig. 5.23 The static potential $V(Q\bar{Q})$ obtained from simply the Wilson loop of Fig. 5.1. N.B. There is no sign of the expected flattening at $r/r_0 \approx 2.4$, where it becomes energetically favourable to create two mesons[149].

in the theory, has caused difficulties in standard nonperturbative methods. Thus string breaking has remained a widely publicized feature of the strong interaction that has never, apart from rough models, been reproduced from the theory.

String breaking can occur in hadronic decays of $Q\bar{Q}$ mesons and is especially relevant when this meson is lying close to a meson–antimeson $(Q\bar{q})(\bar{Q}q)$ threshold. Its effect should be seen most directly by measuring the $Q\bar{Q}$ potential, since the onset of string breaking would change the form of the standard $Q\bar{Q}$ potential in Eq. 5.6 so that the confining term — br for all r — would become br only for $r < r_c$ and have the constant value br_c for $r > r_c$. Unfortunately, this direct approach of trying to see the flattening in the static $Q\bar{Q}$ potential at large separation has only had limited success. An example[149] of the usual outcome is seen in Fig. 5.23. There the potential $V(Q\bar{Q})$ continues to rise linearly way past the value of $r/r_0 \approx 2.4$, where the breaking into two mesons should occur — denoted by the horizontal dotted lines. There is clearly no sign of the expected flattening. The failure of this Wilson loop method seems to be mainly due to the poor overlap of the operator(s) with the $[(Q\bar{q})(\bar{Q}q)]$ state [150, 151]. In three-dimensional SU(2) with staggered fermions an improved action approach has been claimed to be successful with just Wilson loops [152]. However, in full QCD with fermions, effective operators for $[(Q\bar{q})(\bar{Q}q)]$ systems are hard to implement; part of the problem is the exhausting computational effort required to get sufficient statistics for light quark propagators with conventional techniques for fermion matrix inversion. In Ref. [153] this problem was to a large extent overcome by applying the Maximal Variance Reduction method of Subsec. 5.8.3 using SU(3) on a $16^3 \times 24$ lattice with the Wilson gauge action plus the Sheikholeslami–Wohlert quark action and $\beta = 5.2$ (*i.e.* $a \approx 0.14$ fm) with two degenerate flavours of both valence– and sea–quarks.

Since Fig. 5.22 b) is a two-step process it is natural to consider the process as a coupled channels problem between the $[(Q\bar{q})(\bar{Q}q)]$ and $(Q\bar{Q})$ configurations. This leads to the two new processes in Fig. 5.22 d) and c) — the off-diagonal term $[(Q\bar{q})(\bar{Q}q)] \longleftrightarrow (Q\bar{Q})$ and the corresponding diagonal term $(Q\bar{Q}) \longleftrightarrow (Q\bar{Q})$, which is nothing more than the Wilson loop discussed in Subsec. 5.4.1.2.

In Fig. 5.24 the results from a calculation using just the most fuzzed basis states for both $(Q\bar{Q})$ and $[(Q\bar{q})(\bar{Q}q)]$ (a 2×2 matrix) are shown. Unfortunately, the statistics are not sufficient to give accurate plateaux for the energies. However, the authors of Ref. [153] are able to extract a quantity they call the mixing matrix element x, which can be interpreted as an indirect measure of possible string breaking. They find that $x = 48(6)$ MeV. At r_c one would expect the ground and excited state energies to be separated by $2x$. However, in Fig. 5.24 a larger separation is observed, which is presumably again due to insufficient statistics. In Ref. [154], utilizing the SU(3) colour relationship $3 \otimes \bar{3} = 1 \oplus 8$, the $Q\bar{Q}$ at $R = 0$ behave as a singlet(vacuum), so that the remanent $q\bar{q}$ component with $(S_{q\bar{q}}, I_{q\bar{q}}) = (0, 1)$ can be compared successfully to a pion with non-zero momentum.

In the above, the failure of the $Q\bar{Q}$ correlations alone to give a flattening of $V(Q\bar{Q})$ was considered a negative feature. However, in the Conclusion it will be seen that this "failure" could be useful for constructing models.

So far no one has attempted to understand this $B\bar{B}$ data with the extended f-model of Subsec. 5.10.2 for BB states. Such an extension would also have to incorporate $q\bar{q}$ creation and annihilation using some model such as the so-called Quark Pair Creation or $^3P_0$ model [155]. When this $^3P_0$ model was combined with the harmonic oscillator flux tube model of Isgur and Paton [54], it proved successful for describing flux tube breaking and formation [156].

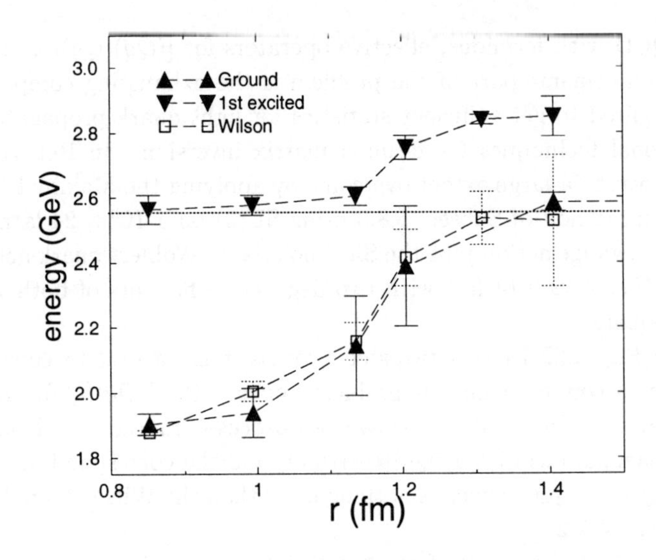

Fig. 5.24 Ground and excited state from a variational calculation including $Q\bar{Q}$ and $Q\bar{q}\bar{Q}q$ operators [153]. The highest fuzzed basis state for both is used here. The ground state of the Wilson loop and $2m_{Q\bar{q}}$ are also shown.

5.12 Conclusions and the Future

In this chapter there has been an attempt to bring together two distinct lines of research:

(1) **Lattice QCD** was applied to the multiquark systems $[(Q\bar{Q})(Q\bar{Q})]$, $(Q\bar{q})$, $[(Q\bar{q})(Q\bar{q})]$ and $[(Q\bar{q})(\bar{Q}q)]$ in Secs. 5.4.2, 5.8.4, 5.10 and 5.11.

(2) **Effective Potential Theories (EPTs)** — based on interquark potentials with four-quark form factors included — were developed in Secs. 5.5–5.7 and 5.10.2 to interpret phenomenologically this lattice data.

However, most of the comparison between these two lines has been devoted to constructing EPTs that give some phenomenological interpretation of the Lattice QCD *energies*. For the $[(Q\bar{Q})(Q\bar{Q})]$ system the latter concentrated on the energies of the six geometries in Fig. 5.10 and this amounted in all to about 100 pieces of data. In spite of four static quarks being a very unphysical system, the Lattice QCD $\leftrightarrow$ EPT comparison showed that a 4-quark potential seemed to be needed and that simply using a sum of 2-quark potentials failed by generating far too much binding for large interquark distances. On the other hand, for small interquark distances the use of only 2-quark potentials was sufficient — the so-called weak coupling limit. However, in Sect. 5.7 it was found that, if the model was extended to explicitly include gluonic excited states, then much of the attraction came from these excited states and that the need for a 4-quark potential was greatly reduced. *This observation for the interaction between four static quarks suggests that the usual approach of simply using 2-quark potentials in multiquark problems could, to some extent, be justified provided gluonic excited states are explicitly included.*

In the more physical case of the $[(Q\bar{q})(Q\bar{q})]$ system the EPTs are complicated by the presence of the light quark kinetic energy and mass. Even so, the Lattice QCD $\leftrightarrow$ EPT comparison in Fig. 5.21 still shows the same effect that simply using a sum of 2-quark potentials fails, if the basis states only contain the gluon field in its ground state.

This apparent need for a 4-quark potential can be viewed as a form factor — a familiar and successful technique in, for example, parametrizations of the NN-potential [27]. However, there the form factor is needed to regularize the potential at *small* values of r, where the potential can have $1/r^n$ singularities. It can, therefore, be simply incorporated as a short-ranged vertex correction modelled from meson fields. In contrast, the form factor needed in the present model is introduced to eliminate problems — essentially the van der Waals effect — at *large* values of r. This form factor is, therefore, modelling a long-ranged effect which could be excited gluon states, which become more effective the larger the system.

Clearly for a better "Lattice QCD $\leftrightarrow$ EPT" comparison more data is needed from a given quark system. This suggests that comparisons should be made using other quantities in addition to the few available lattice energies. The most obvious candidates are radial distributions of the light quark, since — being a function of distance — these introduce many new pieces of data in a way that is more systematic than the earlier choice of simply the six "convenient" geometries of Fig. 5.10. In addition, there can be several types of radial distribution:

- The charge (vector) density where the probe in Eq. 5.65 is $\Theta = \gamma_4$.
- The matter (scalar) density where the probe is $\Theta = 1$.
- The pseudovector density with the probe $(\gamma_\mu \gamma_5)$. This is needed for the $B^*B\pi$ coupling [145] and was exploited in Subsec. 5.10.1.

- In addition to operators that probe the radial distributions of the light quark(s), it is also possible to map out the structure of the properties of the underlying gluon field. This is achieved by using different orientations of the elementary plaquette defined in Eq. 5.16. Using purely spatial plaquettes the radial distributions of the various components of the colour *magnetic* field can be extracted *e.g.* $U_{\Box}^{xy} \to B_z^2$, whereas those plaquettes with a euclidean time superfix give the spatial distributions of the various components of the colour *electric* field *e.g.* $U_{\Box}^{xt} \to E_x^2$. In Ref. [157, 158, 159] these distributions were calculated for two- and four-static SU(2) quark systems, where the latter were restricted to the corners of squares with sides upto 8 lattice spacings (*i.e.*$\approx$ 1 fm). The lattice was $20^3 \times 32$ with $\beta = 2.4$ ($a \approx 0.12$ fm). There is a wealth of information in these flux profiles and they present a formidable challenge to models that attempt to describe them [54, 55]. However, the few comparisons that have so far been made are very encouraging [157].

In principle, all of these distributions can be measured for the light quarks in the 4-quark systems $[(Q\bar{q})(Q\bar{q})]$ and $[(Q\bar{q})(\bar{Q}q)]$ discussed in Secs. 5.10 and 5.11. However, this is probably too ambitious at the present time for meaningful Lattice QCD $\leftrightarrow$ EPT comparisons. First, the radial distributions in the 2-quark $(Q\bar{q})$ system should be better understood — the topic of Sec. 5.9. It is possible that the radial distributions of the $\bar{q}$ in the $[(Q\bar{q})(Q\bar{q})]$ and $[(Q\bar{q})(\bar{Q}q)]$ systems are related to that in the simpler $(Q\bar{q})$ case, since for the corresponding situation in light nuclei the neutron-proton radial correlations in both ^3He and ^4He are very similar to that in the deuteron — see Fig. 5.25. This shows that, although the 3- and 4-nucleon calculations needed to extract such correlations are very complicated, they result in some simplicities. Possibly comparisons of correlations within few quark systems could lead to similar simplifications and so enhance our understanding of such systems. The work in Ref. [61] for the QQq system is a step in this direction.‡

In Sec. 5.9 the charge and matter distributions of the $\bar{q}$ in the $Q\bar{q}$ system ($\approx B$-meson) were measured and attempts made to fit these distributions with simple functions in Subsec. 5.9.3.1 and also by using the Dirac equation

‡Some recent lattice calculations on doubly-charmed baryons have concentrated on their masses[161].

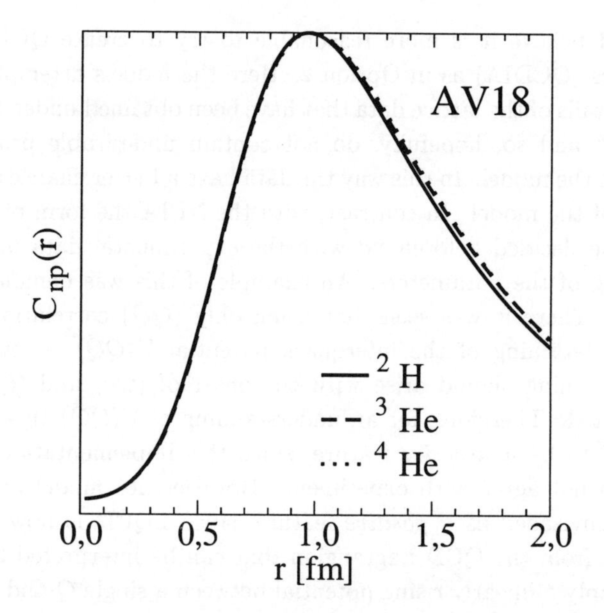

Fig. 5.25 Comparison of the neutron-proton radial correlations in the deuteron, ^3He and ^4He [160].

in Subsec. 5.9.3.2. The latter can be viewed in two ways:

(1) As simply an alternative form of parametrization of the two distributions with no physical interpretation of the parameters needed.
(2) As the construction of an EPT, in which the parameters do have a physical interpretation — albeit phenomenological — on which extensions to multiquark systems can be based.

This leads us to the main problem of how to set up an EPT in order to understand multiquark systems. In the past the key word has been "mimicing", in which models based on potentials have been mimicing the successful models of multi-nucleon systems. Unfortunately, since such models aim directly at the *experimental* data, they may be attempting to describe some mechanisms that are outside the scope of the model. This I have called the Nuclear-Physics-Inspired-Approach (NPIA) and it would correspond to Option 1 in the comparison

$$\text{NPIA} \overset{1}{\longleftrightarrow} \text{Experimental Data } \textit{versus} \text{ Lattice data} \overset{2}{\longleftrightarrow} \text{QCDIA}$$

However, I believe it is more reasonable to try to create QCD-Inspired-Approaches (QCDIA) as in Option 2. Here the models attempt to mimic directly details of the lattice data that have been obtained under "controlled conditions" and so, hopefully, do not contain undesirable processes not included in the model. In this way the data have a better chance of deciding the *form* of the model. In contrast, with the NPIA the form of the model tends to be decided beforehand with the experimental data only leading to a tuning of the parameters. An example of this was demonstrated by Fig. 5.23. There it was seen that using only $(Q\bar{Q})$ correlations did not lead to a flattening of the interquark potential $V(Q\bar{Q})$ — even though such a flattening should arise with the onset of $(Q\bar{q})$ and $(\bar{Q}q)$ mesons being created. Therefore, for an understanding of $V(Q\bar{Q})$ this defect was considered to be a *negative* feature, since this implementation of LQCD clearly did not agree with experiment. However, for model building this could be envisaged as a *positive* feature, since LQCD is now generating *exact* data from the QCD Lagrangian that can be interpreted by a model that is simply a linearly rising potential between a single Q and a single $\bar{Q}$. This intermediate stage model could, hopefully, be more easily extended to the real life situation, when the more complicated correlations in Fig. 5.22 are treated by LQCD. This intermediate stage could be considered as yet another example of the "unphysical worlds" discussed in Sec. 5.2.2 in the context of models with different numbers of colours, spatial dimensions, quark masses *etc.*. This recalls my earlier work in ^{18}O[162, 163]. There the "real" ^{18}O needed to be described by "4-particle 2-hole" states in addition to the usual "2-particle 0-hole" states. This is analogous to the matrix of correlations in Fig. 5.22. It would have been very useful if there had been an "experimental phase" of ^{18}O that only needed 2-particle 0-hole states for its description. In that case the model for "real" ^{18}O would have had less freedom.

So far the QCDIA has only been attempted for the $[(Q\bar{q})(Q\bar{q})]$ configuration in Sec. 5.10. There the $(Q\bar{q})$ lattice data in Fig. 5.18 was first fitted with a Schrödinger equation to give an effective light quark mass $m_{q,\text{ effective}} \approx 400$ MeV. Using a variational method, this Schrödinger approach was then easily extended to the $Q^2\bar{q}^2$ system using the same interquark potential and $m_{q,\text{ effective}}$ determined earlier from the $Q\bar{q}$ data. In principle, this could be extended to any system that can be described in terms of interacting quark clusters.

The problem with the above QCDIA is that it usually results in an $m_{q,\ \text{effective}} \ll 1$ GeV suggesting the need for a more relativistic approach. This can be attempted at different levels:

(1) At the one extreme we can use directly the Dirac equation to describe, say, the $Q\bar{q}$ system as in Subsec. 5.9.3.2. This had some success but we are then confronted with the problem of extending the comparison to multi-quark systems — a second step (2) that is not directly possible for systems containing more than one light quark *i.e.*

$$\text{Lattice data for } Q\bar{q} \xrightarrow{1} \text{1-quark Dirac Eq.} \xrightarrow{2\,??} \text{Multi-quark case}$$

(2) Use some Semirelativistic Schrödinger-like Equation (SRSE) as described in Subsec. 5.9.3.4. This is most easily formulated in momentum space with the kinetic energy $E_{NR} = m_{q,e} + p^2/2m_{q,e}$ being replaced by $E_R = \sqrt{p^2 + m_{q,e}^2}$, where $m_{q,e}$ is an effective quark mass chosen to fit some piece(s) of experimental or lattice data. However, care must be taken to treat the potential terms to the same semirelativistic degree by inserting appropriate factors of $m_{q,e}/E_p$ as in the Blankenbecler–Sugar equation [35]. The second step (2) to multiquark systems is then probably possible *i.e.*

$$\text{Lattice data for } Q\bar{q} \xrightarrow{1} \text{SRSE} \xrightarrow{2} \text{Multi-quark case}$$

(3) Thirdly, a compromise model emerges. First the $Q\bar{q}$ lattice data is fitted by a Dirac equation. This equation then generates other observables that are interpreted in terms of a SRSE, which can be extended to multi-quark systems. This is similar to the philosophy of Bhaduri and Brack [30], who show how the Schrödinger equation with a quark effective mass of $m_{q,\ \text{effective}} \approx 500$ MeV is able to explain some of the results — energies and magnetic moments — of a Dirac equation for a *zero* mass quark *i.e.* they make the comparison

$$\text{Lattice data for } Q\bar{q} \xrightarrow{1} \text{Dirac Eq.} \xrightarrow{2} \text{SRSE} \xrightarrow{3} \text{Multi-quark}$$

So what are the "Bridges from Lattice QCD to Nuclear Physics" as advertized in the title of this chapter? So far, the only clear examples involve mainly static (or heavy) quarks (Q) as in the extraction of the string energy and the lattice spacing from V_{QQ} with a non-relativistic Schrödinger equation (Sec. 5.2) and the energies of the various Q^4 geometries in Secs. 5.3–5.7. The introduction of light quarks as in the $(Q\bar{q})$, $[(Q\bar{q})(Q\bar{q})]$ and $[(Q\bar{q})(\bar{Q}q)]$ systems in Secs. 5.8 – 5.11 leads to major complications and the only partial success is the Schrödinger description of the $(Q\bar{q}) + [(Q\bar{q})(Q\bar{q})]$ *energies* in Sec. 5.10. The real test of whether any "Bridge" exists is only now becoming possible with the advent of the lattice data for radial distributions, since such distributions are also at the centre of much of Nuclear Physics. Of course, the final outcome could be that there is no useful bridge for treating multiquark systems in the way we treat multinucleon systems. This would mean that Lattice QCD would only ever be able to directly address problems involving a few quarks — perhaps upto the six quarks needed for the nucleon-nucleon interaction — but not show a general way for how to deal with multiquark systems. This would result in the two worlds of QCD and Nuclear Physics having little direct connection with each other.

However, before such a pessimistic view is adopted, we should remember that Nuclear Physics earlier had another two-world structure that lasted for many years. Until the 1960's there were essentially two models for the nucleus — the collective liquid-drop-like model that rarely mentioned the nucleon-nucleon potential and, in contrast, the shell model based on this potential. But with the advent of Brueckner theory it was shown how effective interactions in many-nucleon systems could be constructed from the basic nucleon-nucleon potential. This was followed up by the generation of collective effects as interacting particle-hole states [164]. In this way a bridge was made between the basic nucleon-nucleon potential and collectivity — but it took many years.

Acknowledgements

The author wishes to thank Mika Jahma, Jonna Koponen, Timo Lähde and Slawek Wycech for their invaluable help in preparing and commenting on this manuscript. The author, in his rôle as Editor, also wishes to thank the other contributors to this Volume. Without their cooperation and enthusiasm the volume would never have appeared.

Appendix A: Extensions of the f-Model from 2×2 to 6×6

A.1: The 3×3 extension of the unmodified two-body approach of Subsec. 5.5.1

Because the colour group considered so far is SU(2), there is no distinction between the group properties of quarks (Q) and antiquarks ($\bar{Q}$). Four such quarks can then be partitioned as pairs in *three* different ways

$$A = (Q_1Q_3)(Q_2Q_4), \quad B = (Q_1Q_4)(Q_2Q_3) \quad \text{and} \quad C = (Q_1Q_2)(Q_3Q_4), \tag{A.1}$$

where each (Q_iQ_j) is a colour singlet. However, these three basis states are not orthogonal to each other. Also, remembering the fact that the quarks are indeed fermions gives, in the weak coupling limit, the condition in the Appendix of Ref. [52]

$$|A\rangle + |B\rangle + |C\rangle = 0. \tag{A.2}$$

Since $\langle A|A\rangle = \langle B|B\rangle = \langle C|C\rangle = 1$, we get — in this limit — the equalities $\langle A|B\rangle = \langle B|C\rangle = \langle A|C\rangle = -1/2$.

In SU(3) the partitioning problem is different since in that case the three partitions are

$$A = (Q_1\bar{Q}_3)(Q_2\bar{Q}_4), \quad B = (Q_1\bar{Q}_4)(Q_2\bar{Q}_3) \quad \text{and} \quad C = [(Q_1Q_2)^d(\bar{Q}_3\bar{Q}_4)^d], \tag{A.3}$$

where state C is expressed in terms of either colour antitriplet ($d = \bar{3}$) or sextet ($d = 6$) states and so cannot appear asymptotically as two clusters.

If all three basis states in SU(2) are included, then the matrix below is singular for the obvious reason that $|A\rangle + |B\rangle + |C\rangle = 0$ *i.e.*

$$\det \mathbf{N} = \det \begin{pmatrix} 1 & -1/2 & -1/2 \\ -1/2 & 1 & -1/2 \\ -1/2 & -1/2 & 1 \end{pmatrix} = \det \begin{pmatrix} 1 & 1/2 & 1/2 \\ 1/2 & 1 & -1/2 \\ 1/2 & -1/2 & 1 \end{pmatrix} = 0. \tag{A.4}$$

Earlier this was interpreted to mean that it was unnecessary to include all three states and so the symmetry was broken by keeping the two states with the lowest energy, let us say , A and B. A similar effect also occurred in the lattice simulations. There it was found that the energy of the lowest state was always the same in both a 2×2 and 3×3 description, providing

A or B had the lowest energy of the three possible partitions. In addition the energy of the second state was, in most cases, more or less the same – the largest difference occurring with the tetrahedral geometry.

A.2: The 3×3 extension of the f-model of Subsec. 5.5.2

The f-model of Subsec. 5.5.2, by incorporating multiquark interactions, had the good feature that, when fitting the data (E_0, E_1) for a given square, only a single $\bar{f}$ was necessary to get a reasonable fit to both energies — see Fig. 5.7. Of course, $\bar{f}$ was dependent on the size of the square, but a reasonable parametrization was

$$f(Ia) = \exp(-b_s k_f S) \quad \text{(Version Ia)}, \tag{A.5}$$

where S is the area of the square and $k_f \approx 0.5$. Earlier in Eq. 5.43 a slightly different notation was used. The original hope was that, with k_f determined from the squares and nearby rectangles, the model would automatically also fit other geometries with S being the "appropriate" area contained by the four quarks. When the four quarks lie in a plane, the definition of S is clear. However, in non-planar cases the situation is more complicated. One possibility is to simply take S to be the average of the sum of the four triangular areas defined by the positions of the four quarks *i.e.* the faces of the tetrahedon. For example, in the notation of Eq. A.1, the appropriate area $S(AB)$ for f is

$$S(AB) = 0.5[S(431) + S(432) + S(123) + S(124)], \tag{A.6}$$

where $S(ijk)$ is the area of the triangle with corners at i, j and k. For planar geometries this simply reduces to the expected area, but for non-planar cases this is only an approximation to $S(AB)$ – a more correct area being one that is not necessarily a combination of planar areas but of curved surfaces with minimum areas. These possibilities are discussed in Ref. [165]. It would be feasible to incorporate this refinement here, since only a few ($\approx$ 50) such areas are needed for the geometries in Fig. 5.10. However, for a general situation, in which the positions of the quarks are integrated over, it would become impractical to use the exact value of $S(AB)$, since the expression for the minimum area itself involves a double integration. In contrast, the area used in Eq. A.6 is an algebraic expression and is, therefore, more readily evaluated for any geometry. The above model will be referred to as Version Ia.

This model Ia has only the one free parameter k_f in f(Ia) of Eq. A.5. Another possibility with additional parameters f_0, k_P is

$$f(\text{Ib}) = f_0 \exp(-b_s k_f S + \sqrt{b_s} k_P P) \quad \text{(Version Ib)}, \tag{A.7}$$

where P is the perimeter bounding S. This form has been used in Refs. [165]. However, as shown in Ref. [49], this reduces in the continuum limit to the same as Version Ia — the differences at $\beta = 2.4$ being mainly due to lattice artefacts.

Unfortunately, in the 2×2 version both of these models have the feature that, for regular tetrahedra, they are unable to reproduce a degenerate ground state with a *non-zero* energy, since the two eigenvalues are

$$E_0 = -\frac{f/2}{1+f/2}[V_{CC} - V_{AA}] \text{ and } E_1 = \frac{f/2}{1-f/2}[V_{CC} - V_{AA}], \tag{A.8}$$

where, in the notation of Fig. 5.10, $V_{AA} = v_{13} + v_{24}$ and $V_{CC} = v_{14} + v_{23}$ and so for regular tetrahedra $V_{CC} = V_{AA}$, giving $E_0 = E_1 = 0$.

Prior to the work on tetrahedra the geometries considered had, at most, two of the three possible partitions being degenerate in energy (*e.g.* for squares) — see Subsec. 5.5.2. In these cases, it is found that the lattice energies E_0 and E_1 are essentially the same for the three-basis-state calculation $(A + B + C)$ and those two-basis-state calculations $(A + B,\ A + C$ and effectively $B + C)$ which involve the basis state with the *lowest* unperturbed energy. This is one of the reasons why the 2×2 version of the f-model in Eq. 5.38 was quite successful for a qualitative understanding of these cases. However, for tetrahedra and the neighbouring geometries calculated in Subsec. 5.6, it now seems plausible to extend the f-model to the corresponding 3×3 version of

$$[\mathbf{V}(f) - \lambda_i(f)\mathbf{N}(f)]\,\Psi_i = 0, \tag{A.9}$$

in which

$$\mathbf{N}(f) = \begin{pmatrix} 1 & f/2 & f'/2 \\ f/2 & 1 & -f''/2 \\ f'/2 & -f''/2 & 1 \end{pmatrix} \quad \text{and} \tag{A.10}$$

$$\mathbf{V}(f) = \begin{pmatrix} v_{13}+v_{24} & fV_{AB} & f'V_{AC} \\ fV_{BA} & v_{14}+v_{23} & -f''V_{BC} \\ f'V_{CA} & -f''V_{CB} & v_{12}+v_{34} \end{pmatrix}, \tag{A.11}$$

where the negative sign in the BC matrix elements is of the same origin as the one in Eqs. 5.51 and A.4.

This apparently leads to the need for two more factors f', f'' defined by

$$\langle A|C\rangle = -f'/2 \text{ and } \langle B|C\rangle = -f''/2. \tag{A.12}$$

However, with the parametrizations of f as in Eqs. A.5 or A.7 and the definition of S as in Eq. A.6, it is seen that $f' = f'' = f$, since S is simply proportional to the area of the faces of the tetrahedron defined by the four quark positions and is *independent* of the state combination used. Therefore, the 3×3 model has *for all 4-quark geometries* a form where the N and V matrices are

$$\mathbf{N}(f) = \begin{pmatrix} 1 & f/2 & f/2 \\ f/2 & 1 & -f/2 \\ f/2 & -f/2 & 1 \end{pmatrix} \text{ and} \tag{A.13}$$

$$\mathbf{V}(f) = \begin{pmatrix} v_{AA} & fV_{AB} & fV_{AC} \\ fV_{BA} & v_{BB} & -fV_{BC} \\ fV_{CA} & -fV_{CB} & v_{CC} \end{pmatrix}. \tag{A.14}$$

This extension from 2×2 to 3×3 has the good feature that all three basis states are now treated on an equal footing. This is convenient when considering some general four-quark geometry, since it is then not necessary to choose some favoured 2×2 basis, which could well change as the geometry develops from one form to another. In the weak coupling limit (*i.e.* $f\to 1$) the 3×3 matrix in Eq. A.13 becomes the singular matrix in Eq. A.4. However, in this limit, each of the 2×2 matrices corresponding to the three possible partitions A+B, A+C and B+C gives the same results. Away from weak coupling the 3×3 matrix is no longer singular, but now the three possible 2×2 partitions do not necessarily give the same results. In addition to this general problem as $f \to 1$, there are also the following more specific unpleasant features:

(1) For regular tetrahedra all eigenvalues are *zero* as in the 2×2 models. The reason for this is clear. There is only one energy scale in the model, since all the v_{ij} are the same. Therefore, there can not be any excitations.
(2) For a linear geometry, since the "appropriate" area as defined by Eq. A.6 vanishes, we get $f = 1$ *i.e.* we are back to the weak coupling limit and a singular matrix.
(3) For squares the model gives $E_1 = -E_0$, whereas the predictions of the 2×2 version in Eqs. A.8 seem to be nearer the lattice data.
(4) The differences between using the various combinations of the three partitions are often considerably larger than in the corresponding lattice calculation.

The most glaring problem is the fact that the 3×3 model for the tetrahedron gives three degenerate states with zero energy, because there is only one scale in the model. It is, therefore, necessary to introduce a second energy scale. However, any improvements in the model have very limited choices, since there are only two different matrix elements involved — the diagonal ones all equal to $-E$ and the off-diagonal ones all equal to $\pm 0.5fE$. Therefore, the most general modifications are to change the diagonal matrix elements to $d_1 - E$ and the off-diagonal ones to $\pm 0.5f(d_2 - E)$. This results in the eigenvalues

$$E_0 = E_1 = \frac{d_1 + 0.5fd_2}{1 + 0.5f} \quad \text{and} \quad E_2 = \frac{d_1 - fd_2}{1 - f}. \tag{A.15}$$

At first sight it may appear that there is sufficient information to now extract the new parameters $d_{1,2}$, since f can be estimated using the parameters (assumed to be universal) from other geometries — thus leaving two equations for $E_{0,1}$ and E_2 and the two unknowns $d_{1,2}$. However, as said before, this is too much to demand from the f-model, since in the lattice calculation the third basis state in the complete A+B+C basis generally plays a minor rôle in determining the values of $E_{0,1}$ and, therefore, the third eigenvalue is presumably dominated by an excitation of the gluon field. Furthermore, it is of interest to see that a similar feature now arises with E_2 in Eq. A.15, since this third state is removed in the weak coupling limit *i.e.* $E_2 \to \infty$ as $f \to 1$. However, in its present form, the f-model is only expressed in terms of the *lowest* energy gluon configurations, since the gluon field is not explicitly in its formulation, but only appears **implicitly** in the form of the two-quark potentials and the f-factors. But already at this stage we see from the behaviour of E_2 that the effect of *excited* gluon states seem to be playing a rôle — the topic of the next subsection when the model is further extended to a 6×6 version. In view of this, no quantitative attempt should be made to identify the second excited state emerging from the lattice calculation as E_2 in this 3×3 version of the f-model.

In Ref. [102] it was shown that the *two*-state model of Eqs. 5.38 with the overlap factor $f=1$ agreed with perturbation theory up to fourth order in the quark–gluon coupling [*i.e.* to $O(\alpha^2)$] and gave $E_{0,1}$=0 for tetrahedra. Therefore, the non-zero lattice results for small tetrahedra must be of $O(\alpha^3)$ at least. Another aspect of this special situation for tetrahedra is also seen — when extracting or interpreting the value of E_1 — by the need for the third basis state both in the lattice calculation and in the f-model, since in comparison with Eq. A.15 the two-basis-state version gives

$$E_0=\frac{d_1+0.5fd_2}{1+0.5f} \quad\text{and}\quad E_1=\frac{d_1-0.5fd_2}{1-0.5f} \tag{A.16}$$

i.e. both the two- and three- basis-state models have the same ground state, but the former does not show the $E_0=E_1$ degeneracy.

The expressions in Eq. A.15 are not particularly useful unless there is a model for the parameters $d_{1,2}$. However, since it is not the purpose at this stage to make a comprehensive study of models covering all the 4-quark geometries considered in earlier works [52, 86, 87, 88], only a few general remarks will be made here for the tetrahedron geometry. Models for the $d_{1,2}$ need extensions of the potential in Eq. 5.31, so that for the tetrahedron *two* energy scales arise. Here several ways of achieving this goal are suggested:

The effect of an isoscalar two-quark potential.
As discussed in Ref. [87], an isoscalar potential w_{ij} can be introduced into V_{ij} — still ensuring $V_{ij}=v_{ij}$ for a colour singlet two-quark system — by extending the form in Eq. 5.6 to

$$V_{ij}=-\frac{1}{3}\tau_i\cdot\tau_j\,(v_{ij}-w_{ij})+w_{ij}. \tag{A.17}$$

In this case, $d_1=d_2=4w$, since all of the w_{ij} are now equal to w and results in $E_0=E_1=E_2=4w$. Therefore, here w takes on values that range from –0.0035 to –0.0070 *i.e.* they have values much smaller than the corresponding $v_{ij} = v$ in Eq. 5.31. A similar feature was found in Ref. [87], when the form in Eq. A.17 was introduced to improve the model fit for squares and rectangles. However, as shown in Ref. [102], in perturbation theory all terms of $O(\alpha^2)$ are included in the two state model of Eq. 5.38 with $f = 1$. Therefore, in the weak coupling limit w_{ij} must be of $O(\alpha^3)$ at least.

The effect of a three- or four-body potential.
The f factor is itself a four-body operator. However, it is conceivable that additional multiquark effects arise. Some perturbative possibilities are discussed in Ref. [102]. There it is shown that all three-quark terms arising from three gluon vertices always vanish, but that the four-gluon vertex can contribute to 2-, 3- and 4-quark terms at $O(\alpha^3)$. However, in the tetrahedral case ($r = d$), cancellations result in this particular 4-quark term also vanishing.

The effect of non-interacting three gluon exchange processes.
These are also discussed qualitatively in Ref. [102] and contribute at $O(\alpha^3)$ to 2-, 3- and 4-quark potentials.

The effect of two quark potentials where the gluon field is excited.
The first excited state $[V^*(r)]$ of the two-quark potential $V(r)$ is approximately given by $V^*(r) \approx V(r) + \pi/r$ — see for example Refs. [90, 91]. Therefore, if a fourth state, based on such an excited state, is introduced into the model, it will give attraction in the ground state, since it is higher in energy than the three degenerate basis states so far considered. Furthermore, as the size of the tetrahedron increases this fourth state will approach the other three states, so that the attraction felt in the ground state will increase — a trend seen in the tetrahedron results for $E_{0,1}$ in Fig. 5.11. This possibility will be discussed more in the 6×6 extension below.

The above isoscalar potential option now offers a reason for $E_0 = E_1 \neq 0$. But, unfortunately, E_2 is still equal to $E_{0,1}$ since $d_1 = d_2$. However, there is no reason to expect any three or four body forces to also be purely isoscalars. In this case, their contributions to d_1 and d_2 could be different and through the presence of the $(1 - f)$ factor in Eq. A.15 any estimates of E_2 could be very model dependent.

A.3: The 6×6 *extension of the f-model of Subsec. 5.5.2*

The above models both have trouble in describing regular tetrahedra. In Refs. [108, 109] an attempt is made to overcome this problem. An interesting feature of the regular tetrahedron data is that the lowest state becomes *more* bound as the tetrahedron increases in size with the magnitude of E_0 increasing from –0.0202(8) to –0.028(3) as the d^3 cube containing the tetrahedron increases from $d = 2$ to $d = 4$. This is opposite to what happens with squares, where the magnitude of E_0 decreases from $-0.0572(4)$ to $-0.047(3)$ as d increases from 2 to 5. This indicates that there could be coupling to some higher state(s) that becomes more effective as the size increases and suggests that these higher states contain gluon excitation with respect to the A, B, C configurations. Therefore, the 3×3 model in the previous subsection is further extended to a 6×6 model by adding three more states A^*, B^*, C^*, where in analogy with Eq. A.1

$$A^* = (Q_1Q_3)_{E_u}(Q_2Q_4)_{E_u} \quad etc.. \tag{A.18}$$

Here $(Q_1Q_3)_{E_u}$ denotes a state where the gluon field is excited to the lowest state with the symmetry of the E_u representation of the lattice symmetry group D_{4h}. Because it is an odd parity excitation, A^*, B^*, C^* must contain two such states in order to have the same parity as A, B, C. The excitation energy of an E_u state over its ground state (A_{1g}) counterpart is $\approx \pi/r$ for two quarks a distance r apart. As r increases this excitation energy decreases making the effect of the A^*, B^*, C^* states more important, leading to the effect mentioned above. Here we have assumed that these states arise from a combination of excited states with E_u. However, it is possible that they involve other excitations, *e.g.*

$$A^* = (Q_1Q_3)_{A'_{1g}}(Q_2Q_4) \quad etc., \tag{A.19}$$

where the A'_{1g} state is a gluonic excitation with the *same* quantum numbers as the ground state (A_{1g}). For this case the following formalism would be essentially the same. Another possibility, which is not considered here, is that the relevant excitations are flux configurations where all four quarks, instead of two, are involved in forming a colour singlet. In the strong coupling approximation such states would reduce to two-body singlets due to Casimir scaling of the string tensions, namely, the string tension for a higher representation would be more than double the value of the fundamental string tension, thus preventing junctions of two strings in the fundamental and one in the higher representation. This would happen both in SU(2) and SU(3), the only exception being the unexcited C state in SU(3), which would involve an antitriplet string — see Eq. A.3.

For the regular tetrahedral case, in addition to $f = f' = f''$, there are now several new matrix elements that need to be discussed for $\mathbf{N(f)}$ and $\mathbf{V(f)}$ (Eqs. A.9–A.11):

a) With the inclusion of the A^*, B^*, C^* states and the antisymmetry condition $|A^*\rangle + |B^*\rangle + |C^*\rangle = 0$ analogous to Eq. A.2, there are now two more gluon overlap functions $f^{a,c}$ defined as

$$\langle A^*|B^*\rangle = \langle A^*|C^*\rangle = \langle B^*|C^*\rangle = -f^c/2 \text{ and}$$

$$\langle A^*|B\rangle = \langle A^*|C\rangle = .. \text{ etc. } .. = -f^a/2. \tag{A.20}$$

Here it is assumed that $f^{a,c}$ are both dependent on S as defined in Eq. A.6. Since f^c involves only the excited states, it is reasonable to expect it has a form similar to f in Eq. A.5 *i.e.*

$$f^c = \exp(-b_s k_c S). \tag{A.21}$$

b) By orthogonality $\langle A|A^*\rangle = \langle B|B^*\rangle = \langle C|C^*\rangle = 0$

c) In the weak coupling limit, from the $|A^*\rangle + |B^*\rangle + |C^*\rangle = 0$ condition, we expect $\langle A|B^*\rangle = \langle B|C^*\rangle = ... = 0$ at small distances. To take this into account f^a is parametrized as

$$f^a = (f_1^a + b_s f_2^a S)\exp(-b_s k_a S). \tag{A.22}$$

If all three parameters f_1^a, f_2^a, k_a are varied, it is found that f_1^a is always consistent with zero — as expected from the above condition that $\langle A|B^*\rangle = ... = 0$. Therefore, usually f_1^a is fixed at zero.

d) For the potential matrix $\mathbf{V(f)}$ the diagonal matrix elements, after the lowest energy V_{DD} amongst the basis states is removed, are

$$\langle A^*|V - V_{DD}|A^*\rangle = v_{13}^* + v_{24}^* - V_{DD}, \text{ etc.},$$

where $V_{DD} = \min[V_{AA} = v_{13} + v_{24}, V_{BB} = v_{14} + v_{23}, V_{CC} = v_{12} + v_{34}]$ and $v_{ij}^* \approx v_{ij} + \pi/r$ is the potential of the excited E_u state, which is a quantity also measured on the lattice along with the four-quark energies.

In the special case of regular tetrahedra $V_{DD} = V_{AA} = V_{BB} = V_{CC}$ and $\mathbf{V}$ reduces to the form

$$\mathbf{V} = \left[\begin{array}{ccc|ccc} V_{AA} & -fV_{AA}/2 & -fV_{AA}/2 & 0 & -f^aV_a/2 & -f^aV_a/2 \\ -fV_{AA}/2 & V_{AA} & -fV_{AA}/2 & -f^aV_a/2 & 0 & -f^aV_a/2 \\ -fV_{AA}/2 & -fV_{AA}/2 & V_{AA} & -f^aV_a/2 & -f^aV_a/2 & 0 \\ \hline 0 & -f^aV_a/2 & -f^aV_a/2 & V_b & -f^cV_c/2 & -f^cV_c/2 \\ -f^aV_a/2 & 0 & -f^aV_a/2 & -f^cV_c/2 & V_b & -f^cV_c/2 \\ -f^aV_a/2 & -f^aV_a/2 & 0 & -f^cV_c/2 & -f^cV_c/2 & V_b \end{array}\right] \tag{A.23}$$

where V_a , V_b , V_c can be expressed in terms of V_{AA} and $v^*(ij)$ plus some fine tuning parameters. As with all geometries

$$\mathbf{N} = \left[\begin{array}{ccc|ccc} 1 & -f/2 & -f/2 & 0 & -f^a/2 & -f^a/2 \\ -f/2 & 1 & -f/2 & -f^a/2 & 0 & -f^a/2 \\ -f/2 & -f/2 & 1 & -f^a/2 & -f^a/2 & 0 \\ \hline 0 & -f^a/2 & -f^a/2 & 1 & -f^c/2 & -f^c/2 \\ -f^a/2 & 0 & -f^a/2 & -f^c/2 & 1 & -f^c/2 \\ -f^a/2 & -f^a/2 & 0 & -f^c/2 & -f^c/2 & 1 \end{array}\right] . \tag{A.24}$$

The full 6×6 matrix $[\mathbf{V} - (E + V_{AA})\mathbf{N}]$ now factorizes into three 2×2 matrices, two of which are identical – giving the observed degeneracy. These two matrices have determinants of the form

$$\begin{vmatrix} -E(1 + f/2) & -f^a(E - V_a)/2 \\ -f^a(E - V_a)/2 & -E(1 + f^c/2) + V_b + f^cV_c/2 \end{vmatrix} = 0, \tag{A.25}$$

whereas the third 2×2 matrix has the determinant

$$\begin{vmatrix} -E(1 - f) & f^a(E - V_a) \\ f^a(E - V_a) & -E(1 - f^c) + V_b - f^cV_c \end{vmatrix} = 0. \tag{A.26}$$

In this case the problem reduces to solving two quadratic equations for E. However, away from the regular tetrahedron the complete $6{\times}6$ matrix needs to be treated directly.

By fitting simultaneously the energies E_0 and E_1 from the lattice results for the geometries in Fig. 5.10 an interquark potential model can be constructed that is able to explain, on the average, these energies. The full model utilized 6 basis states A, B, C, A^*, B^*, C^* and in its most general form has eight parameters. However, in practice, only 3 of these (k_f, k_a, f_2^a) need be considered as completely free when fitting the data.

The parameters that are of most interest are those connected with the ranges of the various interactions, namely, k_f and k_a. Here "range" is defined as $r_{f,a,c} = \sqrt{1/b_s k_{f,a,c}}$. In the 2×2 version, where k_a is effectively infinite, we get k_f(Ia)=0.57(1) *i.e.* r_f(Ia) = 5.0 in lattice units of 0.12 fm. However, when the excited states A^*, B^*, C^* are introduced, the interaction between the basic states A, B, C decreases by raising k_f to 1.51 giving $r_f = 3.1$. But at the same time this loss of binding by the direct interaction between A, B, C is compensated by their coupling to the A^*, B^*, C^* states. This coupling in Eq. A.22 is found to have about the *same* range $r_a = 5.1$ as r_f(Ia) in Eq. A.5, whereas *the direct interaction between the A^*, B^*, C^* states seems, in all fits, to be satisfied with simply a two-quark description without any four-quark correction (i.e. k_c=0) in Eq. A.21.* The observation that $r_f(\text{Ia}) \approx r_a$ suggests that the energy density has a range dictated by the longest range available — namely r_a. Therefore, when the A^*, B^*, C^* states are not explicitly present, as in Model Ia, the only available range r_f(Ia) has to simulate the rôle of r_a. In the binding energies the contributions from the A^*, B^*, C^* states rapidly dominate over those from the A, B, C states. For example, with squares of side R, the A, B, C states contribute only 85, 40, 10% to the binding energy for R=2,4,6 respectively. Of course, at the largest distances ($\approx$ 0.7 fm) the quenched approximation is expected to break down and the rôle of quark-pair creation to become important.

Appendix B: Extension of the f-Model to $[(Q\bar{q})(Q\bar{q})]$ Systems

When only two of the four quarks are static the corresponding matrices for $Q(\mathbf{r_1})Q(\mathbf{r_2})\bar{q}(\mathbf{r_3})\bar{q}(\mathbf{r_4})$ can be expressed in a similar form but where the matrix elements are now *integrals* over the positions of the two light antiquarks. In the notation of Fig. 5.4 we consider basis state A to be the one realised as two separate heavy-light mesons — $[Q_1\bar{q}_3]$ and $[Q_2\bar{q}_4]$ — when the distance $\mathbf{R} = \mathbf{r_1} - \mathbf{r_2}$ between the two heavy quarks becomes large. In this state the convenient coordinates are then $\mathbf{s_1} = \mathbf{r_3} - \mathbf{r_1}$ and $\mathbf{s_2} = \mathbf{r_4} - \mathbf{r_2}$, whereas for the other partition B the convenient coordinates are $\mathbf{t_1} = \mathbf{r_3} - \mathbf{r_2} = \mathbf{s_1} + \mathbf{R}$ and $\mathbf{t_2} = \mathbf{r_4} - \mathbf{r_1} = \mathbf{s_2} - \mathbf{R}$.

To describe this system in terms of an Effective Potential Theory the three ingredients quoted in Subsec. 5.1.2.2 are needed — see Ref. [60] for more details of the specific example now to be described:

A wave equation.
For simplicity, the system is considered to be non-relativistic resulting in a Schrödinger-like equation

$$|\mathbf{K}'(R) + \mathbf{V}'(R) - E(4, R)\mathbf{N}'(R)|\psi = 0 \tag{B.1}$$

i.e. a Resonating Group equation as discussed by Oka and Yazaki in Chapter 6 of Ref. [36] and also Ref. [37]. This is a generalisation of Eq. 5.37 to non-static quarks and can be solved using a variational wave function taken to have the form [140]

$$\psi(\mathbf{r_i}, f) = f^{1/2}(\mathbf{r_1}, \mathbf{r_2}, \mathbf{r_3}, \mathbf{r_4}) \sum_{i=1}^{N_4} \exp(-\tilde{\mathbf{X}}\mathbf{M}_i\mathbf{X}), \tag{B.2}$$

where $\tilde{\mathbf{X}} = (\mathbf{s_1},\ \mathbf{s_2},\ \mathbf{R})$ and each matrix $\mathbf{M_i}$ has the form

$$\mathbf{M}_i = \frac{1}{2} \begin{pmatrix} a_i & b_i & c_i \\ b_i & d_i & e_i \\ c_i & e_i & g_i \end{pmatrix}. \tag{B.3}$$

Since the present problem considers the masses of the light quarks to be equal, it is sufficient to use a simplified form of $\mathbf{M_i}$ with $b_i = 0,\ d_i = a_i$ and $e_i = c_i$. This is not necessary, but it is expected to be the dominant term in such a symmetric case. Already for $N_4 = 2$, this wave function is indeed adequate for giving sufficiently accurate four-quark binding energies. Even this choice involves five free parameters $(a_1,\ c_1,\ a_2,\ c_2,\ g_2)$ in the variation – with g_1 being fixed at unity to set the overall normalisation. In what follows the positions of the light quarks are integrated over leaving matrix elements that are functions of $\mathbf{R}$. In order to achieve this in any practical way it is necessary to have a form for $f(\mathbf{r_1}, \mathbf{r_2}, \mathbf{r_3}, \mathbf{r_4})$ that has a simple spatial dependence. Here the very symmetrical form in Eq. 5.44, defined by a *single* parameter k_f, is used *i.e.*

$$f = \exp\left[- k_f b_s \sum_{i<j} r_{ij}^2 \right] \tag{B.4}$$

(N.B. Here $k_f = k/6$, where k was defined in Eq. 5.44.) It should be emphasised that this form of f is purely for numerical simplicity leading to analytical expressions for all matrix elements. As in the static case k_f is a free parameter, which should be adjusted to fit the four-quark lattice energies.

The wave function in Eq. B.2 is used for both states A and B. This is an approximation that appears to work well for the $Q^2\bar{q}^2$ system, since A and B are similar in structure for the R values of interest here.

An interquark potential.
This enters in three different contexts:
1) As $v(13)$, $v(24)$, $v(14)$, $v(23)$ in the $V_{Q\bar{q}}$ potential. This is taken to be of the standard form in Eq. 5.6, namely

$$aV(2,r) = -\frac{0.309(38)}{r/a} + 0.1649(36)r/a + 0.629(25), \tag{B.5}$$

which gives a string energy of (445 MeV)^2 for $a = 0.18$ fm. This was obtained by fitting $V_{Q\bar{Q}}$ generated on a $16^3 \times 24$ lattice. Here the emphasis was to get a good fit over the important range of $r \sim (2-4)a$ and is in contrast to the potential in Ref. [51], which was designed to extract the string tension at large values of r.
2) As $v(34)$ for the $V_{\bar{q}\bar{q}}$ potential. Here it is assumed to also be of the form in Eq. B.5.
3) As $v(12)$ for the V_{QQ} potential. This was calculated from the same gauge configurations as the four-quark energies. In this case there was no need to fit V_{QQ} with a function of R, since it is only ever needed at discrete values of R – the ones for which the four-quark energies are calculated.

An effective quark mass $\mathbf{m_q}$.
In this case m_q can be determined beforehand by carrying out an EPT analysis of the correponding *two-body* energies in Fig. 5.13. Using the potential in Eq. B.5, a value of $m_q \approx 400$ MeV is able to give a good fit to the spin-averaged energies with $L = 0$, 1, 2 and 3. However, the results are not strongly dependent on m_q — see Fig. 5.18. A disturbing feature of this

result is that $m_q \ll 1$ GeV indicating the need for a relativistic approach — as discussed in Sec. 5.12.

In Eq. B.1 the normalisation matrix $\mathbf{N}'(R, k_f)$ — a generalisation of $\mathbf{N}(f)$ in Eq. 5.38 to non-static quarks in SU(3) — can now be written as

$$\mathbf{N}'(R,\ k_f) = \begin{pmatrix} N(R,\ 0) & \frac{1}{3}N(R,\ k_f) \\ \frac{1}{3}N(R,\ k_f) & N(R,\ 0) \end{pmatrix}, \tag{B.6}$$

where, after integrating over $\mathbf{s_1}$ and $\mathbf{s_2}$, $N(R,\ k_f)$ can be expressed as a sum of terms of the form

$$\frac{\pi^3}{(aX)^{3/2}} \exp\left[-(Z - \frac{Y^2}{X})R^2\right], \tag{B.7}$$

where $a = 0.5(a_i + a_j) + 3k_f$, $c = 0.5(c_i \pm c_j) + 2k_f$, $d = 0.5(c_i \pm c_j) - 2k_f$, $g = 0.5(g_i + g_j) + 4k_f$, $X = a - k_f^2/a$, $Y = c + k_f d/a$ and $Z = g - d^2/a$.

Since two of the quarks are not static there is now also a kinetic energy matrix in Eq. B.1, namely,

$$\mathbf{K}'(R,\ k_f) = \begin{pmatrix} K_3(R,0) + K_4(R,0) & \frac{1}{3}[K_3(R,k_f) + K_4(R,k_f)] \\ \frac{1}{3}[K_3(R,k_f) + K_4(R,k_f)] & K_3(R,0) + K_4(R,0) \end{pmatrix}, \tag{B.8}$$

where, for example,

$$K_3(R,k_f) = \int d^3 s_1 d^3 s_2 \psi^\star(k_f) \left[-\frac{d^2}{2m_q dr_3^2}\right] \psi(k_f). \tag{B.9}$$

Again these integrals can be expressed in forms similar to that in Eq. B.7.

Finally, the potential matrix — a generalisation of $\mathbf{V}(f)$ in Eq. 5.38 to non-static quarks — has the form

$$\mathbf{V}'(R,\ k_f) = \begin{pmatrix} \langle v(13), 0\rangle + \langle v(24), 0\rangle & \langle V_{AB}, k_f\rangle \\ \langle V_{AB}, k_f\rangle & \langle v(14), 0\rangle + \langle v(23), 0\rangle \end{pmatrix}. \tag{B.10}$$

Here

$$\langle V_{AB}, k_f\rangle = \langle V_{Q\bar{q}}\rangle - \langle V_{\bar{q}\bar{q}}\rangle - \langle V_{QQ}\rangle,$$

where

$$\langle V_{Q\bar{q}}\rangle = \frac{1}{3}\left[\langle v(13),k_f\rangle + \langle v(24),k_f\rangle + \langle v(14),k_f\rangle + \langle v(23),k_f\rangle\right],$$
$$\langle V_{\bar{q}\bar{q}}\rangle = \frac{1}{3}\langle v(34),k_f\rangle \quad \text{and} \quad \langle V_{QQ}\rangle = \frac{1}{3}N(R,k_f)V(2,R). \tag{B.11}$$

Here $N(R, k_f)$ is defined in Eq. B.6, $V(2, R)$ is the potential between the two heavy quarks and, for example,

$$\langle v(13), k_f\rangle = \int d^3s_1 d^3s_2 \psi^*(k_f)V(s_1)\psi(k_f). \tag{B.12}$$

For potentials of the form in Eq. 5.6, these integrals can be expressed in terms of Error functions. The energy $E(4, R, k_f)$ of the two heavy-light meson system is then obtained by diagonalising Eq. B.1. Since this equation is a 2×2 determinant, a prediction can also be made for an excited state $E^*(4, R, k_f)$ and the corresponding binding energy $B^*(4, R)$.

Bibliography

[1] *e.g.* R. K. Ellis, W.J. Stirling and B. R. Webber, "QCD and collider physics", (Cambridge University Press, 1996);
E. Leader, "An introduction to gauge theories and modern particle physics: Vol. 2 CP-violation, QCD and hard processes", (Cambridge University Press, 1996);
T. Muta, "Foundations of QCD: An introduction to Perturbative Methods in Gauge Theories (2nd Edition)", World Scientific Lecture Notes in Physics – Vol. 57 (World Scientific, 1998);
Y. L. Dokshitzer, Phil. Trans. Roy. Soc. Lond. **A359**, 309 (2001) hep-ph/0106348;
P. Hoyer, Nucl. Phys. **A711**, 3 (2002), hep-ph/0208181;
S. Capitani, Phys. Rept. **382**, 113 (2003), hep-lat/0211036

[2] M. Lüscher, Annales Henri Poincare **4**, S197 (2003), hep-ph/0211220;
M. Creutz, Nucl. Phys. Proc. Suppl. **94**, 219 (2001), Latt00 and hep-lat/0010047; "The Early Days of Lattice Gauge Theory", hep-lat/0306024;
C. Davies, "Lattice QCD", Lectures given at 55th Scottish Universities Summer School in Physics: Heavy Flavor Physics, St. Andrews, Scotland, (Institute of Physics 2002, eds. C.T.H. Davies and S.M. Playfer, 2001) p.105, hep-ph/0205181;
G. Münster and M. Walzl, "Lattice Gauge Theory — a Short Primer", Published in Zuoz 2000, Phenomenology of gauge interactions p. 127, hep-lat/0012005;

T. DeGrand, "Lattice QCD at the end of 2003", hep-ph/0312241 to be published in Review for Int. J. Mod. Phys. A. (Worldscience)

[3] S. Aoki *et al.* Nucl. Phys. Proc. Suppl. **106**, 230 (2002)

[4] M. G. Alford, T. R. Klassen, G.P. Lepage, Phys. Rev. **D58**, 034503 (1998), hep-lat/9712005;
G.P. Lepage, Nucl. Phys. Proc. Suppl. **47**, 3 (1996), also Latt95 and hep-lat/9510049

[5] B. Sheikholeslami and R. Wohlert, Nucl. Phys. **B259**, 572 (1985)

[6] M. Lüscher, S. Sint, R. Sommer, P. Weisz and U. Wolff, Nucl. Phys. **B491**, 323 (1997), hep-lat/9609035

[7] A. S. Kronfeld, "Uses of Effective Field Theory in Lattice QCD" in: At the Frontiers of Particle Physics: Handbook of QCD, (ed. M. Shifman) **4** Chp. 39, (World Scientific, Singapore, 2002), hep-lat/0205021

[8] A. S. Kronfeld, eConf C020620:FRBT05 (2002), hep-ph/0209231

[9] A. C. Irving, Nucl. Phys. Proc. Suppl. **119**, 341 (2003), Latt02 and hep-lat/0208065

[10] qq+q Collaboration, F. Farchioni, C. Gebert, I. Montvay and L. Scorzato, "On the price of light quarks", hep-lat/0209142

[11] M. Creutz, "Quarks, Gluons, and Lattices" (Cambridge University, Cambridge, UK, 1983)

[12] I. Montvay and G. Münster, "Quantum Fields on a Lattice" (Cambridge Monographs on Mathematical Physics, CUP 1994)

[13] H. J. Rothe, "Lattice Gauge Theories", World Scientific Lecture Notes in Physics - Vol. 59 (World Scientific Publishing Co., Singapore, 1997)

[14] C. DeTar and S. Gottlieb, Physics Today, February 2004, 45

[15] H. Neuberger, "Lattice Field Theory: past, present and future", hep-ph/0402148

[16] K. Symanzik, "Recent Developments in Gauge Theories", edited by G.'t Hooft *et al.* (Plenum, New York, 1980): Nucl. Phys. **B226**, 187, 205 (1983).

[17] J. Gasser and H. Leutwyler, Ann. Phys. **158**, 142 (1984): Nucl. Phys. **B250**, 465 (1985)

[18] A. W. Thomas, Nucl. Phys. Proc. Suppl. **119**, 50 (2003), hep-lat/0208023.

[19] J. Soto, Quark Confinement and the Hadron Spectrum, Proceedings of the 5th International Conference, Gargnano, Italy, September 2002 (World Scientific 2003) p.227

[20] N. Brambilla, A. Pineda, J. Soto and A. Vario, Nucl. Phys. **B566**, 275 (2000);
A.Vario, Quark Confinement and the Hadron Spectrum, Proceedings of the 5th International Conference, Gargnano, Italy, September 2002 (World Scientific 2003) p.73

[21] U. van Kolck, Nucl. Phys. **A699**, 33 (2002);
U. van Kolck, L. J. Abu-Raddad and D. M. Cardamone, nucl-th/0205058;
D. Phillips, Czech. J. Phys. **52**, B49 (2002), nucl-th/0203040

[22] J. A. Oller, AIP Conf. Proc. **660**, 116 (2003), nucl-th/0207086

[23] S. Weinberg, Physica **A96**, 327 (1979); Nucl. Phys. **B363**, 3 (1991)
[24] S.C. Pieper and R. B. Wiringa, Ann. Rev. Nucl. Part. Sci. **51**, 53 (2001)
[25] N. Glendenning, "Compact Stars" (Springer, 1996)
[26] A. Akmal, V. Pandharipande and D. Ravenhall, Phys. Rev. **C 58**, 1804 (1998)
[27] R. Machleidt and I. Slaus, J. Phys. **G27**, R69 (2001), `nucl-th/0101056`; R. Machleidt, Nucl. Phys. **A689**, 11 (2001) `nucl-th/0009055` : Phys. Rev. **C63**, 024001 (2001), `nucl-th/0006014`
[28] J. L. Richardson, Phys. Lett. **82B**, 272 (1979)
[29] E. Eichten, K. Gottfried, T. Kinoshita, K. D. Lane and T. M. Tan, Phys. Rev. **D21**, 203 (1980);
E. Eichten and F. Feinberg, Phys. Rev. **D23**, 2724 (1981)
[30] R. K. Bhaduri and M. Brack, Phys. Rev. **D25**, 1443 (1982)
[31] M. Baker, J. S. Ball and F. Zachariasen, Phys. Rev. **D51**, 1968 (1995)
[32] P. R. Page, T. Goldman and J. N. Ginocchio, Phys. Rev. Lett. **86**, 204 (2001)
[33] M. Di Pierro and E. Eichten, Phys. Rev. **D64**, 114004 (2001)
[34] H. Crater, B. Liu and P. Van Alstine, "Two-Body Dirac Equations", `hep-ph/0306291`
[35] G. E. Brown and A. D. Jackson, "The Nucleon-Nucleon Interaction", (North–Holland Publishing Co., 1976)
[36] Editor W. Weise "Quarks and Nuclei", International Review of Nuclear Physics - Vol. **1** 1984 (World Scientific Publishing Co. Pte Ltd 1984)
[37] T. Sakai, J. Mori, A. J. Buchmann, K. Shimizu and K. Yazaki, Nucl. Phys. **A625**, 192 (1997), `nucl-th/9709054`
[38] M. Oka, "Baryon-Baryon Interaction in the Quark Cluster Model", `hep-ph/0306173`
[39] F. Coester, "From Light Nuclei to Nuclear Matter — The Rôle of Relativity", `nucl-th/0111025`;
F. Coester and W. N. Polyzou, "Relativistic Quantum Mechanics of Many Body Systems", `nucl-th/0102050`
[40] L. Ya. Glozman and D. O. Riska, Phys. Rep. **268**, 263 (1996), `hep-ph/9505422`: Nucl. Phys. **A603**, 326 (1996), erratum *ibid* **A620**, 510 (1997), `hep-ph/9509269`
[41] S. Chernyshev, M. A. Nowak, and I. Zahed, Phys. Rev. **D53**, 5176 (1996)
[42] H. Heiselberg and M. Hjorth-Jensen, Phys. Repts. **328**, 237 (2000)
[43] E. Shuryak and D. O. Riska, *private communication*
[44] D. H. Perkins, "An Introduction to High Energy Physics" (Addison-Wesley Pub. Co., 1972)
[45] S. Veseli and M.G. Olsson, Phys. Lett. **B383**, 109 (1996), `hep-ph/9606257`
[46] C. Rebbi, Phys. Repts. **12C**, 1 (1974)
[47] K. Wilson, Phys. Rev. **D10**, 2445 (1974)
[48] R. Sommer, Nucl. Phys. **B411**, 839 (1994), `hep-lat/9310022`.
[49] P. Pennanen, Phys. Rev. **D55**, 3958 (1997), `hep-lat/9608147`

[50] UKQCD Collaboration: C. R. Allton *et al.*, Phys. Rev. **D49**, 474 (1994), hep-lat/9309002

[51] R. G. Edwards, U. M. Heller and T. R. Klassen, Nucl. Phys. **B517**, 377 (1998), hep-lat/9711003

[52] A.M. Green, C. Michael and M.E. Sainio, Z. Phys. **C67**, 291 (1995), hep-lat/9404004

[53] A.M. Green, J. Lukkarinen, P. Pennanen and C. Michael, Phys. Rev. **D53**, 261 (1996), hep-lat/9508002

[54] N. Isgur and J.E. Paton, Phys. Rev. **D31**, 2910 (1985)

[55] M. Baker, J.S. Ball and F. Zachariasen, Phys. Rev. **D51**, 1968 (1995); Int. J. Mod. Phys. **A11**, 343 (1996)

[56] UKQCD Collaboration: C. Michael and J. Peisa, Phys. Rev. **D58**, 034506 (1998), hep-lat/9802015

[57] A.M. Green, J. Ignatius, M. Jahma and J. Koponen, work in progress

[58] A.M. Green, J. Koponen, P. Pennanen and C. Michael, Phys. Rev. **D65**, 014512 (2002), hep-lat/0105027

[59] A.M. Green, J. Koponen, P. Pennanen and C. Michael, Eur. Phys. J. **C28**, 79 (2003), hep-lat/0206015

[60] A. M. Green, J. Koponen and P. Pennanen, Phys. Rev. **D61**, 014014 (2000), hep-ph/9902249

[61] J. Koponen, work in progress

[62] C. Michael, Phys. Lett. **B232**, 247 (1989), Nucl. Phys. (Proc. Suppl) **B 17**, 59 (1990)

[63] B. Lucini and M. Teper, JHEP **0106**, 050 (2001);
B. Lucini, M. Teper and U. Wenger, "Features of SU(N) Gauge Theories", hep-lat/0309170

[64] M. Teper, Phys. Rev. **D59**, 014512 (1999);
B. Lucini and M. Teper, Phys. Rev. **D66**, 097502 (2002)

[65] F. Karsch, Nucl. Phys. **B205**, 285 (1982)

[66] G. Burgio *et al.*, Phys. Rev. **D67**, 114502 (2003), hep-lat/0303005 also Latt03 and hep-lat/0309058, hep-lat/0310036

[67] CP-PACS, M. Okamoto *et al.*, Phys. Rev. **D65**, 094508 (2002), hep-lat/0112020

[68] S. Hashimoto and M. Okamoto, Phys. Rev. **D67**, 114503 (2003), hep-lat/0302012;
S. Sakai and A. Nakamura, " Improved gauge action on an anisotropic lattice", hep-lat/0311020

[69] S. Aoki *et al.* (CP-PACS), Phys. Rev. Lett. **84**, 238 (2000), hep-lat/9904012

[70] D. B. Leinweber, A. W. Thomas, K. Tsushima and S. V. Wright, Phys. Rev. **D61**, 074502 (2000), hep-lat/9906027;
R. D. Young, D. B. Leinweber, A. W. Thomas and S. V. Wright, Phys. Rev. **D66**, 094507 (2002), hep-lat/0205017

[71] S. R. Beane and M. J. Savage, Phys. Lett. **B535**, 177 (2002), hep-lat/0202013

[72] S. R. Beane and M. J. Savage, Phys. Rev. **D67**, 054502 (2003), hep-lat/0210046
[73] P. Lacock and C. Michael, Phys. Rev. **D52**, 5213 (1995)
[74] UKQCD Collaboration: M. Foster and C. Michael, Phys. Rev. **D59**, 074503 (1999), hep-lat/9810021;
M. Foster, University of Liverpool PhD thesis 1998
[75] S. R. Beane and M. J. Savage, Nucl. Phys. **A717**, 91 (2003), nucl-th/0208021
[76] E. Epelbaum, Ulf-G. Meissner and W. Glöckle, Nucl. Phys. **A714**, 535 (2003), nucl-th/0207089 and nucl-th/0208040
[77] UKQCD Collaboration: C. Michael and P. Pennanen, Phys. Rev. **D60**, 054012 (1999), hep-lat/9901007
[78] M. Lüscher, Commun. Math. Phys. **105**, 153 (1986) and Nucl. Phys. **B354**, 531 (1991)
[79] S.R. Beane, P.F. Bedaque, A. Parreno and M.J. Savage, "Two nucleons on a Lattice", hep-lat/0312004
[80] P. Bedaque, "Aharonov-Bohm effect and nucleon-nucleon phase shifts on the lattice", nucl-th/0402051
[81] CP-PACS Collaboration: T. Yamazaki *et al.*, "I=2 $\pi\pi$ Scattering Phase Shift with two Flavors of $O(a)$ Improved Dynamical Quarks", hep-lat/0402025
[82] I. Wetzorke and F. Karsch, Nucl. Phys. Proc. Suppl. **119**, 278 (2003), Latt02 and hep-lat/0208029
[83] S. Sasaki, " Lattice study of exotic S=+1 baryon", hep-lat/0310014;
F. X. Lee *et al.*," A search for Pentaquarks on the Lattice", poster at Lattice 2003 in Tsukuba
[84] T. Nakano *et al.*, Phys. Rev. Lett. **91**, 012002 (2003)
[85] G. Bali, Phys. Rept. **343**, 1 (2001), hep-ph/0001312
[86] A.M. Green, C. Michael and J.E. Paton, Phys. Lett. **B280**, 11 (1992)
[87] A.M. Green, C. Michael and J.E. Paton, Nucl. Phys. **A554**, 701 (1993), hep-lat/9209019
[88] A.M. Green, C. Michael, J.E. Paton and M.E. Sainio, Int. J. Mod. Phys. **E2**, 479 (1993), hep-lat/9301006
[89] F. Lenz *et al.*, Ann. Phys. (N.Y.) **170**, 65 (1986);
K. Masutani, Nucl. Phys. **A468**, 593 (1987)
[90] S. Perantonis, A. Huntley and C. Michael, Nucl. Phys. **B326**, 544 (1989)
[91] S. Perantonis and C. Michael, Nucl. Phys. **B347**, 854 (1990)
[92] S.P. Booth *et al.*, Phys. Lett. **B275**, 424 (1992)
[93] M. Della Morte *et al.*, " Static quarks with improved statistical precision", Latt03 and hep-lat/0309080, hep-lat/0307021
[94] S. Hashimoto, Phys. Rev. **D50**, 4639 (1994)
[95] A. Hasenfratz and F. Knechtli, Phys. Rev. **D64**, 034504 (2001);
A. Hasenfratz, R. Hoffmann and F. Knechtli, Nucl. Phys. B (Proc. Suppl.) **106**, 418 (2002)
[96] K. Choi and W. Lee,"Penguin diagrams for the HYP staggered fermions",

Latt03 and hep-lat/0309070;
T. Bhattacharya *et al.*, "Calculating weak matrix elements using HYP staggered fermions", Latt03 and hep-lat/0309105;
S. Bilson-Thompson and W. Lee," Description and comparison of Fat7 and HYP fat links", hep-lat/0310056

[97] S. Ohta, M. Fukugita and A. Ukawa, Phys. Lett. **B173**, 15 (1986).

[98] A.M. Green and J.E. Paton, Nucl. Phys. **A492**, 595 (1989)

[99] C. Michael, Phys. Lett. **B283**, 103 (1992)

[100] M. B. Gavela *et al.*, Phys. Lett. **B82**, 431 (1979).

[101] O. Morimatsu, A. M. Green and J.E. Paton, Phys. Lett. **B258**, 257 (1991)

[102] J. T. A. Lang, J. E. Paton and A. M. Green, Phys. Lett. **B366**, 18 (1996), hep-ph/9508315

[103] O. Morimatsu, Nucl. Phys. **A505**, 655 (1989)

[104] B. Masud, J.E. Paton, A.M. Green and G.Q. Liu, Nucl. Phys. **A528**, 477 (1991)

[105] H. Matsuoka and D. Sivers, Phys. Rev. **D33**, 1441 (1987)

[106] C. Alexandrou, T. Karapiperis and O. Morimatsu, Nucl. Phys. **A518**, 723 (1990)

[107] A.M. Green, G.Q. Liu and S. Wycech, Nucl. Phys. **A509**, 687 (1990)

[108] A.M. Green and P. Pennanen, Phys. Lett. **B426**, 243 (1998), hep-lat/9709124

[109] A.M. Green and P. Pennanen, Phys. Rev. **C57**, 3384 (1998), hep-lat/9804003

[110] C. Bernard *et al.*, Phys. Rev. Lett. **81**, 4812 (1998), hep-ph/9806412

[111] D.E. Groom *et al.*, Review of Particle Physics, Eur. J. Phys. **C15**, 1 (2000)

[112] C. Weiser, Proceedings of the 28th International Conference on High Energy Physics, Warsaw 1996, p.531 (Edited by I. Adjuk and A. Wroblewski, World Scientific 1996) — see also http://www-ekp.physik.uni-karlsruhe.de/ weiser/proc_warsaw/index.html

[113] UKQCD Collaboration: C. McNeile and C. Michael, Phys. Rev. **D63**, 114503 (2001), hep-lat/0010019

[114] A.M. Green, J. Koponen, C. McNeile, C. Michael and G. Thompson, Phys. Rev. **D69**, 094505 (2004), hep-lat/0312007

[115] H. J. Schnitzer, Phys. Rev. **D18**, 3482 (1978)

[116] H. J. Schnitzer, Phys. Lett. **B226**, 171 (1989)

[117] C. Alexandrou, Ph. de Forcrand and A. Tsapalis, Phys. Rev. **D66**, 094503 (2002), hep-lat/0206026

[118] C. Alexandrou, Ph. de Forcrand and A. Tsapalis, Nucl. Phys. Proc. Suppl. **119**, 422 (2003), Latt02 and hep-lat/0209067.

[119] UKQCD Collaboration: C.R.Allton *et al.*, Phys. Rev. **D60**, 034507 (1999), hep-lat/9808016

[120] UKQCD Collaboration: C. McNeile and C. Michael, Phys. Lett. **B491**, 123 (2000), hep-lat/0006020

[121] V. D. Mur, V. S. Popov, Yu. A. Simonov and V. P. Yurov, J. Exp. Theor.

Phys. **78**, 1 (1994), `hep-ph/9401203`

[122] M. Lüscher, K. Symanzik and P. Weisz, Nucl. Phys. **B173**, 465 (1980) M. Lüscher, Nucl. Phys. **B180**, 317 (1981)

[123] M. W. Paris, Phys. Rev. **C68**, 025201 (2003), `nucl-th/0305020`

[124] D. S. Kuzmenko and Yu. A. Simonov, Phys. Atom. Nucl. **66**, 950 (2003), Yad. Fiz. **66**, 983 (2003), `hep-ph/0202277` and `hep-ph/0302071`

[125] H. Ichie, V. Bornyakov, T. Streuer and G. Schierholz, Nucl. Phys. **A721**, 899 (2003), `hep-lat/0212036` also `hep-lat/0304008`;
H. Suganuma, T.T. Takahashi and H. Ichie, "Detailed Lattice-QCD Study for the Three-Quark Potential and Y-type Flux-Tube Formation", `hep-lat/0312031`;
T. Takahashi, H. Suganuma and H. Ichie, "Y-Type Flux tube formation in Baryons", `hep-lat/0401001`

[126] J. Zeng, J. W. Van Orden and W. Roberts, Phys. Rev. **D52**, 5229 (1995), `hep-ph/9412269`

[127] T. A. Lähde, C. J. Nyfält and D. O. Riska, Nucl. Phys. **A674**, 141 (2000), `hep-ph/9908485`

[128] E.E. Salpeter and H. Bethe, Phys. Rev. **84**, 1232 (1951)

[129] E.E. Salpeter, Phys. Rev. **87**, 328 (1952)

[130] G. Breit, Phys. Rev. **34**, 553 (1929)

[131] W. Lucha and F. F. Schöberl, Int. J. Mod. Phys. **A14**, 2309 (1999), `hep-ph/9812368`;
R. L. Hall, W. Lucha and F. F. Schöberl, Int. J. Mod. Phys. **A18**, 2657 (2003), `hep-th/0210149`

[132] K.C. Bowler, L. Del Debbio, J.M. Flynn, G.N. Lacagnina, V.I. Lesk, C.M. Maynard and D.G. Richards, Nucl. Phys. **B619**, 507 (2001), `hep-lat/0007020`;
A.A. Khan *et al.* CP-PACS, Phys. Rev. **D65**, 054505 (2002), Erratum-ibid. **D67**, 059901 (2003), `hep-lat/0105015`;
S. Aoki *et al.* CP-PACS, Nucl. Phys. Proc. Suppl. **106**, 780 (2002), Latt01 and `hep-lat/0110128`

[133] M. B. Voloshin and L. B. Okun, Pisma Zh. Eksp. Teor. Fiz. **23**, 369 (1976)

[134] R. L. Jaffe, Phys. Rev. **D15**, 267 (1976); C. W. Wong and K. F. Liu, Phys. Rev. **D21**, 2039 (1980).

[135] F. Gutbrod, G. Kramer and C. Rumpf, Zeit. Phys. **C1**, 391 (1979)

[136] N. A. Törnqvist, Phys. Rev. Lett. **67**, 556 (1991); Z. Phys. **C61**, 525 (1994), `hep-ph/9310247`

[137] D. Arndt, S. Beane and M. Savage, Nucl. Phys. **A726**, 339 (2003), `hep-lat/0304004`.

[138] M. M. Boyce, "String inspired QCD and E(6) models", Ph.D. thesis, Carleton University, 1996, `hep-ph/9609433`

[139] H. J. Lipkin, Phys. Lett. **B172**, 242 (1986)

[140] S. Zouzou, B. Silvestre-Brac, C. Gignoux and J.-M. Richard, Z. Phys. **C30**, 457 (1986);

J.-M. Richard, "Hadrons with two heavy quarks", Proc. Conf. on future of high sensitivity charm experiments, Batavia (1994), hep-ph/9407224
[141] D. Richards, D. Sinclair and D. Sivers, Phys. Rev. **D42**, 3191 (1990)
[142] C. Stewart and R. Koniuk, Phys. Rev. **D57**, 5581 (1998), hep-lat/9803003
[143] A. Mihaly, H. R. Fiebig, H. Markum and K. Rabitsch, Phys. Rev. **D55**, 3077 (1997);
H. R. Fiebig, H. Markum, A. Mihaly, K. Rabitsch and R. M. Woloshyn, Nucl. Phys. Proc. Suppl. **63**, 188 (1998), hep-lat/9709152
[144] P. Pennanen, A. M. Green and C. Michael, Nucl. Phys. Proc. Suppl. **73**, 351 (1999);
C. Michael, Proceedings of Confinement III, Newport News, VA (1998), hep-ph/9809211
[145] G.M. de Divitiis, L. Del Debbio, M. Di Pierro, J.M. Flynn, C. Michael and J. Peisa, JHEP 9810, 010 (1998), hep-lat/9807032
[146] A. M. Green, J. Koponen and P. Pennanen, Nucl. Phys. Proc. Suppl. **83**, 292 (2000), hep-ph/9908016
[147] UKQCD Collaboration: C. Michael and P. Pennanen, work in progress.
[148] T. Barnes, N. Black, D.J. Dean and E.S. Swanson, Phys. Rev. **C60**, 045202 (1999), nucl-th/9902068;
S. Pepin, F. Stancu , M. Genovese and J.M. Richard, Phys. Lett. **B393**, 119 (1997)
[149] B. Bolder *et al.*, Phys. Rev. **D63**, 074504 (2001)
[150] O. Philipsen and H. Wittig, Phys. Rev. Lett. **81**, 4056 (1998), hep-lat/9807020;
ALPHA Collaboration, F. Knechtli and R. Sommer, Phys. Lett. **B440**, 345 (1998), hep-lat/9807022
[151] P. W. Stephenson, Nucl. Phys. **B550**, 427 (1999), hep-lat/9902002;
O. Philipsen and H. Wittig, Phys. Lett. **B451**, 146 (1999), hep-lat/9902003;
P. de Forcrand and O. Philipsen, Phys. Lett. **B475**, 280 (2000), hep-lat/9912050;
S. Kratochvila and P. de Forcrand, Nucl. Phys. **B671**, 103 (2003), hep-lat/0306011
[152] H. Trottier, Phys. Rev. **D60**, 034506 (1999), hep-lat/0209048
[153] UKQCD Collaboration: P. Pennanen and C. Michael, "String breaking in zero-temperature lattice QCD", hep-lat/0001015
[154] UKQCD Collaboration: P. Pennanen, C. Michael and A. M. Green, Nucl. Phys. Suppl. **83**, 200 (2000), hep-lat/9908032
[155] A. L. Yaouanc, L. Oliver, O. Pene and J. C. Raynal, Phys. Rev. **D9**, 1415 (1974)
[156] N. Isgur, R. Kokoski and J. Paton, Phys. Rev. Lett. **54**, 869 (1985)
[157] A. M. Green, C. Michael and P. S. Spencer, Phys. Rev. **D55**, 1216 (1997), hep-lat/9610011
[158] P. Pennanen, A. M. Green and C. Michael, Phys. Rev. **D56**, 3903 (1997),

hep-lat/9705033

[159] P. Pennanen, A. M. Green and C. Michael, Phys. Rev. **D59**, 014504(1999), hep-lat/9804004

[160] W. Glöckle *et al.*, Acta Phys. Pol. **B32**, 3053 (2001), nucl-th/0109070

[161] UKQCD collaboration, J. M. Flynn, F. Mescia and A. S. B. Tariq, JHEP **0307**, 066 (2003), hep-lat/0307025

[162] G.E. Brown and A. M. Green, Nucl. Phys. **85**, 87 (1966)

[163] A. M. Green, Repts. Prog. Phys., **28**, 113 (1965)

[164] G.E. Brown, "Unified Theory of Nuclear Models and Forces", (North-Holland Publishing, Amsterdam, 1967)

[165] S. Furui, A. M. Green and B. Masud, Nucl. Phys. **A582**, 682 (1995), hep-lat/0006003